1

BIOLOGY PART 1: CELLS, ENERGY, REPLICATION, INHERITANCE, EVOLUTION AND ECOLOGY

HAROLD M. ICKES II, M.D.

Ideal Quill Publishing Group LLC
idealquill.com

Printed in the United States of America

10 9 8 7 6 5 4 3 2 1

ISBN 978-0615841243

Biology Part 1: Cells, Energy, Replication, Inheritance, Evolution and Ecology

by Harold M. Ickes II, M.D.

Table of Contents

Unit 1

The Science of Biology

What is Biology?

Biology is the science that studies life. Unfortunately, that definition uses the word 'life,' which can be very problematic. What constitutes 'life' is not so easily defined, because the processes of life span a huge range of forms. It comes in practically all shapes and sizes. At times, it's hardly recognizable from nonliving things, and nonliving things can appear at times to be alive. At other times, we can recognize it intuitively, from the mailman at your door, to the cat walking across your yard, to the strange thing currently growing in my kitchen sink (put there by my lazy roommate, no doubt). If we are to study and categorize life, clearly we must first specify the difference between 'living' and 'nonliving.' To do this, we have to be very systematic and careful. We need to be scientific. To do that, it is important to understand the meaning of 'science.' While it may seem silly to toil over the meaning of science, it is useful to have a clear definition in mind. All too often, the media is full of debates about what science really is and what it isn't. So, let us talk about this for a little bit.

What is science and why does it exist?

Science is a field of human endeavor that seeks to attain knowledge through a systematic approach of gathering evidence. The knowledge we acquire is then used to predict further outcomes. Why do we do this? Often, we seek out evidence in nature because we are just plain curious. Just as often, we seek out answers because we want to predict something that will make us rich or famous. For example, what natural laws could help me make a better cell phone? What biological components are out there to allow me to help prevent acne? How could I make a floating car? How can I use science to rid me of my neighbor's terrible music? All of these things benefit mankind and help to make the scientist rich. However, in order to gain benefit or wealth from your research, your theories have to be proven to work. Theories are tested and tested and tested and tested for the slightest weakness. When that process is done, it's tested again. If something about a theory appears to falter, millions of people stand ready to point out the weakness and propose a better theory, especially if the new theory leads to better cell phones or acne preventions or floating cars. That is the business of science, and no accepted scientific theory exists by virtue of it being cool or trendy. Our theories are in place because we can't find holes in them, but rest assured we test them all the time. As long as people are curious, in need of improving the human condition, or simply want to get rich, our theories will be rigorously tested.

Three Assumptions about Reality

Ok, how do we do science? We do science using the **scientific method**. This method is quite simple and surprisingly easy to do, but we first have to understand three assumptions about reality. They may seem obvious to you, but it's important that these assumptions be made. The assumptions are: 1) **Natural Causality,** 2) **Uniformity of Space-Time**, and 3) the assumption of **Common Sensation**. These three principles, together with the scientific method, allow us to slowly unravel the causes of Nature.

First, let's talk about **Natural Causality**. What does it mean? Well, in history, there were two ways of trying to explain Nature. The first was to suppose a divine origin. The thing in my kitchen sink exists, one might say, because a divine being such as God put it there. The second way to explain nature was through Natural Causality, **a principle that assumes that all things occur as a consequence of some other natural event that can be understood** (although not necessarily *easily* understood). Natural Causality is basically a way of saying: "let's see how much we can understand about our reality without assuming anything supernatural." It is not a statement about the existence of God nor an attempt to disprove Him (Her?). **Disproving God is *not* the goal of science**. We are simply trying our best to understand Nature before shrugging and saying that God did it. Therein lies a subtlety, however, for when do we ever know if something cannot be explained by natural events? Perhaps we haven't looked deep enough, after all. The answer may be around the corner, so in effect, Natural Causality stops us from *ever* making any statement about a supernatural origin of any phenomena. We assume an answer exists and keep digging.

The second assumption is the **Uniformity of Space-Time**. What is that? Is it some mad and convoluted monstrosity from the mind of Einstein? No, not at all. It just means that **we assume the laws of physics work the same everywhere in the Universe and have worked the same throughout all time**. Imagine how incredibly impossible it would be to try to figure anything out if the laws of physics were completely different depending on the era or the place. Now THAT would be mad and convoluted. For example, suppose your car stops working, and you're trying to figure out what is wrong with it. Now suppose the laws of physics completely change every 3rd second. Nothing would ever make sense, and you'd never get your car working. (I doubt cars would ever have been invented in that case, actually). What if the laws of physics change when you go from one state or country to another? Your car would stop altogether because the physics that runs it would change.

The third principle is the assumption of "**Common Sensation**." No, I didn't say common sense, which is anything but common. "Common Sensation" **means that we all observe the world the same**. We see, hear, feel, smell, and taste things in the same fashion. In other words, if we are both looking at a square, we will both say it is a square. It won't mystically look like a circle to you for some reason. If we are both looking at the color blue, we will both comment it is blue. What is hot to me is hot to you, and so forth. This principle speaks about things we can measure physically; we must be able to agree that 5 meters is 5 meters or the very nature of measurement is sketchy, and then no one can get rich doing science.

The assumption of common sensation is not referring to how we feel about those things. So, for example, you and I could watch the same movie, but I might like the movie while you might not. We may disagree about the quality of the movie, but we can readily agree that we saw the same movie, with the same actors, scenes, action, and so forth. How we feel about things, such as movies, politics, religion, etc., has no formal place in science. We all may have opinions about scientific theories, but unless we can *show* that these theories are wrong, our opinions are just opinions. Your opinion cannot be made into a theory, unless it is well formulated with mountains of rock-solid evidence behind it.

Note that the assumption of common sensation requires a healthy and normally functioning mind. There are a number of unfortunate illnesses that cause people to be unable to accurately sense and decipher the physical world. As a result, they often suffer from being unable to understand what is real and what is not.

Important Founding Principles of Science

1) Natural Causality – phenomena have natural explanations that can be understood.
2) Uniformity of Space-Time – the laws of physics work the same throughout the Universe and throughout all time.
3) Common Sensation – we all observe the physical world in the same ways.

The Scientific Method

The scientific method is a systematic approach to investigating reality. It begins with the **observation**. From that observation, a **hypothesis** is made. A hypothesis is an educated guess, (unless you are my housemate, in which case it would just be a guess). The hypothesis is only useful if it can be tested. For example, suppose I declare that I have an untouchable, unseeable, undetectable Doobadooboo monster in my closet. If it is impossible to detect it in any way, then it is silly to even argue about whether or not it is there. Science deals with the testable, and if your theories cannot be tested, it is not science.

Once we have a testable hypothesis, the next thing to happen is to **experiment** to verify or disprove the hypothesis. When we do this, we do physical measurements, once under one condition and then again under a different condition. When changing the conditions, it is best to change only one thing at a time. For example, suppose you want to understand the effects of pressure on the boiling point of water. In one situation, you boil the water under conditions of standard room pressure. In the second situation, you do *exactly the same thing*, changing *only* the pressure in the room. Assuming nothing else changed, the change in pressure can then be blamed for whatever change happens to the boiling point. Suppose, on the other hand, you *also* changed the pot as well. Could you be certain that the change in boiling point was due to the change in pressure and not the change in the pot? Sadly, you couldn't. So, to avoid that, we change *only one thing at a time*. This is an important point to understand. Before your idea will become an accepted theory, it will be attacked from all directions by people who will argue that more than one thing changed. You must be able to demonstrate that only one thing changed between the two conditions. Sometimes it is trickier than it sounds, especially in biology and related fields like medicine.

Lastly, after the experiment, we make a **conclusion** by accepting, altering, or rejecting our hypothesis. If the hypothesis fails experimentation, we come up with a new hypothesis and start over.

Summary

So, to summarize this nifty little chapter, we learned about the science of biology, the nature of science, and the intractable laziness of my housemate. In future sections, we'll build upon these ideas.

- What is biology?
  - It is the science that studies life.
- What are the three assumptions before studying any science?
  - Natural Causality – phenomena have natural explanations that can be understood.
  - Uniformity of Space-Time – the laws of physics work the same throughout the Universe and throughout all time.
  - Common Sensation – we all observe the physical world in the same ways.
- What is the scientific method?
  - 1) Observation – we observe something that we don't understand.
  - 2) Hypothesis – we make an educated and testable guess about our observation
  - 3) Experiment – we test the hypothesis by changing one parameter of the experiment at a time.
  - 4) Conclusion – we accept alter, or reject the original hypothesis, returning to (2) if need be.

In this section we cover:

1) The definition of life, focusing on an overview of its layered structure, from particles on up to the biosphere.

2) Taxonomy

Unit 2

What is Life?

All right, enough grumbling about what science is. Now that we have made those introductory statements, let us clarify what we mean by the difference between living and nonliving. In other words, let's define what life is. All living things have eight traits in common: **complex structure**, **response to stimuli**, **homeostasis**, **metabolism**, **growth**, **reproduction**, **inheritable genetics** and **evolution**. There were a couple unfriendly words in there, so let's discuss these a little more in depth. Once we've done that, we'll briefly discuss how life is categorized.

Complex Structure

Living organisms are incredibly complex things, made up of very complex structure. Biology divides these structures into various levels of organization, which we will discuss in much more detail in the next few sections. Here is a glance of that structure from smallest to largest.

The smallest entities in the known Universe are **particles**. Particles are (probably) indivisible – meaning that we cannot break them down into smaller things, as far as we know. Particles come together to make **atoms**.

Atoms, in turn, clump together in various fashions to make **molecules**. Molecules that have carbon in them are called organic molecules or organic compounds.[1] Organic compounds are important in biology, because they are the predominant molecules in living systems. We will be discussing them in future units until we are blue in the face. The next level of biological organization is the **organelle**. These are collections of organic molecules put together in a very specific way. Organelles have many purposes in life, but we will put off their functions until later. For the moment, think of them as little chemical factories.

---

[1] It is very unfortunate that society has chosen the word "organic" to denote a food that was naturally grown. All food is, by its very nature, organic since it has carbon in it. If it were not made with carbon it would be inorganic and probably poisonous to you. Actually – come to think of it – it is also unfortunate that we say something is "naturally" grown. After all, I can't remember the last time I saw something that was supernaturally grown.

One more level up is the **cell**. The cell is the basic unit of life. Cells contain organelles, water with organic molecules floating in it, and a whole lot of other important things needed for life's processes. Think of cells as little water balloons with tiny things floating inside. The boundary of the cell is comprised of the cell membrane, which is like its skin, in a sense. Cells come in all shapes and sizes. Some are amorphous (like blobs, oozes, or my roommate). Some are long and slender, like nerve cells. Some are like cubes. The shape of the cell depends on what its purpose is, as will become evident in future units. Most life forms have only one cell, like bacteria for example. Bacteria are just simple cells living on their own in water. They can be rod-shaped, long and branched, or ball-shaped. Though they are often living on their own, many can clump together into little colonies.

For those organisms that are made up of two or more kinds of cells, we must go further in our categorization of the levels of organization. A collection of different types of cells (that cannot survive on their own) makes up a **tissue**. There are many types of tissues, such as nervous tissue, muscle tissue and connective tissue, to name a few.

Next on the list of organization levels is the organ. **Organs** are made up of two or more types of tissues. The heart, the brain, the pancreas and the liver are examples.

Organ systems come next. The brain connected to the spinal cord (another organ) make up the nervous system; the heart connected to all the blood vessels make up the circulatory system, etc.

A collection of organ systems makes up an individual, or the **multicellular organism**. This is the level we readily recognize in everyday life. When we greet each other, we see each other as complete individuals. We never see a brain moseying in to the post office by itself, or spot a collection of connective tissue out for a day's fishing. On the other hand, not all organisms necessarily have all organs. Trees do not have hearts, squids do not have knee caps, and my roommate has managed very well without a brain.

One might suppose that our discussion of the structure of life would end here, but it does not. Biology concerns much more than this. A collection of individuals living within the same area and capable of interbreeding is called a **population**. These individuals tend to share very similar traits, since they interbreed with each other. The humans living in a city on an isolated island (say before boats and planes) would be a population, because they can interbreed with each other and tend to be very similar. They might, for example, all have blond hair with many of them having green eyes. Who knows?

A collection of populations, all capable of interbreeding with each other, is known as a **species**. Two populations could be on the opposite side of the earth from each other, but if they are capable of breeding with each other, they are part of the same species. For example, the indigenous peoples of South America are one population, and the people of Japan are another, but since they are capable of breeding and producing viable reproductive offspring, the two populations are considered part of the same species. If a *healthy male* and a *healthy female* are *not* capable of breeding together, they are *not* part of the same species. If those two organisms *can* breed together, then they *are* part of the same species. For a quick and rough definition: Interbreeding potential = same species.

Next on the list is the **community**. A community is defined as two or more species, such as a species of frog, a species of fish and a species of mosquito, living in the same area.

The **ecosystem** is one level higher. It is the community with the nonliving elements considered as well. So, the frogs, the fish, the mosquitoes, along with the rocks, the streams, etc. make up the ecosystem in which those species live.

Last on the list (phew!) is the **biosphere**. It basically means the part of the Earth in which any form of life can be found. There is only one biosphere until we find another planet with life on it (or put life on it), or until we mess up everything and kill the one we're on.

Though we have discussed the structure of life from bottom to top, the first semester of biology typically does not involve a great deal of topics in multicellular organisms (except perhaps closer to the end). The topics of community, ecosystem and biosphere will be encountered again near the end of this book when we talk about ecology and biogeography.

Quick list of biological organization:

Particle – The tiniest known pieces of matter in the Universe

Atom – A collection of particles put together in a very specific way.

Molecule – A collection of atoms put together in a very specific way.

Organelle – A collection of molecules that have specific purposes in living reactions

Cell – The basic unit of life, containing many organelles

Tissue – A collection of two or more types of cells, (which could not survive without each other)

Organ – A collection of two or more types of tissues

Organ System – A collection of two or more organs working in concert

Multicellular Organism – An individual made up of multiple cells, often arranged as a collection of organ systems. (The generic term "organism" is often used instead. The term organism can, however, refer to any kind of critter, multicellular or not.)

Population – A collection of interbreeding organisms.

Species – A collection of populations, all capable of interbreeding if the opportunity arises.

Community – A collection of different species living in a geographic area.

Ecosystem – The community plus the nonliving elements of the geographic area.

Biosphere – The surface and atmosphere of the Earth, as well as a small amount of subterranean region where life can be found.

Responding to Stimuli

Now we know that living things are complex and made up of a tremendous number of carbon-containing molecules known as organic compounds. The next quality that life has is the ability to respond to environmental circumstances that could alter the continued survival of the organism or its offspring. For example, if a cat gets hit with a stick, the cat will jump away. Why? The impact of the stick upon the cat causes pain to the creature, which is a built-in response that tells the cat "this is causing damage that could threaten your life and/or ability to reproduce." So, the cat jumps away to avoid being hit again.

Now suppose I hit a small rock with a stick. It appears to "jump" away as well. Does that mean that the rock is fleeing to save its life? No, of course not. The rock flew away because I imparted momentum upon it. It's a simple physics phenomenon. After all, suppose I hit a *big* rock with a stick. The rock will just sit there with absolutely no response, even though continued strikes could (in theory) "threaten" the continued existence of the rock.

That's all well and good, but what if the cat is drugged so that it is incapable of responding to being hit by a stick? Can we then say that the cat is no longer alive because it cannot respond to stimuli? No. We cannot make such a conclusion, because we know the cat would respond under standard conditions. Just because the cat is *stoned* doesn't mean it's the same as the *rock*. . .

Yeah, that's right. I actually went all that way just to set up a terrible joke. As it turns out, sticks and stones break monotony.

Incidentally, I do not, under any circumstance, advocate that you hit cats. Animal cruelty is never advisable, nor acceptable. You are more than welcome to have a go at the thing in my kitchen sink, though. I'm quite certain it will respond to the stimuli, however, and you probably won't like the response. You've been warned.

Homeostasis

Homeostasis is a very important quality of life, and we hinted at it with our example of the cat. The word is derived from Greek and means roughly "stay the same." All living organisms maintain their internal structure to the best of their ability for as long as they can. If they are low on fluids, they get thirsty and seek out water. If they are low on a nutrient, they tend to seek out specific types of food that contain that nutrient. Homeostasis is not just a conscious drive, however, for it happens at the level of the cell as well. If a certain protein (which we will discuss later) is running low within the cell, the cell will gather up the necessary materials and make more of the protein. This quality of maintaining balance is vital to the success of the organism. Without it, life would be like the cat that just sits there – doomed.

Metabolism

Metabolism is the acquisition of material and energy from the environment. These things are manipulated at a chemical level inside the organism and converted into energy and material forms that help it to sustain homeostasis and the other qualities that are required for survival. Think of metabolism as internally controlled chemistry that releases energy and converts scavenged chemicals into more useful forms.

Growth & Reproduction

All living creatures grow. They start small and grow to a bigger size before reproducing. Reproduction takes many forms. Bacteria multiply in number by dividing in half, a means known as **asexual reproduction**. They basically grow until they are twice their original size, then split in half in an orderly fashion to start the cycle anew. Some organisms, such as water hydra, reproduce by **budding** new individuals directly from their sides. In contrast, sexual reproduction occurs when the **gametes** (small genetic packets, if you will) from a male and female combine. Depending on the species, this can occur inside or outside the organism.

Inheritable Genetics

When living organisms reproduce, genetic material is passed from the parent to the child. That genetic material is the DNA,[2] a complex molecule that we will discuss in detail later. This molecule stores biological information in it, acting like a blueprint for the new offspring. It dictates such things as the color of your hair (assuming you haven't dyed it), the color of your eyes (minus contacts), your height (to a large extent), and facial features (i.e. family resemblance), to name a few. What we learn (i.e. the skills and most behaviors) is *not* stored in the DNA and is therefore *not* inherited by our children. You are free to learn whatever subject you like, but that

---

[2] Some viruses use a related molecule called RNA as its genetic material. Whether or not viruses are considered living, however, is hotly debated.

particular subject will not be passed to your children. There seems to be plenty of anecdotal evidence that we inherit the *tendency* to learn some subjects quicker than others. In other words, suppose music comes "natural" to your family. One might predict that your daughter would tend to learn piano quicker than the neighbor's rotten kid. It's possible but hard to prove. The problem is that if your daughter is surrounded by music from birth, she'll have picked up much more of it than children of non-musical families, genetic tendency or no. In addition, your experience in music might make you a superior music teacher to her, versus a family with no contacts with musicians. The point is that it's hard to prove that someone is necessarily natural at something, since all the factors may not be evident. Remember, the strongest arguments about cause are those in which only *one* thing is different between two examples. If multiple things in the environment are different, it's hard to pinpoint which is to blame for your family's musical nature.

Either way, there are plenty of things that are definitely NOT inherited, including (and not limited to) scars and most behaviors. These "markings" are not encoded in the DNA. They are acquired during life, and once the organism carrying them dies the markings go with it. If you acquire some psychological injury due to a traumatic event, your children will not be born with the psychological injury, because the details of the psychological injury are not recorded in the DNA.

Basically, the rule of thumb is that *nothing* is inherited unless it is recorded in the DNA, and practically nothing is deliberately written into your DNA as a consequence of your life's experience. Learn to play piano all you want, but your kids won't inherit a scratch of it.

Evolution

Evolution is a complex topic that causes a lot of fuss. Unfortunately, the general populace does not understand the concept very well, because it is poorly taught. The best understanding of the theory of evolution (and what it tells us) comes after we understand a great deal about basic biology, so we will wait until near the end of this book to cover it in detail. By that point, you will have all the prerequisite knowledge necessary to fully grasp the nature of the theory. For the moment, however, let's have a peek at evolution, just for a preview. . .

What is evolution, and why is it so important to understand? It's important because modern biology makes *little sense* without it, and *every bit of sense* with it. The basic idea is simple: inherited information from the DNA contributes to the odds of a creature's survival. Individuals with poor genetics (genes that give them unfavorable physical and basic mental characteristics) have a low chance of survival. Over generations, they fail to reproduce as much as individuals with good genetics and are eventually removed from the species. Individuals with superior genetics have increased odds of survival. They tend to reproduce more successfully, and over the course of generations they increase in percentage.

DNA regularly undergoes **mutations**, which changes the genetic information encoded within an individual. When that individual passes this mutated DNA to an offspring, that offspring may have a new trait that was previously unseen in the species, altering its odds of survival. If this trait is advantageous, the offspring will tend to leave more offspring of its own, and over time the trait will become common. Through the introduction of mutations, new traits appear, and the species slowly changes from one generation to the next. *Individuals do not evolve* (in the scientific meaning of the word). It is a process that takes place in (usually) tiny steps from one generation to the next, sometimes requiring *millions* of years to be obvious. Monkeys did not suddenly climb out of a tree, put on a suit, and decide to walk upright. Well, maybe my housemate did, but he's the only exception.

Anyway, let's discuss a quick example. Suppose a species of bird relies on its ability to break open seeds in order to eat the nutrients inside. Birds born with tiny beaks will have trouble with this task and will have a disadvantage in harvesting nutrients and energy from the seeds. Birds born with larger beaks will be more successful. Each time a mutation appears that makes the beaks a little larger, the birds with these larger beaks tend to do much better than those without. So, they tend to be more likely to survive. Better survival tends to increase the opportunities for mating. Therefore, they tend to mate more. The trait slowly spreads throughout the species. After thousands of years, (or hundreds of thousands), the birds slowly evolve bigger and bigger beaks. The genetics for small beaks are lost because those birds couldn't get enough of the seeds opened.

Biological Classifications or Taxonomy

Having discussed the qualities that all living organisms have in common, and having skimmed the idea of evolution, let us now talk briefly about how we categorize organisms by biological type. Biological classification is called **taxonomy** and is divided into levels. At the top level are the domains. These are divided into kingdoms, which are in turn divided into phyla. These are then divided into classes, then orders, then families, then genera (singular is genus), and finally species.

So, to summarize the categories:

Domain -> Kingdom -> Phylum -> Class -> Order -> Family -> Genus -> Species

There are butt-loads of mnemonics out there for remembering this. My favorite is:

Deadly killer poo can obviously fumigate gathering spots.

Silly, but effective . . . well, it is to me anyway. Think of biological classifications as a kind of filing system like you find on computers. It organizes creatures into groups by similarity. The closer they are grouped along the right side of the categories (same family or genus, say), the closer they are related to each other. We categorize organisms based on similarity and our best guess of evolutionary relatedness. For example, organisms belonging to the same family are thought to be far more related to each other than organisms that belong to an entirely different kingdom. Below, this chart shows that the species "B" is much more related to species "C" than it is to species "A." I've left out names for simplicity.

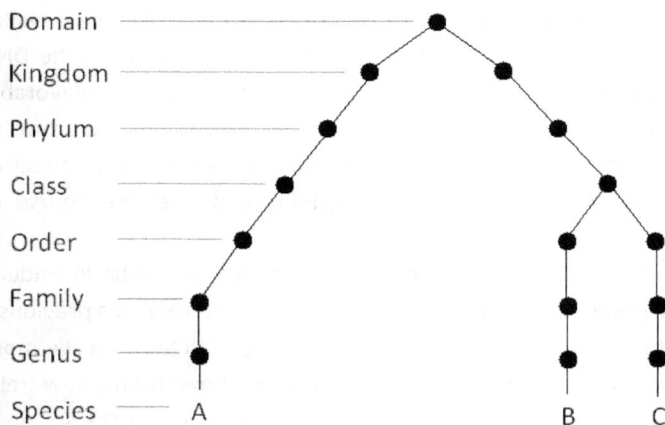

16

The concept of the **domain** is relatively new. There are three: **Archaea**, **Bacteria** and **Eukarya**. The Archaea and Bacteria domains contain single-celled organisms that do not have a cell nucleus. What is a cell nucleus? The cell nucleus, (which we will be discussing in depth later), is basically a part of the cell that has its own membrane. For the Eukarya, the DNA is in this inner compartment. For the Archaea and Bacteria, the DNA is just floating free in the cell.

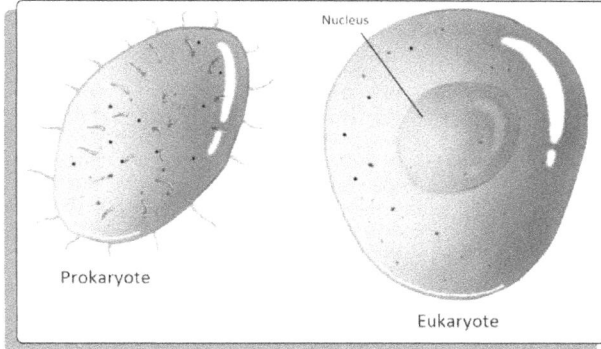

Nucleus

Prokaryote

Eukaryote

The cell nucleus is believed to be a more advanced characteristic, having evolved much later than the Archaea and Bacteria organisms.

You will frequently see the words **eukaryote** and **prokaryote** used to describe the topic of cell nucleus. Eukaryote means roughly "good nut" and refers to cells that have cell nuclei. Prokaryote refers to cells that do not have cell nuclei. Technically, 'pro' means "before," so 'prokaryote' literally means "before nuts." In the same vein, "eu" means true, so I guess 'eukaryote' literally means "truly nuts." Humans are eukaryotes, so the name fits well.

The Archaea and Bacteria domains each have only one kingdom, named the same as the domain. The Eukarya domain is divided into a variable number of kingdoms, depending on the textbook you use, or the country you're in. Sadly, there is no consensus. One of the more popular systems divides Eukarya into Animalia, Plantae, Fungi, and Protista. If you are reading along with a class, follow the classification system your professor uses. Here is how humans compare to a number of other organisms (chimpanzee, sockeye salmon and the French field elm).

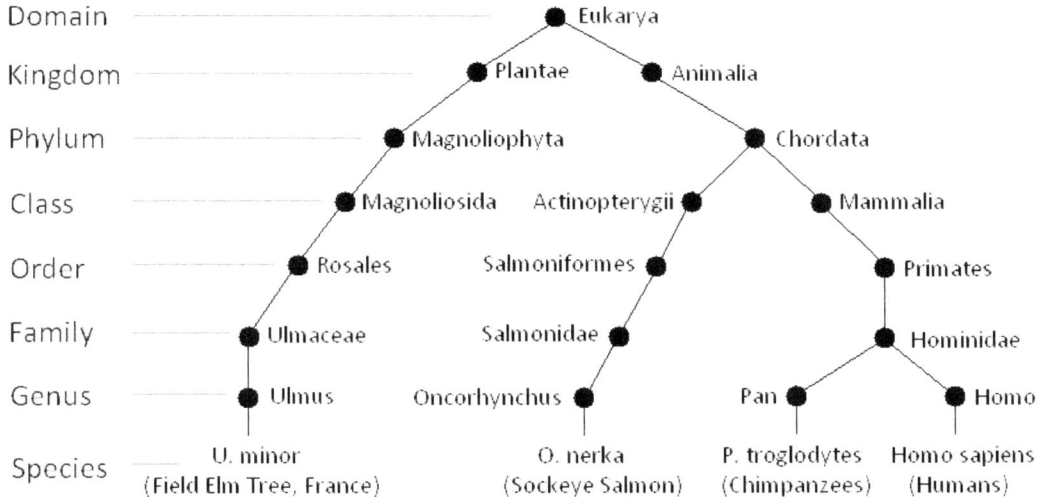

That brings our preliminary discussion about biology to a close. Next we dive head first into the topic. Put on your helmet, because here we go.

Summary

In the unit, we studied nature and categorization of life:

- The eight traits all living things have in common are:
    - complex structure – built up from particles to biosphere, as follows:
        - particles
        - atoms
        - molecules
        - organelles
        - cells   (unicellular organisms)
        - tissues
        - organ
        - organ systems
        - multicellular organism
        - population
        - species
        - community
        - ecosystem
        - biosphere
    - response to stimuli – altering behavior or state in response to a change in the environment
    - homeostasis – the ability to remain relatively unchanged over long periods of time as a result of internal chemical maintenance.
    - metabolism – the ability to chemically alter gathered resources into chemicals not found in the environment or for energy release.
    - growth - increase in mass and usually volume in preparation for reproduction.
    - reproduction – the creation of an offspring, either asexually (fission or budding) or sexually (male and female gamete fusing).
    - inheritable genetics – traits recorded in the genes that get passed from one generation to the next. Scars, injuries (physical or mental), and most behaviors do not get inherited.
    - evolution – the slow change from one generation to the next due to new mutations and changing environmental pressures. Usually requires thousands of years to notice.
- Taxonomy is the categorization of living organisms into a hierarchy of eight levels:
Domain -> Kingdom -> Phylum -> Class -> Order -> Family -> Genus -> Species

    The domains are very broad. The species are very specific.

Unit 3

Nonliving Building-blocks of Life

Part 1

The processes of life are comprised of physical materials that obey the laws of physics. Here we discuss what those basic materials are, how they look and some simple characteristics that result from how they come together.

The Particle

The tiniest pieces of the Universe, as far as we can tell, are particles. They are incredibly small things, smaller than anything else in existence (except, of course, my housemate's chances of doing the dishes).

To give you an idea of their size: a *tiny* shred of paper on a table has an unfathomable number of particles in it. It *easily* has more particles than the number of people on Earth. You can reasonably think of particles as being *almost* infinitesimally small dots that have *essentially* no volume at all, approximated by theoretical things we call **point masses**.

There are many types of particles in the Universe, but the most important ones in the processes of life are the **proton**, the **neutron**, the **electron** and the **photon**. Protons are positively charged, and electrons are negatively charged, whereas neutrons and photons have no charge. By charge, I'm referring to the same concept as you encounter with batteries, and at a basic level the physics behind batteries is the same as the physics behind the processes of life: plusses and minuses interacting. "Like" charges repel each other. Opposite charges attract. So, protons and electrons attract each other. As for photons, let's put off that discussion until a later section, except to say that photons make up light.

Protons and neutrons have roughly the same mass, whereas electrons are about 1000 times *lighter*. In fact, electrons have so very little mass that their behavior is unlike anything you would expect. Why would this be? In physics (don't panic) there is an important theory known as the **Uncertainty Principle**. Although the details of this topic are far outside the range of this book, it suffices to say that the Uncertainty Principle makes it practically impossible to pinpoint exactly where *anything* is, especially the tiniest entities in our Universe. In a way, it's a matter of practicality, for how do you capture something so ridiculously small? It's like catching one specific grain of dust in a category 5 hurricane. The darn thing could be anywhere! The best you can do is say that the grain of dust in question is most likely within a fifty square mile radius. The analogy isn't perfect, of course, but when applied to electrons, it becomes necessary to draw them as fuzzy "probability" clouds, where the clouds represent the most likely space where the electrons will be found. For the sake of simplicity, electrons are commonly drawn as dots within little balloons (which are usually attached to atoms and molecules – but we'll get to that later). Sometimes, when we don't care about the actual physical position of the electron, we draw them as just simple dots. Don't be deceived, however, for electrons are not just little dots. Their exact nature is still the subject of a great deal of theory. Here are important things to remember about the little buggers: 1) they are super tiny, 2) negatively charged and 3) impossible to pinpoint exactly, so we typically draw them as clouds or balloons.

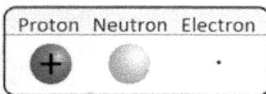

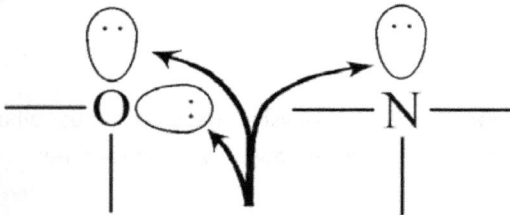

Proton Neutron Electron

Sample electron clouds (two dots = two electrons in each). The "O" is an oxygen atom. The "N" is a nitrogen atom. Atoms are covered a little later.

Here are a few pictures for your brain to gnaw on. I've drawn the electron as a dot to give you an idea of their relative mass compared to protons or neutrons.

Protons and neutrons clump together in various combinations to form the nuclei of atoms. Electrons, in turn, are clouds around the atomic nuclei, attracted by the positive charge. Most biology textbooks, (and many introductory chemistry books as well), portray electrons as "orbiting" around the nucleus in neat little circular paths. For reasons I explained above, this is not actually realistic, but for the purposes of most biological considerations, it works.

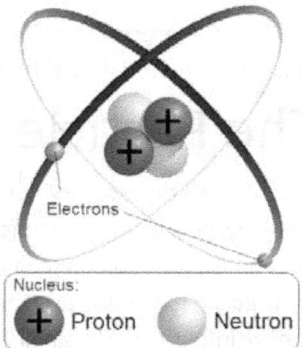

Electrons

Nucleus:
+ Proton   Neutron

The Atom

The number of protons that an atom has in its nucleus is called its **atomic number**, and it determines the type of **element** that the atom is. For example, carbon atoms have six protons in their nuclei. If you find an atom with six protons, it is carbon, and never nitrogen, zinc, iron, etc. If you find a carbon atom, it *always* has six protons, and never five or seven or some other number. The cartoonish image of the atom above is a helium atom, because it has two protons in the middle – helium atoms never have more than two or less than two protons.

Unlike protons, the number of neutrons in atoms can vary, depending on the element, but the variation is usually not much. For example, most carbon atoms have six neutrons in their nuclei, while some have seven neutrons instead. Atoms that have the same number of protons but different numbers of neutrons are called **isotopes** of one another. So, the carbon with 6 neutrons is one isotope, while the carbon with seven neutrons is another isotope. The total mass of an atom is measured in **atomic mass units** (amu).[3] It is easily estimated by simply adding the number of protons and neutrons in the nucleus, because protons and neutrons each have a mass of about one amu. The mass of electrons is so incredibly small that we usually don't include them in the calculation. We indicate what isotope we are talking about by writing its atomic mass to the upper left side of the element's symbol.

$$^{12}\text{C}$$ $$^{13}\text{C}$$

Carbon 12   Carbon 13
6 protons   6 protons
6 neutrons   7 neutrons
mass: 12 amu   mass: 13 amu

Most atomic nuclei are *very* stable things, resisting all kinds of energy: heat, vibration, light, my ex-girlfriend's constant nagging, you name it. Electrons, on the other hand, can be ripped away from their orbits without much trouble – we'll get back to that. Some isotopes, however, are *not* stable and spontaneously eject various forms of radiation from the nucleus, without any external reason. These isotopes are said to be **radioactive**. When radiation emits from the nucleus, the atom changes isotope and atomic number. This is basically what happens to uranium and plutonium as they give off radiation to slowly transform into lead. For example, the first step in the breakdown of uranium-238 into lead is:

$$^{238}_{92}U \rightarrow {}^{234}_{90}Th + {}^{4}_{2}He^{2+} + 2e^-$$

This reaction basically means that the uranium-238 nucleus boots out a quartet of two neutrons and two protons, written as the $^{4}_{2}He^{2+}$. The nucleus left over is a thorium nucleus, which itself is unstable and eventually makes its own adjustments to become something else in time. Anyway, that's enough of that for the moment . . .

If you look at the periodic table you will see over one hundred different elements, each identified by their atomic number. Many of the elements with atomic number over ninety are radioactive. Since they are unstable, you cannot find much of them in Nature anymore. They've mostly broken down by now.[4] Anyway, there are 92 naturally occurring more-or-less stable elements, each having unique properties. Here is a table of the common elements that are found in living systems:

Element name	Atomic Symbol	Atomic Number
Hydrogen	H	1
Carbon	C	6
Nitrogen	N	7
Oxygen	O	8
Sodium	Na	11

---

[3] One amu equals approximately 1.66053886 x 10-24 grams. One amu = 0.00000000000000000000000166053886 grams. The amu is like one person living amongst 90,000,000,000,000 populated Earths.

[4] In case you're wondering, most, if not all, of these atoms came from stars. There's a lot of quantum mumbo-jumbo and mathematical hubbub, but basically stars are so incredibly hot that they smash atomic nuclei together, combining little nuclei into bigger nuclei. At some point in the distant past, stars emitted these atoms into space, which gradually formed together into planets . . . That or it all came out of a big galactic magic hat. I like to keep an open mind; it leaves room in the event my sanity returns.

Phosphorus	P	15
Sulfur	S	16
Chlorine	Cl	17
Potassium	K	19
Calcium	Ca	20

Let's move on to discuss the electrons around the atom – those sketchy little cloudlike thingies. An atom will have an equal number of electrons around it as it has in its nucleus. So, a carbon atom will have six electrons around it, since it has six protons in its nucleus. The six protons in the middle will give the atom a plus six charge, while its six electrons contribute a negative six charge, making it neutral (+6 – 6 = 0). Atoms that have unequal ratios of electrons to protons will have an overall charge and are called **ions**. We'll talk more about those later.

As we said earlier, electrons gather in clouds around atoms. These clouds are called **orbitals**, even though they are *not* circular in shape (despite simplified pictures). Some orbitals are spherical in shape, while others have stranger shapes, such as bowties and clovers. Below is an image of a few sample electron clouds. The image is not comprehensive by any measure, for there are lots of other possible shapes available. The first two drawn are the ones that are encountered the most in basic biology.

Simplified nucleus

"s" orbital

"p" orbital

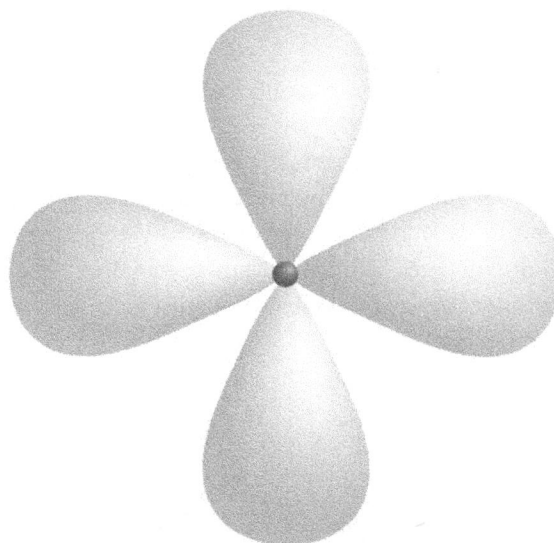

"d" orbital

So what do these cloudlike orbitals represent? They represent the statistically most likely place where you would find an electron that is in that orbital. For example, consider the "p" orbital from the picture. The electrons in that orbital are most likely to be found somewhere within that bowtie shape at any given point in time. The bowtie squeezes down to zero in the middle, so that says the electron shouldn't be found there. However, the shape of the orbital is only a general guideline for where the electron could be found. *Technically* the electron can be found

22

outside the orbital as well, but the odds are *overwhelming* that at any given moment it is somewhere *within* the bowtie.

I wouldn't worry too much if the electron orbital shapes seem alien. It is generally covered in detail in courses on chemistry, and it's not necessary to understand the concept in its fullness at this moment. Just be aware that the actual physical shape of the orbitals ultimately affects the physical shape of the atoms, and that ultimately affects how atoms interact and fit together to make molecules. Those molecules, in turn, have definite shapes that are critical in the reactions of life. The molecules inside you recognize and react with each other because of their characteristic shapes. So, by extension, the fact that electrons orbit around atoms in weird ways makes life possible.

Leaving their shapes behind, let's look at a few other characteristics of electron orbitals.

Each electron orbital can hold two electrons, which are said to be "paired." Electrons *like* to be paired,[5] despite the fact that they both have negative charges. In fact, as a hard and fast rule, unpaired electrons tend to be rather unhappy, and their attempts to get paired leads to quite a bit of chemistry. It's a bit like dating in that sense. Don't get me wrong, molecules and compounds (and metals) *do* exist that have unpaired electrons, so the rule isn't perfect, but these compounds also tend to be unstable in the long run, breaking down in the direction that allows their electrons to become paired.

Multiple orbitals around atoms are divided into sets called **electron shells**. The first shell is closest to the nucleus and has only one orbital (which looks like a sphere – the "s" orbital in the picture above). Since the first shell has only one orbital, only two electrons can be in it. The second shell is a bit farther from the nucleus, is bigger, and has four orbitals in it (one sphere, and three bowties – one on each of the three-dimensonal axes). Each of these orbitals can hold two electrons. So, with four orbitals in the second shell the total number of electrons that can fit there is eight. The third shell has nine orbitals in it (of a variety of weird shapes), which means the third shell can hold up to eighteen electrons. Most of the time, only four of the orbitals are filled, however, meaning that usually only eight electrons are found in the outermost shell of an atom in a stable form (the **octet rule** we see in chemistry classes). There are technically an infinite number of possible shells: the higher the shell number, the higher the energy of the electrons that are in it. The lowest shells fill up first, though, so we rarely deal with the higher level shells. The row on the periodic table that the element is in determines the base number of shells it will have. For example, hydrogen is in the first row, so it has one shell. Carbon is in the second row, so it has two shells. Chlorine is in the third row, so it has three shells. Most biological chemicals are made from elements in the first three rows, so understanding of the first three shells will suffice to allow you to study the vast majority of topics in biology. These shells are summarized here.

Electron shell #	Number of orbitals	Max number of electrons
1	1	2
2	4	8
3	9	18 (8 is commonly found)

---

[5] It's actually due to a strange quantum mechanics principle known as the "spin number," as well as minimizing magnetic fields. I once tried to explain this to a student I was tutoring, and his head exploded, so I won't go into detail about spin numbers. All joking aside, just know that electrons like to be paired because of quantum mechanics.

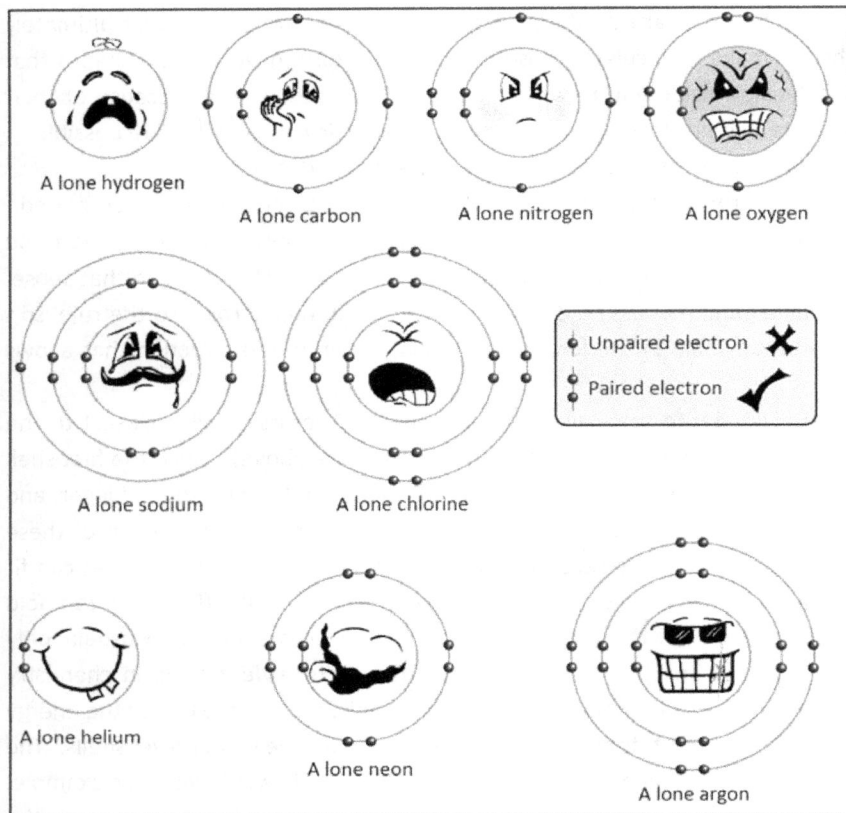

A lone hydrogen

A lone carbon

A lone nitrogen

A lone oxygen

Unpaired electron	✖
Paired electron	✔

A lone sodium

A lone chlorine

A lone helium

A lone neon

A lone argon

An atom's outermost shell is called the **valence shell**. Atoms generally interact with each other exclusively through their valence shells. That, in essence, is where chemistry happens. If we ignore the actual shape of the individual orbits, we can use the popular strategy of depicting the shells as just circles. To the left are a few sample atoms with their electron shells. Some are happy. Some are not. Can you figure out why?

The image shows lone atoms. The hydrogen and helium have only one electron shell. Carbon, nitrogen, and neon have two shells, while sodium, chlorine, and argon have three shells. Notice that the first two shells of sodium, chlorine, neon, and argon have the maximum electrons in them (two and eight, respectively). The third shell of argon has eight electrons (which is typical, even though the third shell can technically have as many as eighteen). So why are some happy and some sad? Here is a golden rule of chemistry: **Atoms are happiest when their valence shells are full.** The reasons are, as always, due to quantum mechanics, but you can think of it like this: if the shells are not full, the atom will be lopsided in one way or another.[6] Nature doesn't like that. Further, since electrons like being paired, the electrons in the full valence shells are all paired off.

The pattern is simple:

2, 8, 8 make atoms great.

---

[6] It will be lopsided not just physically, but magnetically as well. Also, the third shell doesn't need 18 electrons to avoid being lopsided. Eight electrons will suffice. For the purpose of discussion we will consider 8 electrons as making the third shell "full."

Helium, neon and argon already have full outer shells. These atoms are completely happy by themselves. They don't need to worry about anything and rarely interact chemically with the world around them. Since they don't interact much with other atoms, (and are typically found as gases), we call them the **noble gases**.

Look at the other atoms in the picture. Their outer shells are not full. They are unhappy, unstable and lopsided. How can they fill their outer shells? One option, of course, is to simply add or subtract electrons. This makes an ion, and some atoms do indeed take this pathway. Here's what hydrogen, sodium, and chlorine can do, (leaving carbon and nitrogen for the moment):

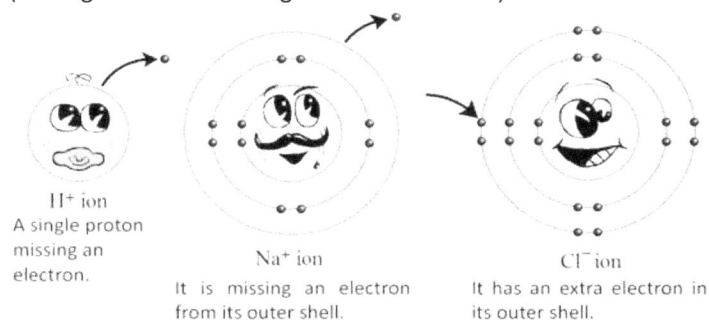

H$^+$ ion
A single proton missing an electron.

Na$^+$ ion
It is missing an electron from its outer shell.

Cl$^-$ ion
It has an extra electron in its outer shell.

In the image to the left, we see three different ions: H$^+$, Cl$^-$ and Na$^+$. The first one might surprise you. Hydrogen is probably the only atom that has the luxury of just getting rid of its electron all together – it can become a simple proton, a wanderer, a vagrant, living carefree without the worries of all this orbital business. That must be nice. It's not necessarily the best option, and it doesn't always do this though, as we will see, for most hydrogen atoms would rather bond to another atom. The sodium takes a similar strategy of abandoning an electron. By doing so, the shell below it now becomes its valence shell, full and *not* lopsided. The chlorine simply gains an electron from somewhere (perhaps an atom, like sodium, that is stabilized by losing an electron) and becomes stabilized for the same reason.

The important thing to understand about these atoms is that they now have an imbalance in their proton to electron ratio. They are ions. The new hydrogen has one proton but no electrons, so it has a positive charge. The new chlorine has seventeen protons and eighteen electrons. That means it has a -1 overall charge. The new sodium has eleven protons and ten electrons. That gives it an overall charge of +1.

Many atoms form ions in this fashion. Here is a quick list of those that are relevant to biology: K$^+$, Ca^{2+}, Fe^{2+}, Fe^{3+}, Mg^{2+}, and Zn^{2+}. The "2+" above an ion, (such as for calcium Ca), means that it is missing two electrons. Also, not all ions are made from single atoms. Some ions are actually groups of connected atoms that collectively have extra or missing electron(s). These are called **polyatomic ions**. dihydrogenphosphate (H$_2$PO$_4^-$), ammonium (NH$_4^+$) and bicarbonate (HCO$_3^-$) are examples. You'll see many of these over and over, so don't sweat what they are at the moment.

Getting back to nitrogen and carbon, not all atoms can pull off this trick of gaining or losing electrons. So, don't think of the ion as the default cure for getting full electron shells. Atoms such as nitrogen and carbon generally do not use this strategy in stable compounds because it results in too much charge built up in too little space. They must therefore interact with other atoms through bonds in order to get full valence shells. This forms molecules. Although hydrogen has the option of abandoning its electron, that strategy is the exception not the rule. Most hydrogen atoms also participate in bonds as a result of the need to have a full valence shell. Bond and molecule formation will be the focus of the next unit.

Summary

In this section, we discussed the basic particles that are important in biology and how they come together to make atoms.

The basic particles are:
- Proton – positively charged mass found in the nucleus of atoms.
- Neutron – neutral mass found in the nucleus of atoms.
- Electron – negatively charged smaller mass that "orbits" around the nucleus in funky shaped clouds called orbitals.
- Photon – neutral particles that make up light.

Here are the basics we covered about atoms:
- Positively charged nucleus has protons and neutrons in it. Negatively charged electrons "orbit" in funky shaped probability clouds around the nucleus. Each orbital can hold two electrons.
- The number of protons in the nucleus is the atomic number and determines what kind of element the atom is.
- The nuclei of elements in the 90's and above tend to be unstable and break down by radiation, spitting pieces out of the nucleus to make a smaller nucleus, repeating the process until the nucleus is finally stable. The atomic number (equal to the number of protons in the nucleus) changes when this happens.
- Electron orbitals are divided into sets called shells. The first shell has only one orbital (an s orbital), shaped like a sphere. The first shell can hold only two electrons. The second shell has four orbitals of various shapes. The second shell holds up to eight electrons. The third shell has nine orbitals and can hold up to eighteen electrons, but most atoms are happy having eight in the outer shell. There are more than three shells, but the first three are the most relevant to the study of biology. The outermost shell is called the valence shell, and it's where chemistry between atoms takes place.
- Electrons like to be paired and atoms don't like to be lopsided, so atoms prefer to have their valence shell full of paired electrons. 2, 8, 8 makes atoms great. Hydrogen and helium have only the first shell, so they only need two electrons to have a full valence shell. All others have two or more electron shells, so they need at least eight electrons in their outer shell. Elements in the third row of the periodic table (and below) can have more than eight electrons in their outer shell, but most atoms are happy having just eight.
- Some atoms acquire full outer shells by gaining or losing an electron. These atoms have an imbalance in their proton to electron ratio. So, they are charged. We call them ions. The word ion can also refer to molecules that have imbalances in proton to electron ratio. We call these molecules polyatomic ions. Atoms that are unable to make full valence shells by becoming ions must interact with other atoms to become molecules.

In this section we cover:

1) Atomic bonding – nonpolar and polar

2) Basic molecular formation and shape

3) Electron lone pairs

4) Hydrogen bonding

5) Polar molecules

Unit 4

Nonliving Building-blocks of Life
Part 2

Atomic Bonding

Atoms tend to be most stable when their outer shells are full and each electron is paired. Once again, the third shell can actually hold up to eighteen, but most atoms are happy with eight. Atoms that have full valence shells due to their inherent number of electrons (the hermit-like helium column in the periodic table) are **inert**. Unfortunately, most of the 'unhappy' atoms in other parts of the periodic table cannot simply add or subtract spare electrons to fill their shells (like chlorine and sodium), because most of them cannot handle the charge that results. For example, an isolated carbon with eight electrons in its outer shell would have an overall −4 charge all packed within one atom's space. Electrons like pairing, but when too much negative charge piles up in one place the repulsion of all the charge starts to overpower the electron pairing, breaking up the bunch. Similarly, too much positive charge tends to attract electrons back to the positively charged atom. So, what is an atom to do when it wants more electrons but cannot do it on its own? The answer is to share. Atoms share electrons through bonds with other atoms that similarly want their valence shell plump-full of paired electrons. When atoms share electrons in this way, they are said to form a **bond** between them. This type of union is called a **covalent bond**, and the formation and breaking of such bonds are called **chemical reactions**. The existence of two or more atoms bound together through

27

covalent bonds makes up a **molecule**, a concept we will return to shortly. For the moment, a simple covalent bond would look like two atomic nuclei sharing electron clouds, shown here.

The more electrons that are shared between atoms, the stronger the overall bond will be. Covalent bonds found in living systems can have as many as six electrons shared between neighboring atoms, but two and four are

3D image of a hydrogen-hydrogen bond

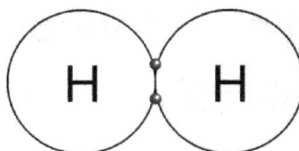

Simplified valence shell image of a hydrogen-hydrogen bond

H–H

The lazy chemist's shorthand stick verson of a hydrogen-hydrogen bond. Crude by affective.

the most common. Neighboring atoms that share two electrons (like the H-H in the image above) are said to have a **single bond**. Sharing four electrons is called a **double bond**. Sharing six electrons is called a **triple bond**. These are the strongest. Sharing more electrons than this is a phenomenon that only takes place in atoms not typically seen in living systems.

Molecular Oxygen

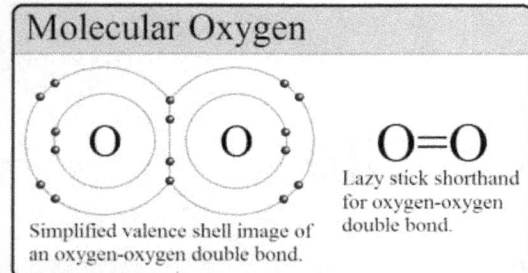

O=O

Lazy stick shorthand for oxygen-oxygen double bond.

Simplified valence shell image of an oxygen-oxygen double bond.

Molecular Nitrogen

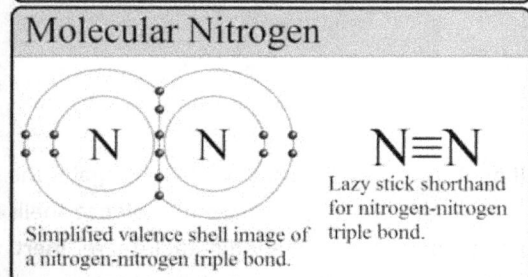

N≡N

Lazy stick shorthand for nitrogen-nitrogen triple bond.

Simplified valence shell image of a nitrogen-nitrogen triple bond.

When atoms share electrons, they don't necessarily share them fairly. Some atoms tend to hog the electrons more than others. When this happens, you get a bond that is said to be **polar**. In a polar bond, the shared electrons are pulled closer to one atom than to the other. This typically happens when the atoms that are sharing electrons are not of the same element, such as oxygen bonding with carbon. The elements in the top right part of the periodic table (excluding the weird hermits in the far-right "helium" column) are more selfish than the other elements, a characteristic called **electronegativity**. The more electronegative an element is, the more it will tend to try to hog electrons. Curiously, carbon and hydrogen have almost identical electronegativities, and so when these two elements get together in bonds, they share electrons equally. When electrons are shared equally, you get a **nonpolar** bond. All the bonds in the images above and to the left are nonpolar bonds, because they involve elements of the same type bonding with each other. For example, even though oxygen likes to hog electrons, when two oxygen atoms are bound together, they try to hog the electrons equally, cancelling the effect entirely.

Let's take a closer look at polar bonds. Polar bonds look like electron clouds that are not positioned evenly over the atoms: one atom will have more of the electron cloud over it, while the other will have less. Since electrons are negatively charged, the more-covered atom will have a sort-of negative charge (a "partial negative," represented by the funny little symbol δ^-), while the other less-

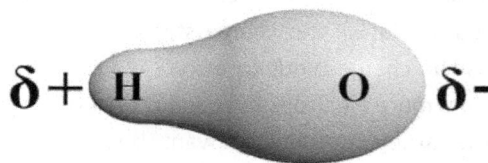

$\delta +$ H   O $\delta -$

covered atom will have a sort-of positive charge (a "partial positive," δ^+). Overall, the two-atom combination is still neutral, because there are an equal number of electrons to protons, but in this situation, the electrons are not evenly distributed, so the effect is that one end has a negatively charged pole, while the other end has a positively charged pole.

This last image is a hydrogen-oxygen bond – a *very* important polar bond in biology – which for some reason came out looking more like a bowling pin or perhaps one of those mutated peanuts. I hate those things, because the little end always has some gnarly looking nut runt in it. Anyway, the main thing to understand about polar bonds is that it causes neighboring polar molecules to get attracted. They tend to want to line up in various geometries. For example, H-Cl also looks like the mutated peanut image above, and in really *cold* samples of H-Cl,[7] they tend to line up as shown so that the positive poles attract the negative poles:

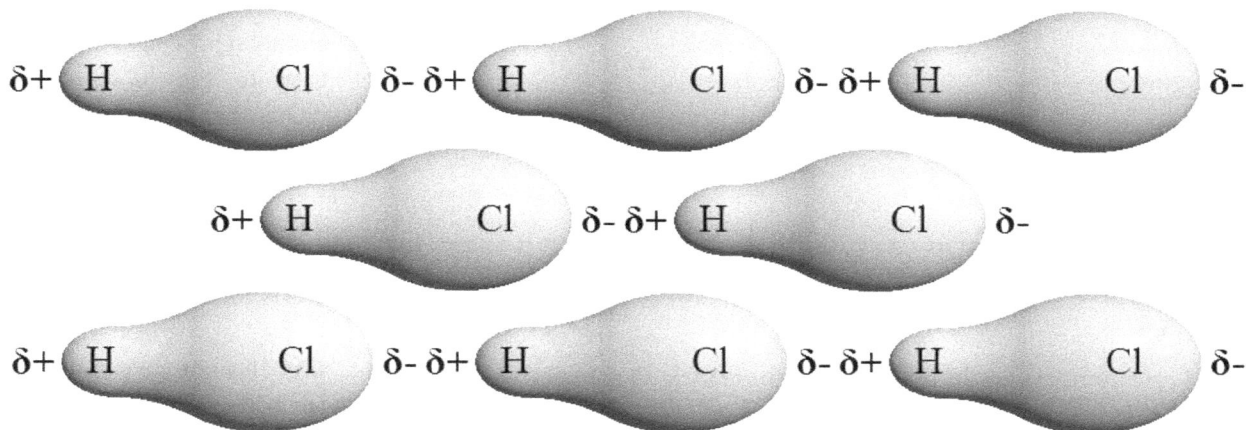

In the previous unit about atoms, we talked about ions. A simplified way of looking at ions is to consider them to be extreme examples of polar bonds. Instead of the electron clouds being off-center, the electron is ripped away from one atom entirely and hovers around the other atom instead, creating a positively-charged ion and a negative-charged ion. The two ions are then held closely to one another by the attraction of opposite charges.

The attractive force that holds ions together is called an **ionic bond**. This occurs, for example, when one atom only needs one electron to fill its valence shell with paired electrons, whereas the other can have a valence shell full of paired electrons if one electron is *removed* from its outermost shell. So, one atom gives up its electron and acquires a +1 charge, and the other gains the electron and subsequently has a -1 charge.

Extra electron

The Molecule

Molecules are collections of atoms that are covalently bonded with each other. The number of bonds that an atom will engage in is determined by the number of electrons it wants to pair up in its outer valence shell. Hydrogen has only one electron, but its valence shell has room for two, so it participates in simple, single bonds. Oxygen has

---

[7] Heat makes molecules move around a lot, so the effect wouldn't be very noticeable in HCl unless the sample is really cold. The point is still the same though.

room for two more electrons in its outer shell, so it engages in two bonds. This may come in the form of two separate single bonds, or one double bond. Carbon has four electrons in its outer shell, but would like to have four more, so it participates in four bonds. It might make these four bonds as four single bonds, or as two single bonds and one double bond, or as two double bonds, or one triple bond and one single bond. Nitrogen needs three electrons to fill its shell so it participates in three bonds, and as with oxygen and carbon, these bonds can be through any combination of single, double, and triple bonds. Here is an image of those bonds:

Hydrogen

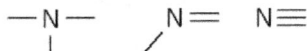

Oxygen

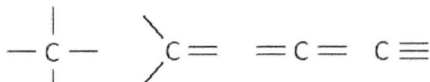

In this image you'll notice that some of the bonds are at weird angles. This goes back to the funky electron orbital shapes we talked about in the previous unit. That's why I made such a fuss about it. Take the nitrogen that participates in three single bonds for example. It forms a shape like a trigonal pyramid, with the nitrogen at the top. The carbons that participate in four single bonds form a shape called the **tetrahedron**, a four-sided geometric thingamajig with triangular sides. The carbon is deep in the center of the shape. While this gets a little too far afield for the present study, it should definitely be kept in mind that the true shape of the molecules differ from the way they are often drawn.

Nitrogen

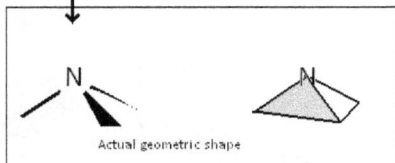

Carbon

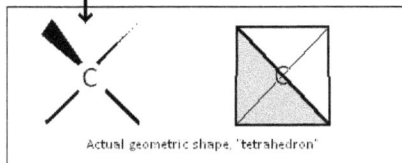

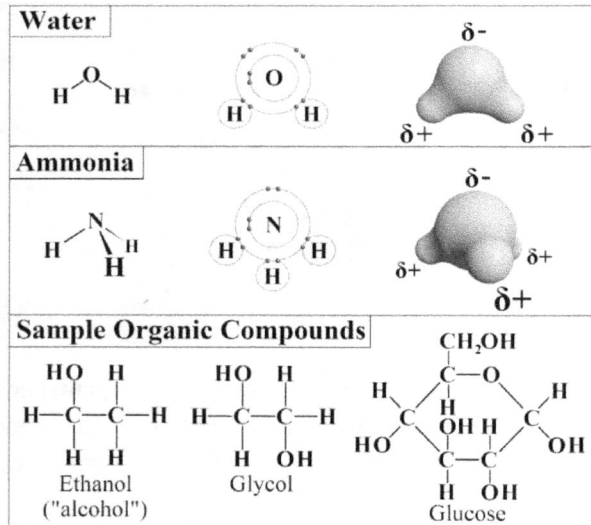

No matter the weird shapes, since carbon atoms can have up to four bonds, they can connect a lot of atoms together. Because of this, carbon is often found as the backbone for biological compounds. These molecules are called **organic molecules**, and are the basis for the incomprehensible study of organic chemistry. They exist in Nature because 1) they can and 2) to terrorize chemistry students the world-over. Organic molecules, along with water (H_2O), ammonia (NH_3), and a few others form the primary ingredients in the processes of life. The image to the left is of water, ammonia, and a few sample organic molecules. Don't worry about the details of the organic molecules shown. We will encounter them again as we need them. They are here just to give you an idea of what they would look like and how they are drawn.

30

Lone Pairs

Electrons within the outer shell of atoms that are already paired do not participate in bonding and instead gather in bulges that stick out from the atom (not shown in previous images for simplicity). These are called **lone pairs**, and are very characteristic of atoms like oxygen and nitrogen.

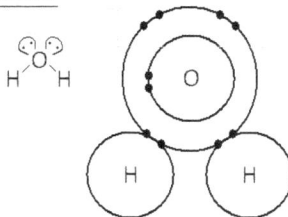
Water

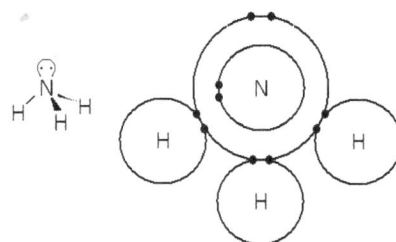
Ammonia

Note that in the abbreviated Lewis structures on the previous page, only the valence electrons are shown. The fact that lone pairs exist in atoms like oxygen and nitrogen is very important. The main consequence is that they represent negatively charged clouds poking out the side of atoms. This means that if any other positively charged thing comes nearby, it will tend to be attracted to the lone pair. One VERY important example of this is the hydrogen bond.

The Hydrogen Bond

As we saw earlier, hydrogen-oxygen bonds are very polar because the electrons in the bond are hogged by the oxygen. As a result, hydrogen tends to have a partial positive charge while oxygen gets a partial negative charge. The hydrogen's partial positive charge makes it tend to stick to negatively charged objects in the immediate environment. Further, the oxygen has two rabbit-ear-like lone pairs bulging out from its side, which are both very negatively charged. These act to attract other hydrogen from nearby hydrogen-oxygen bonds, creating a network of sticky molecules. Hydrogen-nitrogen bonds behave in the same way. As we will see in future sections, this unique "bond" gives water and many other biological molecules very unique characteristics that are crucial for the processes of life.

Polar Molecules

When we discussed chemical bonds, we brought up the topic of polar covalent bonds, an example of which you just saw in the water molecule. The concept of polarity can be extended to molecules at large. If molecules have polar bonds that cause the overall charge distribution in the molecule to be unevenly spread out, the molecule itself is said to be polarized.

One important aspect of this topic is that polar molecules do not mix well with non-polar molecules. Why is this so? Well, polar molecules tend to stick to each other because their partially charged ends attract each other. In the attempt to stick to each

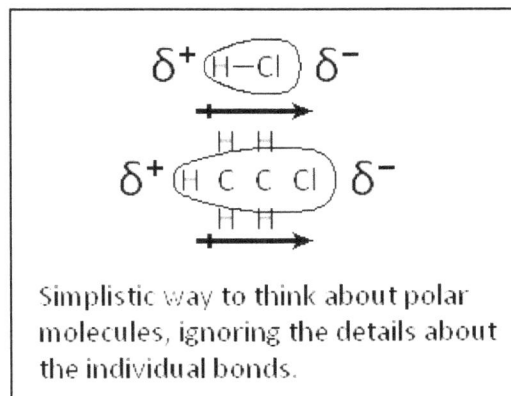
Simplistic way to think about polar molecules, ignoring the details about the individual bonds.

other, they push nonpolar molecules out of the way. So, it's not that polar molecules dislike nonpolar molecules, they just like polar molecules *much* more.

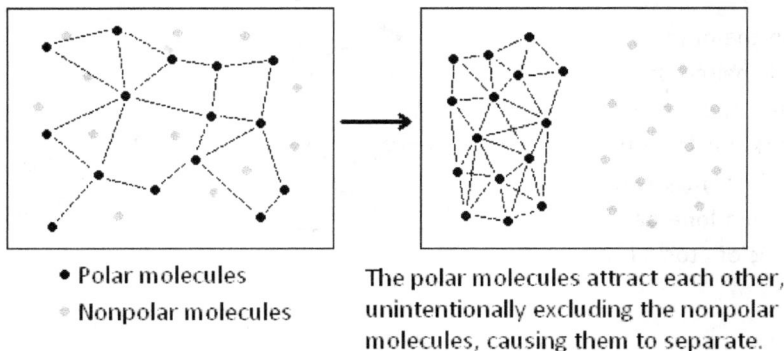

- Polar molecules
- Nonpolar molecules

The polar molecules attract each other, unintentionally excluding the nonpolar molecules, causing them to separate.

Since water is polar it mixes well with other polar molecules. We call molecules that mix well with water **hydrophilic** ("water loving"). Compounds that are non-polar, like oil and wax, do not mix well in water. These we call **hydrophobic** ("water fearing"). This concept will be important when we talk about cell membranes in the near future.

With that, we bring our discussion of the building blocks of life to a close. In the next section we will look closer at water and the qualities that make it important for the existence of life.

Summary

In this section, we covered the basics of atomic bonding.

- Atoms share electrons in an effort to fill their valence shell with paired electrons. The bond that forms is called a covalent bond. The collection of atoms participating in covalent bonds is called a molecule.
- Since electrons like pairing, bonds almost always involve electron pairs. A single bond has two paired electrons in it. A double bond has two sets of paired electrons (four electrons total). A triple bond has three pairs of electrons in it (six electrons total).
- Elements in the upper-right part of the periodic table, (excluding the "helium" column, noble gases), are more electronegative that the rest. These atoms hog electron, so bonds they make with atoms in the other parts of the periodic table are polar.
- Polar bonds involve unevenly shared electrons in a covalent bond, and cause the participating atoms to have partial electronic charges, δ^- and δ^+. These partial charges attract other partial and/or full charges from the environment, affecting how nearby molecules behave.
- An ionic bond is an extreme form of a polar bond, where the electron from one atom is ripped away and goes to the other atom entirely. This makes an ionic bond.
- Carbon can participate in four bonds total. It is commonly strung together to make the backbone of big molecules.
- Nitrogen can participate in three bonds total.
- Oxygen can participate in two bonds total.

- Hydrogen can only participate in one bond.
- Covalent bonds are usually at strange angles, so that the molecules are difficult to draw on paper. These strange bond angles are due to quantum mechanics. The bond angles give molecules specific shapes.
- Electrons that are paired prior to covalent bonding form "lone pairs," sticking out of the side of the atom. These are negatively charged balloon that attract partial and full positive charges nearby.
- Hydrogen bonds involve the partial positive charge of the hydrogen bonded to either nitrogen or oxygen (H-N, H-O) with the lone pairs of oxygen or nitrogen. It is a very important bond in biology.
- Polar molecules have uneven distribution of electron clouds over the entire collections of atoms. This causes the entire atom to have regions of partial negative and partial positive charge. Nonpolar molecules have electron clouds that are evenly spread over the entire collection. So, they do not have partial charges. Polar molecules attract each other, excluding nonpolar molecules. This is why they do not mix. Water is polar. Oils are nonpolar.

Unit 5

Water and Life

Water's Hydrogen Bonds

In the previous chapter, we discussed the nature of lone pair electrons and how they bulge out of the sides of atoms. Each lone pair forms a cloud of negative charge to which positive things nearby are attracted. Let's quickly rehash the example of water. The image to the right is a water molecule.

The image shows the two lone pairs as well as the two polar bonds. Taken together, the entire structure is tetrahedral, just like carbon when it has four single bonds. Each of the hydrogen atoms in water has its electron drawn toward the oxygen, so that H carries a partial positive charge. These positive charges get attracted to nearby lone pairs from other water molecules to form

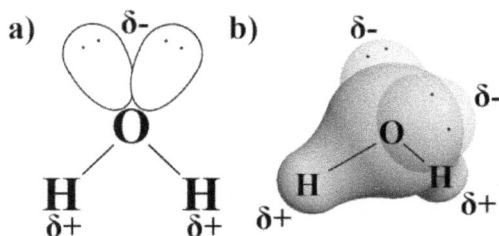

a) Simple stick representation of water with lone electron pairs. b) 3-D water molecule with lone pairs shown (omitted in previous 3-D image of water). Notice that it is similar to carbon's tetrahedral bonding structure.

hydrogen bonds. Although these are called "bonds," they can be broken and reformed much more easily than ionic or covalent bonds. At room temperature, for example, water molecules are moving around sufficiently quickly that hydrogen bonds are constantly being formed and broken. When the temperature drops, so does the rate at which the molecules tumble. The slower movement allows the water molecules to begin to line up based on their hydrogen bonding. The water slowly expands in volume and gradually freezes. When this happens, the water (now ice) becomes less dense and floats to the surface.

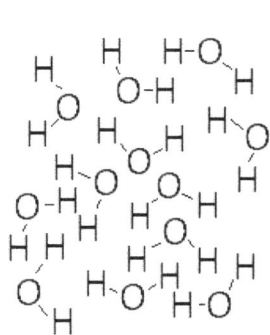

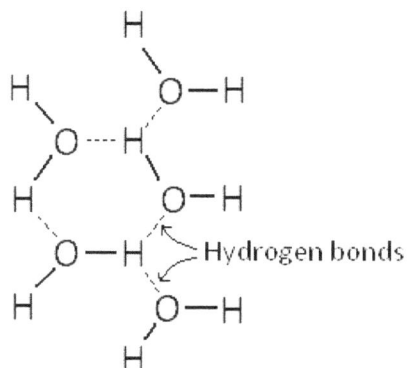

Liquid water   Solid water (ice)

Since ice floats, it forms a thin barrier at the top of lakes and streams, helping to trap in the remaining warmth underneath. This slows the freezing of the water at the bottom and prevents fish and other organisms from freezing during the winter. Otherwise, the population of fish and other critters in the inland water systems would be very small (if not zero). Since a lot of land ecology depends on fish for nutrients, life as we know it would probably be very different if water froze from the bottom up. Keep in mind that fish have been a major food source for mankind throughout its history, so without it who knows where we would be?

Water is a Good Solvent

Water is an excellent **solvent**. Solvents are generally liquids that dissolve other molecules. So, for example, if you take a small spoonful of table salt (Na^+Cl^-) and pour it into a cup of water, you will begin to notice the salt dissolve into the water. This is because water is polar, and the sodium chloride is made of positive and negative charges. These charges interact with the partial positive and negative charges of the water molecules. Each ion gets ripped away and carried off by a number of water molecules, until the ions are so spread out in solution that you can't see them anymore.

Water also dissolves polar molecules by the same mechanism. Since polar molecules have positive and negative ends, water is attracted to these ends and pulls these molecules away from the other molecules. Sugar, for example, is a polar compound (not an ionic compound like salt), but it dissolves in water like salt does.

The fact that water is such a good solvent is important because it allows many of the molecules that are involved in the reactions of life to be dissolved so they can float around and find each other for reaction to do

Salt dissolved in liquid water

important biological things. Otherwise, they would just sit there, like the salt in the spoon, or like my housemate on the couch. Come to think of it, he would also do well to be introduced to the concept of water.

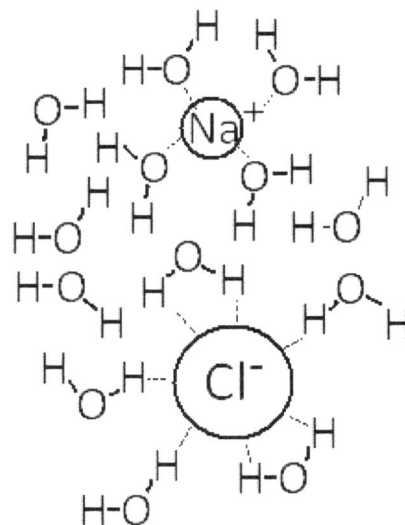

Water as a Temperature Moderator

You may have noticed that on incredibly hot days, rocks and other solids tend to heat up rather quickly. Water, on the other hand, heats up much slower. Why is this? Temperature is basically a rough estimate of how quickly the molecules in a material are moving around and/or vibrating. The hotter it is the more vigorously they move about and/or vibrate. In liquids, this means they tumble and roll passed each other, continuing on through the crowd of similarly moving molecules. In solids, it means that the atoms are vibrating energetically, but relatively locked in place. Water heats up slowly because the energy that it takes to get them moving, rolling and tumbling is high. This is because it takes energy to disrupt all those hydrogen bonds that are in water. The net result is that water tends to maintain relatively stable temperatures while the surrounding land gets hot or cold quickly. This is one reason that people enjoy living near large bodies of water; the slow change of water temperature helps keep the surrounding land cooler in the summer and warmer in the winter versus areas with little water. Deserts, for example, are quite low in water, and they are often extremely hot during the day, but quite cold at night. Living organisms have a lot of water on the inside, and this helps to keep their bodies in relatively constant temperature despite fluctuations in the temperatures outside.

Summary

In this section, we covered the importance of water in biological systems.

- Its hydrogen bonding makes it expand when frozen, forming a layer that protects the underwater communities from freezing in the winter.
- Hydrogen bonding makes water a very good solvent. It is easy to dissolve many types of salts and polar molecules in water. A lot of biological molecules are polar, so water helps to keep them floating around, maximizing their ability to interact with other biological molecules. This maximization speeds the rate of biological reactions, which would otherwise be too slow to support life.
- Water is an excellent temperature mediator and helps to keep our body temperatures relatively constant despite fluctuations in outside temperature.

Unit 6

Acids and Bases

As we have seen, water is made of one oxygen atom and two hydrogen atoms. The bonds between these atoms are very strong, but every once in a rare while, a hydrogen snaps off. The stray hydrogen, (which is literally just a proton, H^+), floats away, but leaves its electron behind with the ^-OH. This remaining part is called **hydroxide** (^-OH). Here is a simplified view:

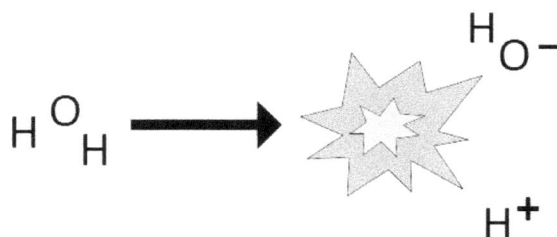

Water isn't the only molecule that can do this, however. In fact, a *lot* of chemicals are capable of breaking into pieces, releasing H^+. Molecules that release H^+ are called **acids**. HCl is an excellent example. When you put it in water, essentially every molecule breaks in half (whereas only a tiny fraction of water does).

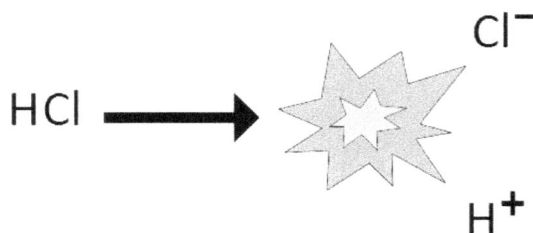

Since HCl breaks apart so easily and releases lots of H^+, it is a very strong acid.

Conversely, any molecule that breaks apart to release OH^- is called a **base**. NaOH is an excellent example. Like HCl, when you put it into water, it practically all breaks apart (dissociates) into ions:

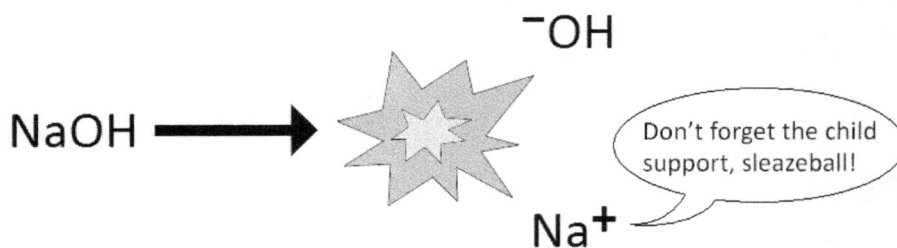

Since NaOH releases a lot of OH⁻, we call it a very strong base.

If a solution of water is pure, there will be equal numbers of H⁺ and OH⁻. However, what if we put something else in the water that easily gives up H⁺? Suppose we put HCl into the water. When this happens, the concentration of H⁺ far outweighs the concentration of OH⁻, not simply because H⁺ is being added to the solution, but also because the H⁺ rejoins with spare OH⁻ to recreate water. Water has a peculiar nature, however, for if you know the concentration of H⁺, you can figure out the OH⁻ concentration by using the following equation:

$$[H^+][OH^-] = 10^{-14} \tag{1}$$

The brackets "[ ]" around a chemical name means the "concentration of" and is typically measured in units of molarity (M).[8] So, from this equation, we can gather that as the concentration of H⁺ increases, the concentration of OH⁻ drops, and vice versa. When there is more H⁺ than OH⁻, we say the solution is acidic. When there is more OH⁻ than H⁺, we say the solution is basic or **alkaline**.

The way we measure the acidity or basicity of a solution is by the **pH** scale. That is determined by knowing the concentration of H⁺ and using the following formula:

$$pH = -\log[H^+] \tag{2}$$

This formula demonstrates that solutions that have high H⁺ concentrations have low pH values. This is sometimes tricky to understand at first, so I've included a series of practice problems at the end of this section. It includes the answers as well. There is a shortcut you can use to approximate the pH of a system. If you are given the H⁺ concentration in scientific notation (e.g. 1.234×10^{-5}), the pH will be approximately equal to the opposite value of the exponent (the little number above the 10). There is a little bit of error using this shortcut, but the actual pH will be between that guess value and the value one unit lower. So, here are some quick examples:

---

[8] "M" stands for "molar," which is short for "moles per liter." Moles are a unit of quantity. One mole equals 6.22×10^{23} molecules. So 1 M, would be 6.22×10^{23} molecules in 1 liter of solution.

Example 1) [H⁺] 5.4 x 10⁻⁵ M   Guess: pH 4-5   Actual pH = 4.27

Example 2) [H⁺] 3.8 x 10⁻¹² M   Guess: pH 11-12   Actual pH = 11.42

Example 3) [H⁺] 1.3 x 10⁻³ M   Guess: pH 2-3   Actual pH = 2.89

Example 4) [H⁺] 0.3 M   Guess: pH 0-1   Actual pH = 0.52

(Same as 3.0 x 10⁻¹)

Be warned though! The only way this trick works is if the concentration is listed in scientific notation, specifically as $X.YZ \times 10^{-W}$, where X, Y, Z, and W are just digits. The units of measure *also have to be in M*. If there is something *other* than M after the concentration number, you have to do some mathemagic on the number to convert it. That is the focus of work done in chemistry classes. The odds are small you would need to do this in a typical introductory course in biology. At any rate, with a little practice you can get pretty good at eyeballing what the pH will be using this shortcut.

Anyway, **acids have low pH values** and **bases have high pH values**. Pure water has a pH of 7, and this value is considered **neutral**. Theoretically, the most acidic that a solution in water can get is 0. Conversely, the most basic that a solution in water can get is fourteen. The table below gives some examples of chemicals and their pH values.

pH	Example	
14	Drain cleaner	Most basic (alkaline)
13	Bleach	
12	Soapy water	
11	Ammonia	
10	Milk of Magnesia	
9	Baking Soda	
8	Sea Water	
7	Pure water	
6	Milk	
5	Vinegar water	
4	Tomato juice and beer	
3	Orange juice and grapefruit	
2	Soft drink (soda, pop)	
1	Stomach acid	
0	Battery acid	Most acidic

Humans and most other animals have an internal pH of around 7.3 to 7.4, but the internal pH of other organisms can vary greatly depending on the species. This value *cannot* change very much without having disastrous effects on the delicate living system. On the other hand, living organisms are choked-full of all kinds of reactions that eject or soak up H⁺. If these went unchecked, the pH in a living organism would fluctuate wildly (usually dropping, such as in muscle tissue during exercise), resulting in the untimely demise of the creature from the inside out. To counteract this, living systems have chemicals known as **buffers**. Buffers are just molecules and ions that act to soak up extra H⁺ and OH⁻ floating around, keeping the H⁺ concentration stable. Think of buffers as pH police, catching any stray H⁺ that tries to get away. There are many important buffers in living systems, including HCO_3^- (bicarbonate), HPO_4^{2-} (mono-hydrogen phosphate), and many proteins and amino acids. These things are generally covered more in detail during the second semester of biology.

Practice pH Problems

1) What is the pH of a solution with a $[H+] = 4.67 \times 10^{-5}$ M?

Since the little number above the ten is "-5," we guess the pH will be between 4 and 5. Here is the calculation:

pH = -log(4.67×10^{-5}) = 4.33

In a graphing calculator:

1) (-) button
2) LOG button
3) 4.67
4) EE button (you may have to hit the 2ND button first)
5) (-) button then 5
4) ENTER button

In a simpler calculator, you'll have to experiment around to figure out how to get it to work. You often have to do the steps backwards though.

2) What is the pH of a solution with $[H+] = 2.15 \times 10^{-10}$ M?

Guess: pH 9-10.

pH = -log(2.15×10^{-10}) = 9.67

3) pH for $[H+] = 1.334 \times 10^{-2}$ M?

Guess: pH 1-2.

pH = -log(1.334×10^{-2}) = 1.87

4) pH for $[H+] = 0.003244$ M?

Guess: ? (If you convert this scientific notation, you'll get 3.244×10^{-3}, so we would guess the pH to be 2 to 3.)

The calculator doesn't care if it's scientific notation, so we type it in as is.

pH = -log(.003244) = 2.49

5) pH for $[H+] = 9.88 \times 10^{-11}$ M?

Guess: 10 to 11

pH = -log(9.88×10^{-11}) = 10.01

Summary

In this section, we covered the basic concepts of acids and bases.

- Acids produce H^+ in water (or take away OH^-).
- Bases give OH^- in water (or take away H^+).
- Acidic solutions have high H^+ concentrations and low OH^- concentrations.
- Basic solutions have high OH^- concentrations and low H^+ concentrations.
- Basic solutions are also called alkaline.
- The H^+ and OH^- concentrations in water are related by the equation $[H^+][OH^-] = 10^{-14}$. So, if you know the concentration of one, you can find the concentration of the other.
- The pH of a solution is given by the equation: $pH = -\log([H^+])$. So, if you know the H^+ concentration, you can find the pH.
- Acidic solutions have low pH (< 7).
- Basic solutions have high pH (>7).
- Fluctuating pH values are dangerous to living organisms, so they have buffers inside them that keep the H^+ concentration stable.

In this section we cover:
In this section we cover:
1) Functional groups in organic chemistry
2) Basics of organic chemistry stick models
3) The basic concept of metabolism
4) Monomers, polymers, dehydration synthesis, and hydrolysis

Unit 7

Some Simple Organic Chemistry

So far we've talked about particles, the atoms they make, and how the resulting atoms come together to form molecules. Then we talked about water and looked at how it behaves in the context of acids and bases. Now let's return to our discussion of molecules, specifically those that have carbon in them (i.e., **organic molecules**). Carbon can have up to four single bonds, and in this situation, the carbon's bonds are in a four-sided pyramid-like shape called a **tetrahedron**. Carbon is especially good at connecting long chains of carbon-based molecules together. Ultimately, these molecules can have some complex overall shapes and structures if enough carbons are strung together. For example, here are a few of the shapes you'll encounter in a jaunt through the chemicals of life (see graphic, following page).

As you can see from the image on the next page, these molecules can get *incredibly* complex. Chemists are wild and crazy people though, so instead of drawing each "C" and "H" atom, they represent organic molecules as knobby little stick thingies. It's how they roll. In simplified notation, the carbons are the intersections of lines, or corners if you will. An end of a line is also a carbon. All carbons have four bonds, and so any intersection that is not met by four lines is assumed to have enough hydrogen attached at that point to make up the remaining bonds. For example, an intersection with two lines is assumed to have two hydrogens attached. An intersection with three lines is assumed to have one hydrogen attached.

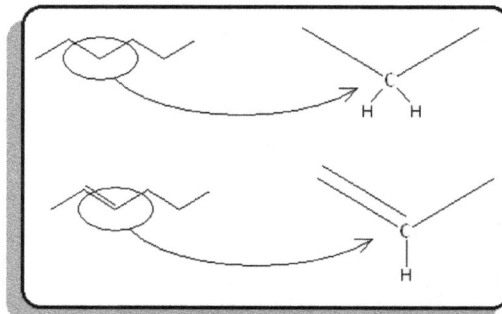

42

Cholesterol ("Lipids")

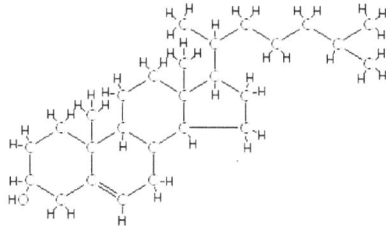 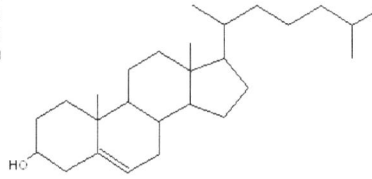

All details shown   Simple shorthand stick notation

A Fatty Acid ("Lipids")

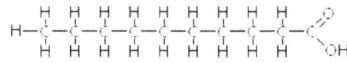 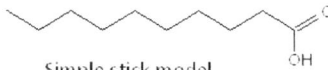

Detailed molecule   Simple stick model

Glucose ("Carbohydrates")

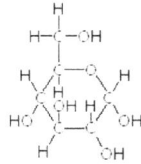 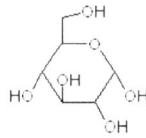

Detailed molecule   Simplified stick model

A Nucleotide, Adenosine ("Nucleic acids")

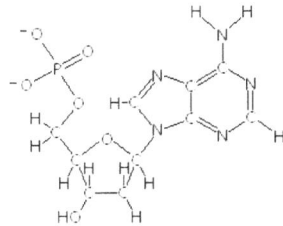 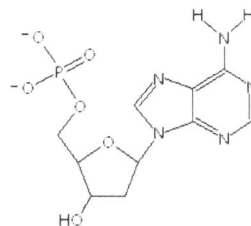

All details shown   Simple stick model

Some Amino Acids ("Proteins")

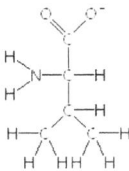 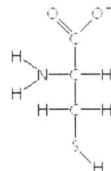 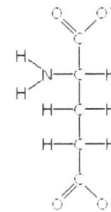

Valine

Glycine

Cysteine

Glutamic acid

Stick models or not, carbon is the backbone of life, and from this backbone are attached an assortment of other atoms and groups of atoms that we call **functional groups**. Functional groups are basically where *most* of the chemical action takes place, so it's important to have a fundamental understanding of what they are. Examples of the essential biological functional groups are listed in the figure below.

That's all well and good, but how do these molecules get made inside living organisms such as you and me? Do we literally take individual carbon, hydrogen, oxygen and nitrogen atoms from the environment and put molecules together one atom at a time? No. I can't remember the last time I found an individual carbon atom sitting all by itself waiting for someone to pick it up. For all practical purposes, carbon is found in the environment *already* bound into all manner of molecules. So, if a living system needs carbon, it will have to process away all the other atoms that are tagging along. That's not to say that we can eat just *any* carbon-based molecule, strip away the junk and get the carbon from it. Sometimes a molecule can have a structure that makes it poisonous. So, we stick to the things our bodies know how to process: **sugars, fats, proteins**, etc. We have the chemical machinery to break these down.

Hydroxyl (-OH)	Amino ($-NH_3$)
Carboxyl (-COOH)	Phosphate ($-PO_4^{2-}$)
Methane ($-CH_3$)	Sulfide (-SH)

And break them down we do! The food we take in gets split into smaller and smaller pieces, releasing both the atoms and the energy locked in the bonds, until eventually all the carbon that was there gets turned into CO_2, and the hydrogen and oxygen gets turned into H_2O. The nitrogen gets packaged into ammonia (NH_3). This is the process of **metabolism**, which we'll talk about in later sections. Here's an overview image to wet your whistle.

Complex oganic molecules with carbon, hydrogen, oxygen and nitrogen. → **Metabolism** → CO_2 H_2O NH_3

→ ENERGY!

Often, big complex molecules are broken down into a discreet set of specific molecules, such as **glucose, amino acids, fatty acids**, and so on. These smaller molecules, which we are very adept at using, get strung together one at a time in an organized fashion to make bigger molecules. It's like the bricks of a house; you put them together to make a grander thing. In the jargon of stuffy scientists, these molecular 'bricks' are called **monomers**, and the bigger molecules they make are called **polymers**. In other words, polymers are just long chains of monomers put together.

Monomers

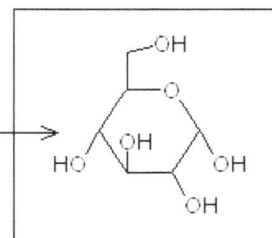

The molecules to the left are incredibly lazy symbolic representations of glucose.

Polymer

The single most important type of reaction for stringing monomers together is called **dehydration synthesis**. You are probably familiar with the idea of dehydration from the common experience of exercise. It means, in a manner of speaking, "removing water," only here it does not occur because organic molecules go for a 5K run. Dehydration synthesis occurs when two molecules that contain hydroxyl groups, (or one hydroxyl and one amino group), are paired together and water is ejected, leaving the two parent molecules joined where the hydroxyl groups used to be. Here is a picture to help illustrate what happens:

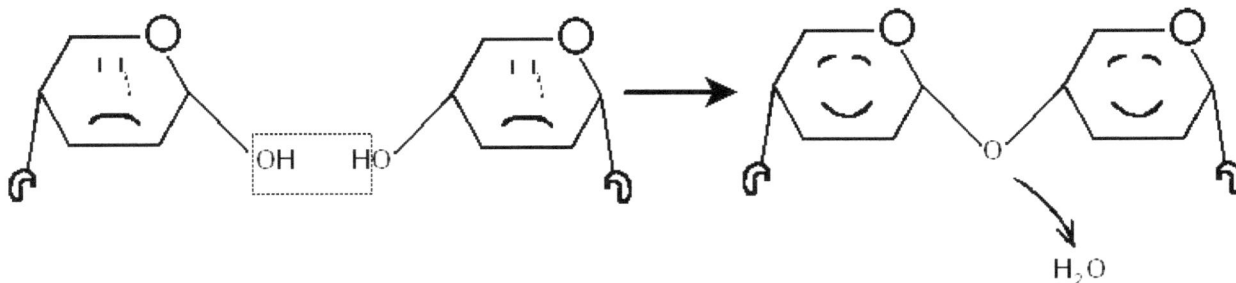

H_2O

This reaction is the basis for almost all biological polymer synthesis, and it is well worth knowing. The way these polymers are broken down is in the opposite fashion, a reaction called **hydrolysis**, which means roughly "to break with water." In that case, it looks basically the same, only going in the reverse direction.

So, now that we know this little bit of organic chemistry, we are set to talk about the four main biological molecule types: **carbohydrates**, **lipids**, **proteins** and **nucleic acids**. That will be the focus of the next several sections.

Summary

In this section, we covered the basics of organic chemistry.

- Organic molecules are carbon-based molecules.
- Carbon is the backbone of biological molecules, since it can connect up to four atoms at a time.
- When chemists draw organic molecules, they represent the carbons as the corners in a stick model and leave out the hydrogen atoms. The number of hydrogen atoms present is four minus the number of lines attached to the corner.
- Functional groups are specific collections of atoms that typically appear attached to organic molecules. They are usually where the chemistry takes place.
- Common functional groups in biology are –OH (hydroxyl), -NH$_2$ (amino), -COOH (carboxyl), PO$_4^{2-}$ (phosphate), -CH$_3$ (methyl), and –SH (sulfide).
- Carbon, oxygen, hydrogen, nitrogen and other important elements come already attached to other atoms in the environment. In order for living systems to use them, they have to be able to deal with the other stuff that is attached. Unfortunately, no living system can process all the possible molecules that are out there, and many molecules can be poisonous.
- Living systems extract energy from organic molecules using metabolism. CO_2, H_2O and NH_3 are typical products of that process.
- Certain molecules found as monomers (or made into monomers from other molecules) are strung together to make chains of monomers called polymers. This process occurs through dehydration synthesis, in which water is removed each time two monomers are connected to each other.
- Hydrolysis is basically the reverse of dehydration synthesis, in which water is added to break two monomers apart from each other.

Unit 8

Carbohydrates

Monosaccharides

Carbohydrates (a word that means "carbon and water") are biological molecules that are comprised of repeats of carbon, hydrogen and oxygen in a ratio of 1:2:1, or in other words, multiples of H-C-OH. The molecules that result are called **monosaccharides**, meaning "single sugars" in Greek. Monosaccharides typically have three to six H-C-OH repeats in them. Most monosaccharides are used in simple energy reactions, while the larger ones, (those with five or six H-C-OH repeats), can also be strung together into large polymers to serve the roles of energy storage, structural supports, or even information storage (such as in DNA). Of the monosaccharides, the 6-carbon molecules are the most numerous and important, so we'll focus on those during this unit. Before we do, however, let's have a brief look at a few of the smaller molecules to get our feet wet on the topic.

This image shows three molecules, each produced by H-C-OH repeats. It may not look like repeats of H-C-OH since the carbons at the top are double-bonded to oxygen, but if you count the carbon, hydrogen and oxygen atoms, the ratio is indeed 1:2:1. Glyceraldehyde is an intermediate in energy metabolism. We'll see it again in future units. Erythrose has a number of minor roles in living systems, but they are outside the scope of this book. Ribose is used to make **DNA**. We'll see it again too.

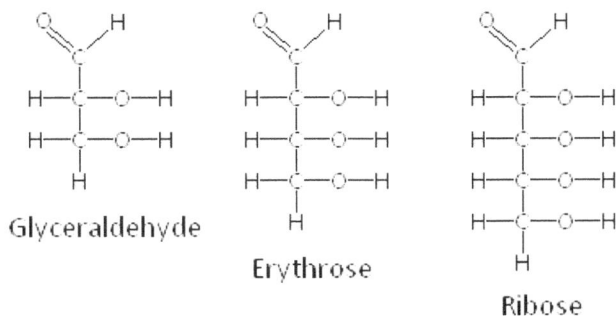

Glyceraldehyde

Erythrose

Ribose

Having touched upon the general structure of sugars, let's zoom in on a few of the specifics of the 6-carbon monosaccharides. These molecules can be ring-shaped or straight, switching back and forth spontaneously in living

47

systems. Most of the time, they will be in the ring-shape, so I will draw them in that form for the remainder of this book. Here are some examples of several 6-carbon monosaccharides:

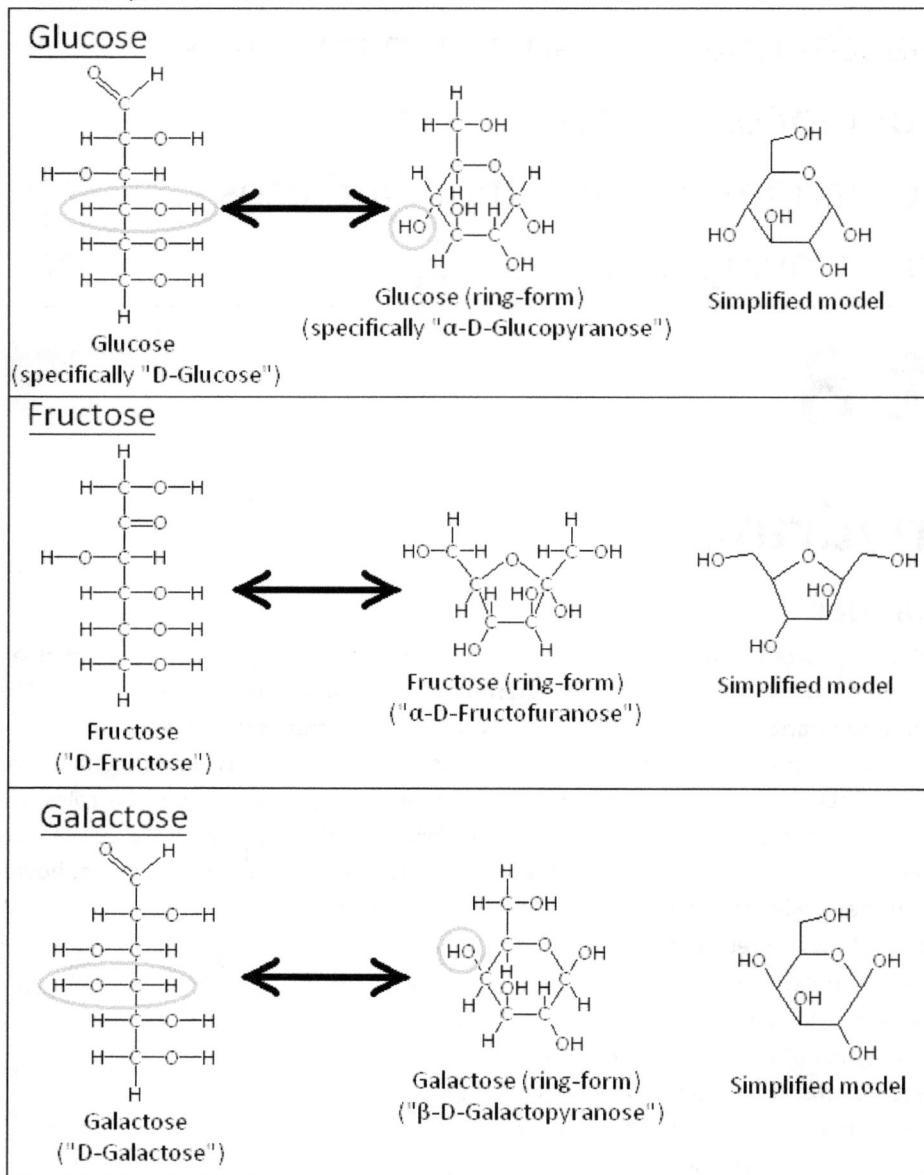

Glucose

Glucose
(specifically "D-Glucose")

Glucose (ring-form)
(specifically "α-D-Glucopyranose")

Simplified model

Fructose

Fructose
("D-Fructose")

Fructose (ring-form)
("α-D-Fructofuranose")

Simplified model

Galactose

Galactose
("D-Galactose")

Galactose (ring-form)
("β-D-Galactopyranose")

Simplified model

Notice in this figure that simple sugars can differ from one another in the orientation of the side groups. For example, when you switch the direction of the third H-C-OH in glucose from right to left, it changes the molecule to galactose. It may not seem like it should matter, but this subtle switch gives the molecule completely different chemistry inside the living system. Molecules that have the same chemical formula but different orientation of side groups are called **isomers**. So, glucose and galactose are isomers of each other. There are many other possible isomers within the group of 6-carbon monosaccharides,[9] but we won't delve too far into the topic except to say that some types of isomers can be divided into two groups: "D" and "L."[10] The D's are mirror-images of the L's. For example, what we have been calling "glucose" is really "D-glucose," and its mirror image is "L-glucose."

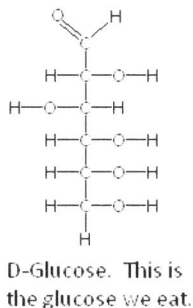

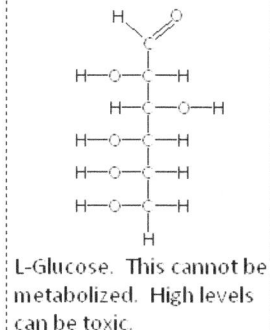

D-Glucose. This is the glucose we eat.

L-Glucose. This cannot be metabolized. High levels can be toxic.

The two molecules certainly look like they could be the same thing, but they aren't! If you were to make a physical model of these two chemicals (an exercise you would do in a course like organic chemistry) you would see that it is impossible to superimpose them. They are definitely not the same. D-Galactose has a mirror image too: L-galactose. A picture of it would be displayed in a similar manner with a little mirror-like reflection showing a molecule that has oppositely positioned side groups.

So far, the six-carbon molecules that we've looked at are only a few isomers out of the total. We've barely scraped the surface! It is very easy to get bogged down by the details and lose our way, however. The main point to understand is that there are many ways to put together H-C-OH repeats, leading to many different possible molecular structures. Each one is chemically unique. The more of the H-C-OH repeats you put together, the greater the number of possible isomers that can be made. The existence of these isomers makes the carbohydrate class of biomolecules a very detailed class. Most basic biology courses do not go in depth in the discussion of monosaccharide isomers, however, as this topic is more properly pursued in courses on organic chemistry and biochemistry. If these courses are in your future, you can put off the torture until that time. If you do not intend to take those courses, you can put off the torture indefinitely. Either way, it is not necessary to know all the isomers in order to understand general biology. Presently, just be aware that the isomers are there, and it creates a *large* number of possible sugar molecules for living systems to use. That said, we're now going to see how these six-carbon sugars get strung together to make polymers called **polysaccharides**.

Disaccharides

---

[9] It's a bit sketchy as to how many isomers actually exist for six-carbon monosaccharides. Theoretically, it should be 32. However, the number of naturally occurring isomers is probably somewhere between twelve and 24. Once in a while, a strange bacteria, plant, or critter will be found to use an unusual monosaccharide isomer, but by and large, most organisms seem to use the same basic set of twelve.

[10] Most of the natural monosaccharide isomers are in the D group. D is short for "dextro" and L is short for "levo." Dextro monosaccharides twist polarized light to the right, while levo monosaccharides twist polarized light to the left. If this sounds Greek to you, don't sweat it. It's way outside the scope of general biology. You learn more about polarized light in courses on physics.

The smallest polysaccharides that can be made from simple sugars are the **disaccharides**, which are just two monosaccharides linked together. These are made with dehydration synthesis. Some examples are shown here.

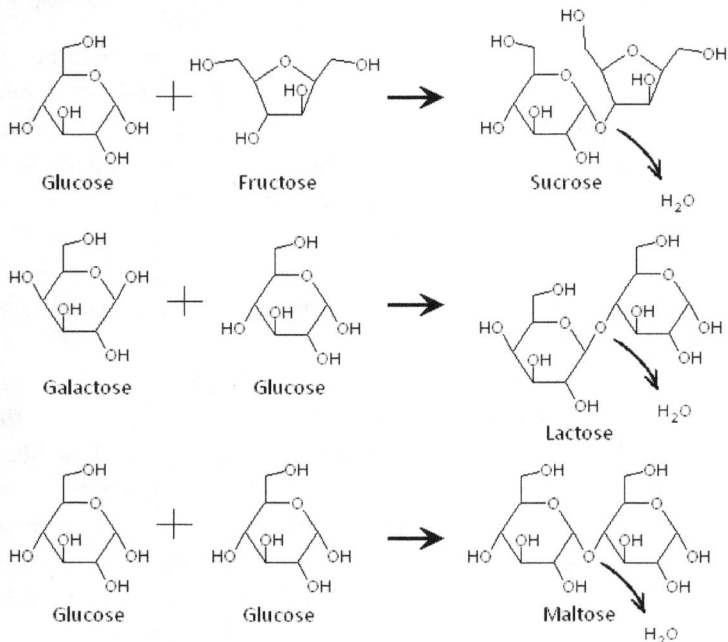

Glucose + Fructose → Sucrose + H_2O

Galactose + Glucose → Lactose + H_2O

Glucose + Glucose → Maltose + H_2O

Like many monosaccharides, most disaccharides are also used as energy sources, and many plants use them for longer-term energy storage. Lactose is found in breast milk and is an important source of energy for young developing mammals. Sucrose is the common table sugar. Maltose is obtained from grain, such as barley, and is used to make malt beverages. Animals break these disaccharides down into their individual monosaccharides by hydrolysis at the -O- bond between the two monomers.

Oligosaccharides

As we string more monosaccharides together, we arrive at polysaccharides known as oligosaccharides, or "many sugars." These have a variety of uses in Nature, from energy storage to forming the basic structure of many antibiotics, to being the source of alcohol produced in some beverages.[11]

Long Chain Polysaccharides

Long chain polysaccharides have two general uses in living systems. Some polysaccharides serve as long-term fuel sources (i.e., starch in plants and glycogen in animals). Other polysaccharides, such as cellulose and chitin serve as structural molecules. Cellulose is made in plants and gives their stems rigid strength. Chitin, a polysaccharide that also has nitrogen groups in it, is made by insects and crabs to form their tough outer skeletons (**exoskeletons**). Here is an image of what glycogen,[12] cellulose and chitin look like (see graphic, next page).

---

[11] The main component of alcoholic beverages is ethanol, CH_3CH_2OH, known commonly as "alcohol." It is not a monosaccharide, but is made from monosaccharides by microscopic organisms.
[12] Starch looks very similar to glycogen.

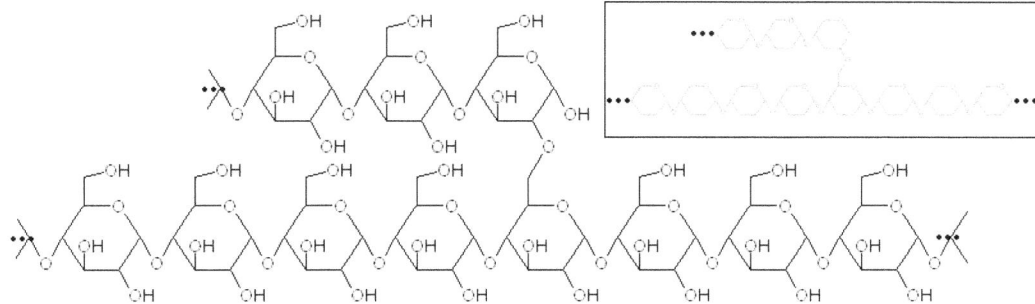

Glycogen

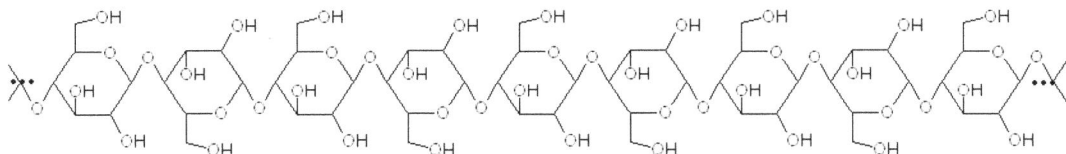

Cellulose

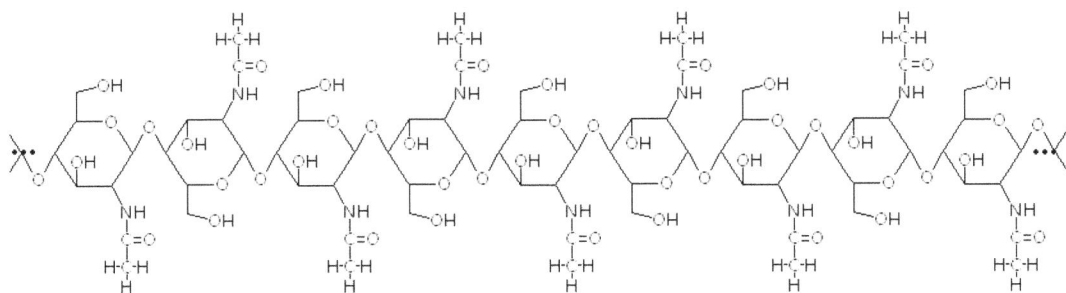

Chitin

One should remember that although the basic unit of polysaccharides is the "simple sugar," this does not imply that all polysaccharides can be eaten and digested. The reality is quite the opposite. Apart from starch and glycogen, very few polysaccharides are actually digestible. In fact, only a handful of organisms (mostly microscopic ones) are capable of breaking down polysaccharides such as cellulose and chitin. Though you can eat lettuce, the cellulose in it does not break down into its respective sugar pieces, but instead gets passed through the gut. You can be thankful for that, because if we COULD digest cellulose, the actual number of calories in a salad would be horrendous.

The Water Loving Nature of Carbohydrates

If you take another look at any of the molecules we studied in this section, you'll notice that they all have a bunch of -O- and -OH groups. Each oxygen has lone pairs sticking from its side, (although they were not drawn explicitly). In addition, the -OH groups are notorious for participating in hydrogen bonding. You may remember that in an earlier unit we noted that water also engages in a lot of hydrogen bonding. Taken together, these qualities make carbohydrates attract a LOT of water, so much so that large polysaccharides are like molecular sponges. Long chains of cellulose, for example, are typically laid down next to each other and interconnected, and when the water infiltrates these water-loving strands, the entire structure swells and becomes incredibly rigid. It is this behavior that gives cellulose part of its structural abilities. Glycogen attracts a lot of water as well, but it is not cross-linked in the same manner. It's connected together in the middle and allowed to spread out, branching here and there. The result is that it swells up like a big molecular ball, with each individual polysaccharide strand sticking out into the surrounding water. Since there are a large number of strands sticking outward, there are a lot of possible sugar monomers that can be broken

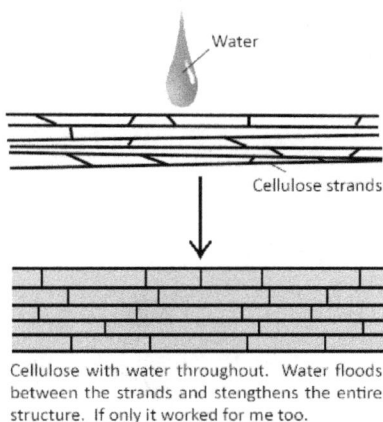

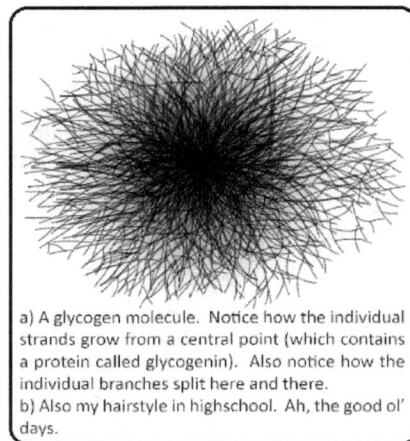

Water

Cellulose strands

Cellulose with water throughout. Water floods between the strands and stengthens the entire structure. If only it worked for me too.

a) A glycogen molecule. Notice how the individual strands grow from a central point (which contains a protein called glycogenin). Also notice how the individual branches split here and there.
b) Also my hairstyle in highschool. Ah, the good ol' days.

off and used for energy at any given point on the ball's surface. While this means that energy can be released from glycogen very quickly, it also means that glycogen takes up a LOT of room. If a living system needs to store a LOT of energy, it would quickly run out of room if it chose to store all of it as glycogen. So how does a living system store a lot of long-term energy? The answer to this dilemma lies in a different class of molecules called lipids, which we will explore in the next section.

Summary

To summarize this chapter:

- Monosaccharides are a class of molecules made by putting together repeats of H-C-OH. Their overall structure has a carbon to hydrogen to oxygen ratio of 1:2:1.
- Monosaccharides are largely used in energy management. Most of the small monosaccharides are used primarily for this, though they have some other minor uses as well.
- Monosaccharides have a lot a different ways they can be put together, creating isomers of each other. Monosaccharides are broadly divided into "D" and "L" isomers. D and L isomers are mirror images of each other. Living systems use mostly D isomers.
- Six-carbon monosaccharides have a lot of isomers, also broadly divided into the naturally occurring "D" isomers, and the rare "L" isomers.
- The six-carbon monosaccharides are often linked together by dehydration synthesis to make polymers called polysaccharides.
- Disaccharides are polysaccharides with only two monosaccharides linked together. They are used primarily in energy processes.
- Oligosaccharides are a little bigger than disaccharides and are used primarily for energy. They also have smaller roles as the structural basis for some antibiotics, as well as a few other miscellaneous uses.
- Long chain polysaccharides are used for long-term energy storage, but some have structural roles.
- Glycogen is an important energy-storing polysaccharide found in animals.
- Starch is an important energy-storing polysaccharide found in plants.
- Cellulose is a structural polysaccharide made by plants. It helps gives them their rigid stems and leaves.
- Chitin is a nitrogen-containing polysaccharide derivative that is found in bug and crab shells. It gives the shells their hardness.
- Carbohydrates attract water. This is why cellulose networks swell and get rigid, and why blobs of glycogen spread out and take up lots of space.

Unit 9

Lipids

Lipid Subcategories

Lipids are mostly non-polar molecules that are made up of primarily carbon and hydrogen, and are constructed from repeats of $-CH_2-$ linked together. There are three broad categories of lipids: 1) fats, oils and waxes, 2) phospholipids and 3) steroids. Each category tends to have specific uses.

Fats, Oils, and Waxes

Fats, oils, and waxes are made up of two general parts: 1) a **glycerol** molecule and 2) one or more **fatty acid** chains. A fatty acid chain is a long molecule that has numerous $-CH_2-$ repeating units with a carboxyl group on one end. These are combined with the glycerol molecule through dehydration synthesis. Three such fatty acids and a glycerol make a **triglyceride** molecule, as shown to the right:

When we discussed carbohydrates, we talked about how storing energy as glycogen has the unfortunate drawback of taking up lots of room. The sugars in polysaccharides like to stretch out into the water, soaking up the water molecules like a sponge. Lipids do the opposite. They hate water and shrink

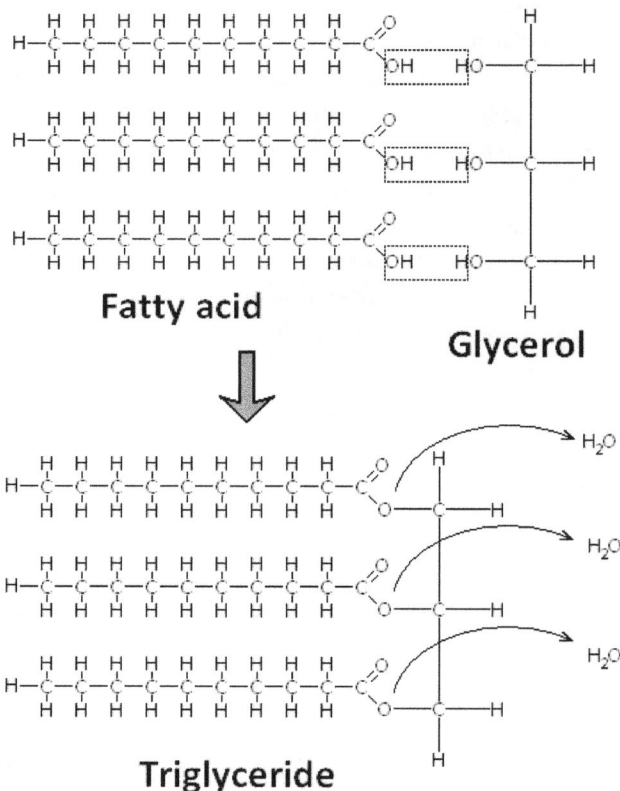

Fatty acid

Glycerol

Triglyceride

54

away from it. Also, the C-H bonds in lipids are very energetic (meaning they can store a lot of energy to be used later by the living organism). With these two qualities together, lipids serve as a very efficient form of biological energy storage. Triglycerides like that shown on the previous page are a very common molecule used to store energy.

Let's look a little closer at fatty acids. In the first image, we saw little fatty acid chains that were ten carbons long, but fatty acids can have variable numbers of carbons in them. The majority have an even number, however, (i.e. 10, 12, 14, 16 and 18 "–CH_2–" repeats, etc.). Fatty acids that have double bonds in them are called **unsaturated**. Those that have only single bonds linking all the carbons together are called **saturated**. This is the origin of the terms "saturated fat" and "unsaturated fat."

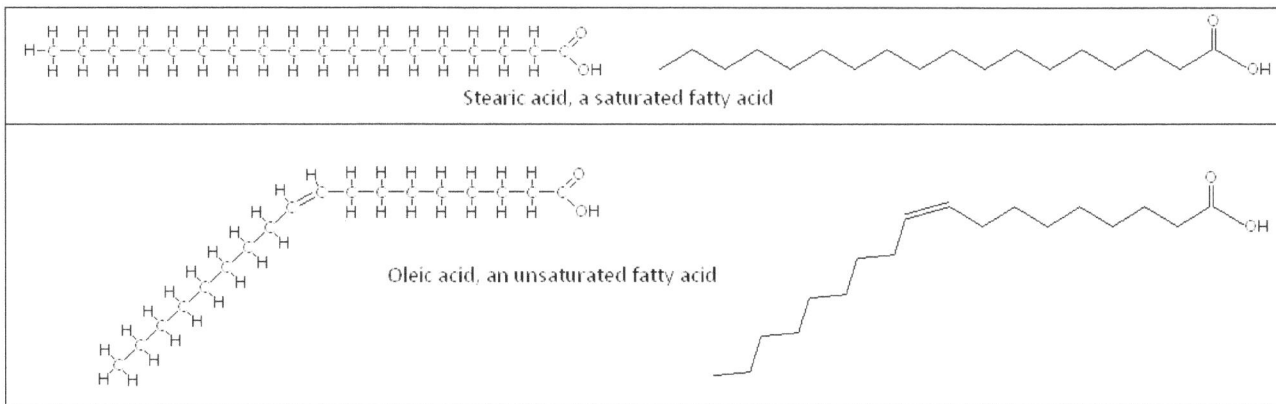

Stearic acid, a saturated fatty acid

Oleic acid, an unsaturated fatty acid

There are two consequences to fatty acids being saturated. First, their non-polar chains can straighten out and nestle against each other, packing tightly. This not only means more efficient energy storage, but it also means they tend to be more solid or semi-solid at room temperature, like lard or shortening. Second, saturated fatty acids have more energy than unsaturated fatty acids, because C=C double bonds are not as energetic as the C-C single bonds.

Dude! Seriously?!

A bunch of saturated fatty acids packing very closely to each other. One unsaturated fatty acid stands out in the middle, unable to pack as closely because of the double-bonded kink in its tail.

to haul our big butts around.

Fats (such as the kind stored in our rumps) have mostly saturated medium-length fatty acid chains (16, 18, or 20 carbons, etc.). Oils, on the other hand, are shorter, mostly *un*saturated fatty acids or are short hydrocarbon chains (linear molecules with nothing but carbon and hydrogen). Since they are very small and many do not pack very efficiently with each other, they are liquid at room temperature. Oils have a wide variety of uses in biology, from energy storage to surface waterproofing to chemical defense. For example, the oil on skin traps water inside the body, helping to keep us from drying out. The crude oil that is found deep in the ground is a collection of all the biological lipids that have deposited over the billions of years of life on the planet. All that fat from poor T-Rex's love-handles has become buried, and now we use the unspent biological energy to run our vehicles. T-Rex's lizard butt now helps

Waxes tend to have *very* long fatty acid chains. This makes them stick and tangle together with each other more tightly than fats and oils. Consequently, they form solids at room temperature. Waxes are not edible by most animals, but serve instead as waterproofing and structural support in things like plants, animal furs and beehives. Earwax, for example, provides a basic defense against bacteria, fungi and insects. It also does an incredibly good job at annoying us on a regular basis. Precisely how cavemen managed without cotton swabs is beyond me.

Phospholipids

Phospholipids are very similar to triglycerides in that they have a glycerol backbone. Instead of three fatty acid subunits, however, they have only one or two. Like triglycerides, these subunits are connected by dehydration synthesis. In place of the third fatty acid, it has a very polar side group that often contains phosphate and nitrogen. They come in many forms, but here is a typical example:

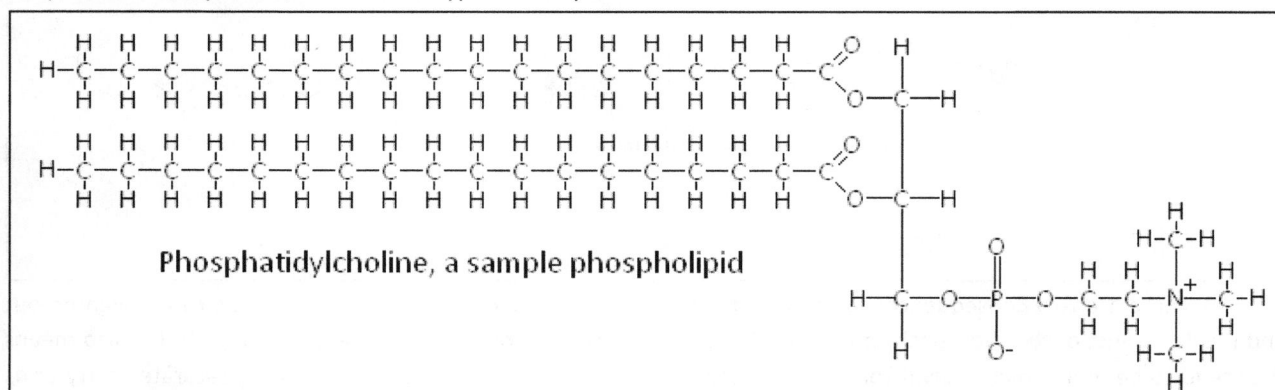

Phosphatidylcholine, a sample phospholipid

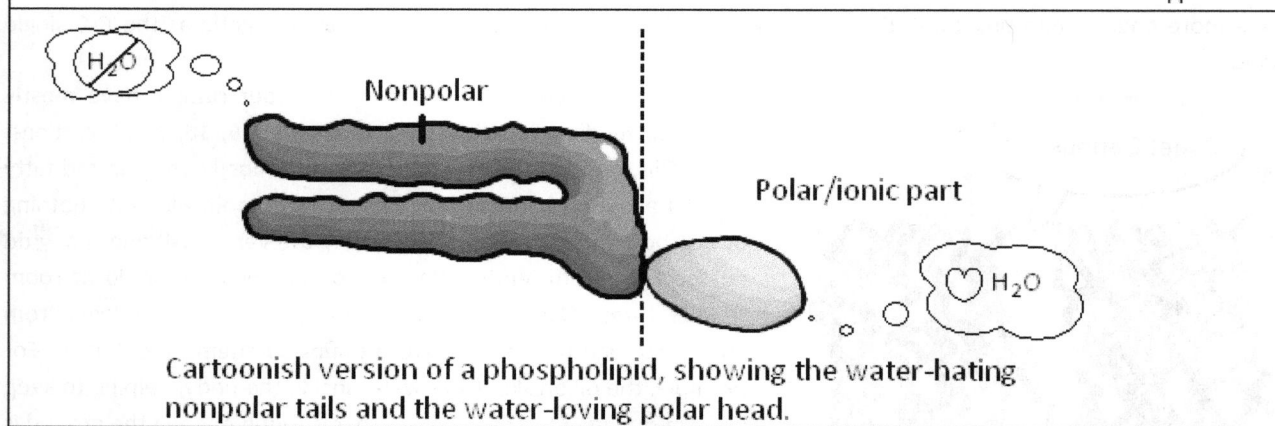

Cartoonish version of a phospholipid, showing the water-hating nonpolar tails and the water-loving polar head.

This unique structure has hydrocarbon tails that are hydrophobic and head groups that are polar (sometimes ionic) and thus hydrophilic. This causes phospholipids to line up side by side, with their polar heads pointing into the water, and their non-polar tails tucked tightly against each other away from the water. This is a crucial feature for the development of the cell membrane, which we shall discuss in a later section.

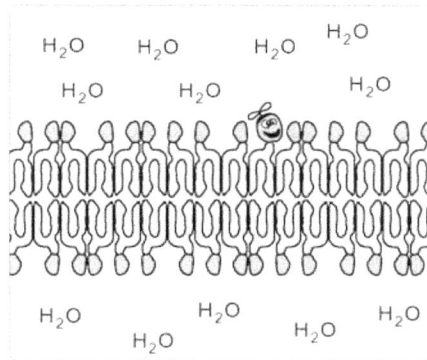

Steroids

Steroids are unlike the other two categories of lipids in that they are not made from glycerol backbone and fatty acids. Rather, they consist of series of $(-CH_2-)$-containing rings based on the general design for cholesterol (shown in the graphic below).

Steroids have a very diverse role in biology. For example, cholesterol can be found in cell membranes where it helps to keep phospholipid hydrophobic tails from packing too tightly. This keeps the membrane from becoming too solid. Other steroids are used as chemical messengers called hormones (**testosterone**, **estrogen** and **cortisol** are examples), which signal distant parts of the body about changes in body conditions. Steroids, like cholesterol, can also serve as the starting point for the creation of several important molecules. These topics are generally the focus of later courses, however, such as second semester general biology, anatomy and physiology or biochemistry. We will not discuss them much further in this book.

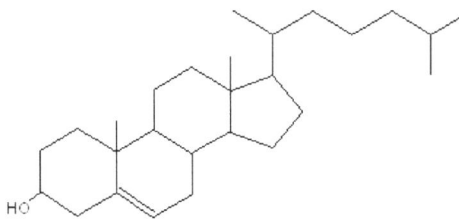
Cholesterol

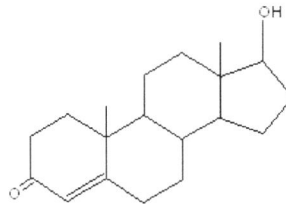
Testosterone

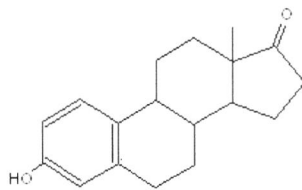
Estrogen

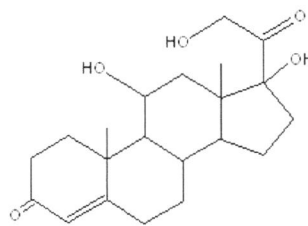
Cortisol

57

Summary

In this section, we covered the basics of lipids.

- Lipids are mostly hydrophobic compounds divided into three main categories:

  1) fats, oils and waxes

      - fats are semi-solid compounds used primarily for compact biological energy storage.

      - oils are short fatty acid (or hydrocarbon) chains that are liquids at room temperature. They typically serve as energy storage and waterproofing.

      - waxes are very long fatty acid chains that pack so closely to each other that they form solids at room temperature. They are used to protect/waterproof surfaces and to create structures such as beehives.

  2) phospholipids

      - Two fatty acid chains + 1 glycerol molecule + 1 hydrophilic head. These line up with each other so that their hydrophobic tails pack together, while their hydrophilic heads interact with water. This creates a double layer of molecules within the water, and becomes the important basis for cell membranes.

  3) steroids

      - These are multi-ring chemicals based off of the cholesterol molecule. They have a diverse number of uses in biology, such as helping to keep the cell membrane more fluid, or serving as chemical messengers called hormones. Cholesterol is the starting point for the creation of several other important molecules.

Unit 10

Proteins

The topic of proteins is a very complex one. Here we discuss the basic concepts of the structure and uses. We will encounter proteins in a number of places throughout the remainder of the book and will pick up more detail as those topics arise.

Protein Roles

Proteins are biological polymers that have very important roles in living systems. First and foremost, they can act as **enzymes**. Enzymes are catalysts - they speed the rate of other reactions. Enzymes are specific and selective in this task, each particular enzyme speeding only one particular reaction or a set of similar reactions. Second, proteins can serve as microscopic structural supports. Third, some proteins can be **contractile** (changing shape upon stimulation), serving as the means by which muscles shorten (we'll discuss the specifics in a later unit). Fourth, they

59

can serve to transport chemicals rapidly from one place to another inside living systems. Fifth, they can be used as sources of nutrition for the young, such as might be found in breast milk. Sixth, some proteins are used as **hormones**. Hormones are chemical messengers that link together physiological systems. We talked very briefly about this topic when we looked at the cholesterol-based steroids. Last, and certainly not least, proteins make up **antibodies** and most of the other components of the immune system, which fights off infections and helps to keep us healthy. Hormones and antibodies are usually covered in the second semester of biology, or in advanced biology classes.

Protein Roles - Overview
1) Enzymes – Speeds the rate of specific reactions.
2) Structural units – microscopic framework upon which other molecules can attach, or upon which reactions can take place.
3) Contractile units – the primary mechanism that acts to shorten muscle.
4) Transportation – efficiently carry materials through the body.
5) Nutrition in milk – helps to provide nutrients to offspring.
6) Some Hormones – chemical messenger that acts to inform physiological systems of external stimuli.
7) Antibodies – helps fight off infection.

Protein Synthesis, the Basics

Proteins are polymers. The monomer building blocks used to make these polymers are called **amino acids** (commonly abbreviated aa's). All amino acids are related by a common backbone chain of three atoms: one nitrogen and two carbons, as shown here:

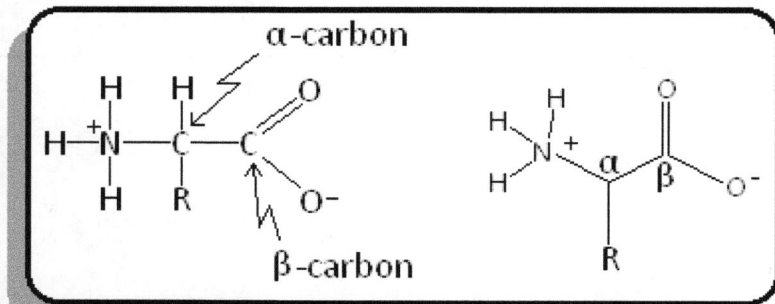

The middle carbon is called the **alpha-carbon**. The end carbon is called the **beta-carbon**. Attached to the alpha-carbon is a side group, labeled "R" in the picture. Each amino acid has a unique R group, and it is this R group that determines the unique chemical behavior of that amino acid. There are twenty fundamental **amino acids** that living systems have in common. In the image below you can see that the R groups range from very simple (just an H for glycine) to very complex, such as the rings in tryptophan and the long nitrogen chain in arginine.

Polar Amino Acids

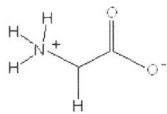
Glycine

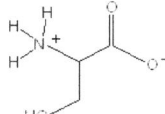
Serine

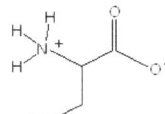
Cysteine

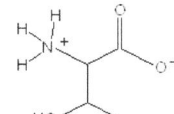
Threonine

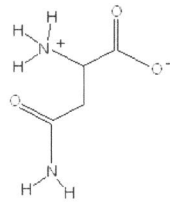
Asparagine

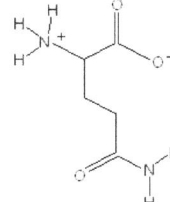
Glutamine

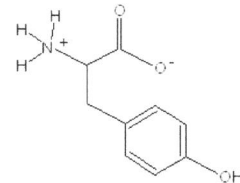
Tyrosine

Nonpolar Amino acids

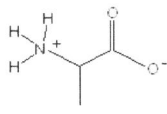
Alanine

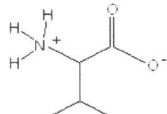
Valine

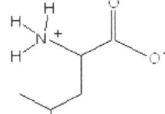
Leucine

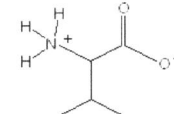
Isoleucine

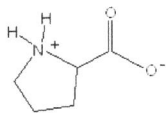
Proline

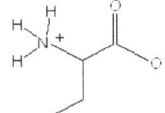
Methionine

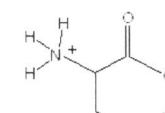
Phenylalanine

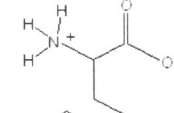
Tryptophan

Acidic Amino Acids

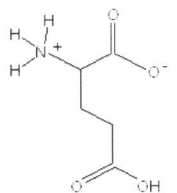
Glutamic acid

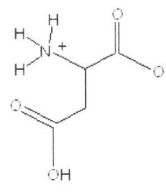
Aspartic acid

Basic Amino Acids

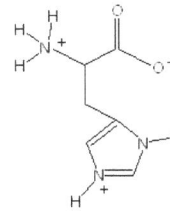
Histidine

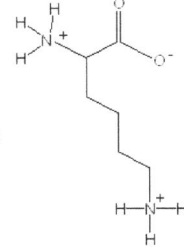
Lysine

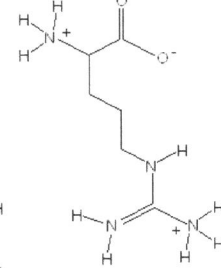
Arginine

61

You'll notice in this image that each amino acid backbone has a positively charged nitrogen on the left and a negatively charged oxygen on the right. The reason for this is that the COOH functional group is acidic (tending to release its H^+ to become COO⁻, as shown), while the NH_2 group is basic (tending to attract H^+ to become NH_3^+). So, individual amino acids are typically drawn with COO⁻ and NH_3^+.

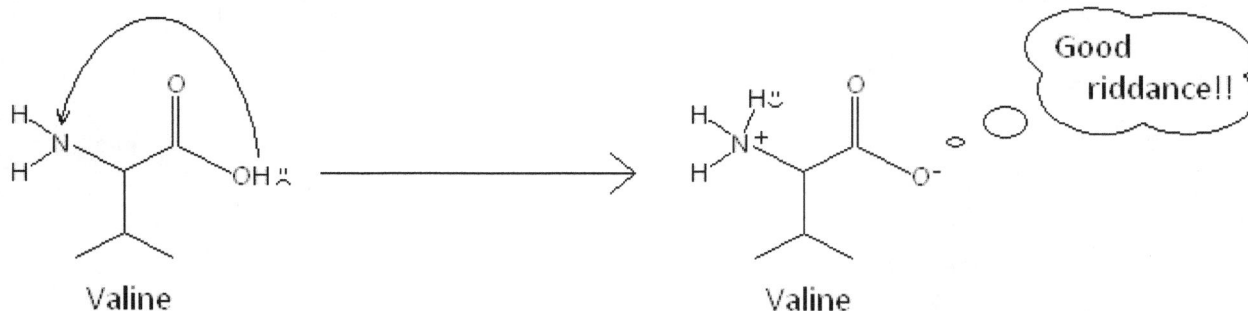

Valine   Valine

Since the hydrogen that is released from the COO⁻ on the right is actually H^+, you might be wondering why I put the "+" next to the nitrogen in the drawing. This is because of the way the electrons are shared between the nitrogen and H^+; the positive charge tends to gather over the nitrogen.

The next thing to notice in the big image of all the amino acids is that they are divided into four broad categories: nonpolar, polar, acidic and basic. These categories refer to the characteristics of the amino acids when they are bound together into proteins. To shed light on that, let's look at how amino acids are connected together.

Like the previous polymers we studied, amino acids are strung together using dehydration synthesis. The easiest way to see this is to draw the amino acids the same as the valine in the last image – with the H still attached to the oxygen on the right. The reason is to avoid confusion, since the dehydration synthesis reaction is otherwise not very obvious. The real reaction turns out to be the same regardless of how the amino acids are drawn, but it involves some otherwise unimportant details that will only serve to confuse rather than to educate. At any rate, the result is the usual: water is released when the monomers are connected together. The following picture shows three amino acids getting connected into a small protein polymer:

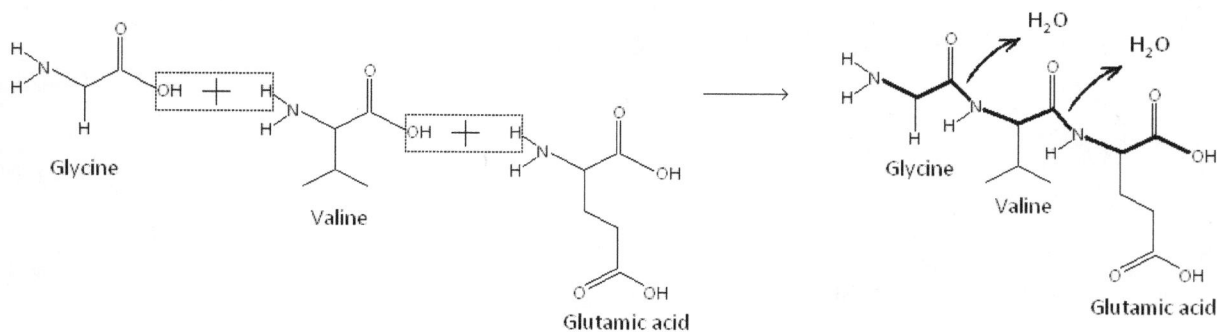

This little protein (also called a **peptide**)[13] is only three amino acids long, with a knobby, crooked backbone connecting each monomer (drawn as dark thick lines). We could easily extend this to any length we like, of course. For example, here is a cartoonish protein with thirteen generic amino acids, (i.e., side group "R"), drawn with the backbone as a simplified straight line.

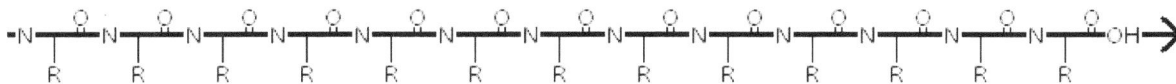

Peptide chains sometimes have a directional arrow attached to one end to make it clear which direction is the end with the COOH. In this last picture, the arrow points to the right, because the right end of the protein has the COOH and the left end has the nitrogen.

As we saw in the picture of the three amino acids, the N-C-C backbones of proteins are *not* straight, but very crooked, bending at each atom because of the weird bonding angles they make. The strange shape of the protein backbone isn't the only determining factor in the overall shape of the protein, however. As it turns out, the backbone is very flexible, allowing the protein chain to fold into all kinds of bizarre shapes. This, in turn, leads to some interesting structures. . .

Protein Structure

The order of amino acids in a protein is called its **primary structure**. The backbone of the primary structure, as we learned, is an alternating pattern of N-C-C-N-C-C. . . and so forth. The nitrogen-hydrogen bonds in this backbone are polar, with a partial positive charge positioned over the H. In addition, the oxygen atoms of the repeating C=O bond have two lone pairs poking out from them. This makes them partially negative. So, if we zoom in on our cartoonish protein from the last image, we would see a whole bunch of partial plusses and minuses everywhere.

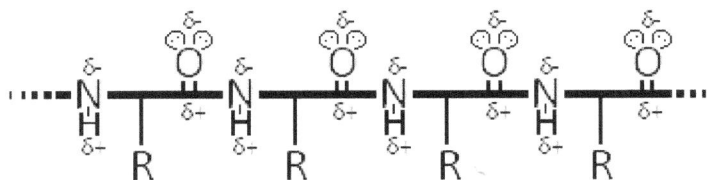

The result of all these polar bonds is that the protein chain is very sticky. It's like a shoe string that has gooey stuff all over it. Before too long it gets all knotted up, sticking to itself due to the partially positive sites being attracted to the partially negative sites. As it turns out, there are two common ways that protein chains bond in response to these attractive forces. The first is the **alpha-helix**. In this process, the chain starts to spiral around in such a way that a hydrogen from farther along the chain forms a **hydrogen bond** with an oxygen farther back, causing a corkscrew shape.

---

[13] Peptide = one or more amino acids stuck together by this mechanism.

Here is roughly what it looks like:

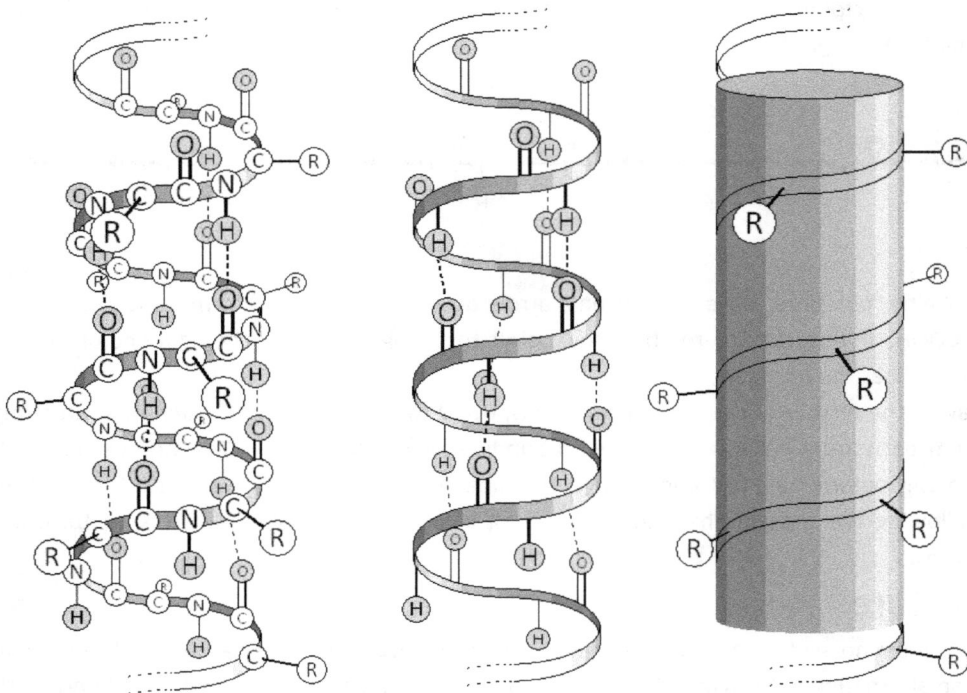

The far left shows all the atoms in their raw complexity. The ribbon that connects them is the backbone that runs along the chain. This, of course, is a cluttered picture, so have a look at the central image. This image shows the backbone ribbon with only the hydrogen bonding between the sticky hydrogen and oxygen shown as individual atoms. The cylinder on the far right is a stylized version of the alpha-helix; this is how it is typically drawn in many biology books. In that simplified version, you can see that the alpha-helix is basically like a tube where the side groups stick outward.

The second way that protein chains often organize is through the **pleated-sheet** or **beta-sheet**. In this process, the chain bends backwards and overlaps itself, like a person touching their toes.

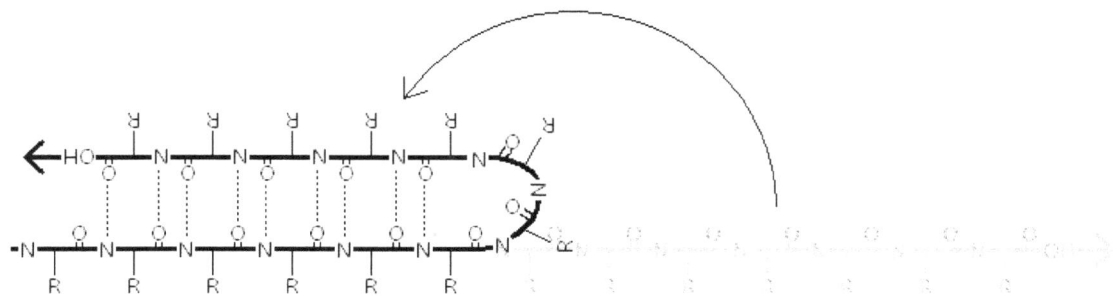

It looks fairly simple in this picture, but the actual 3-D structure is like a ruffled potato chip, or a wavy roof, with the "side" group sticking up and down.

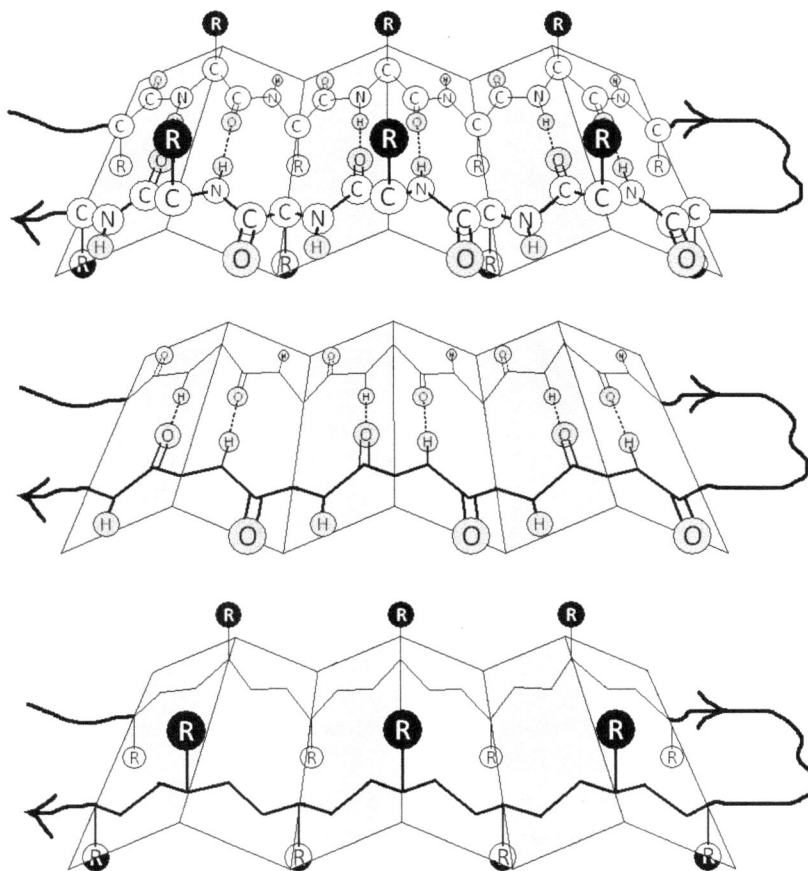

The bottom version of the beta sheet is typically drawn with flat arrow symbols, like the image to the right.

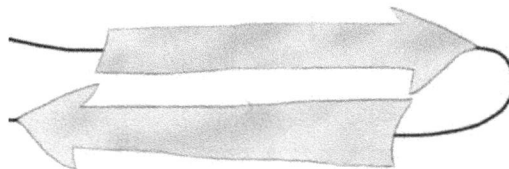

   Whereas the order of the amino-acids is called the primary structure, the alpha-helix and beta-sheet structures are referred to as the **secondary structure** of the protein. These structures, in turn, tend to rest against each other to form the **tertiary structure** of the protein. The precise method in which all of this occurs is mainly dependent on the side groups that are sticking out (the R's). Amino acids that have non-polar side groups are hydrophobic and will cause the entire thing to arrange so that they are tucked inside the protein and away from water. Amino acids that are acidic, basic or polar side groups are hydrophilic and will cause the structures to arrange so that they are pointing out into the surrounding aqueous (water) environment.

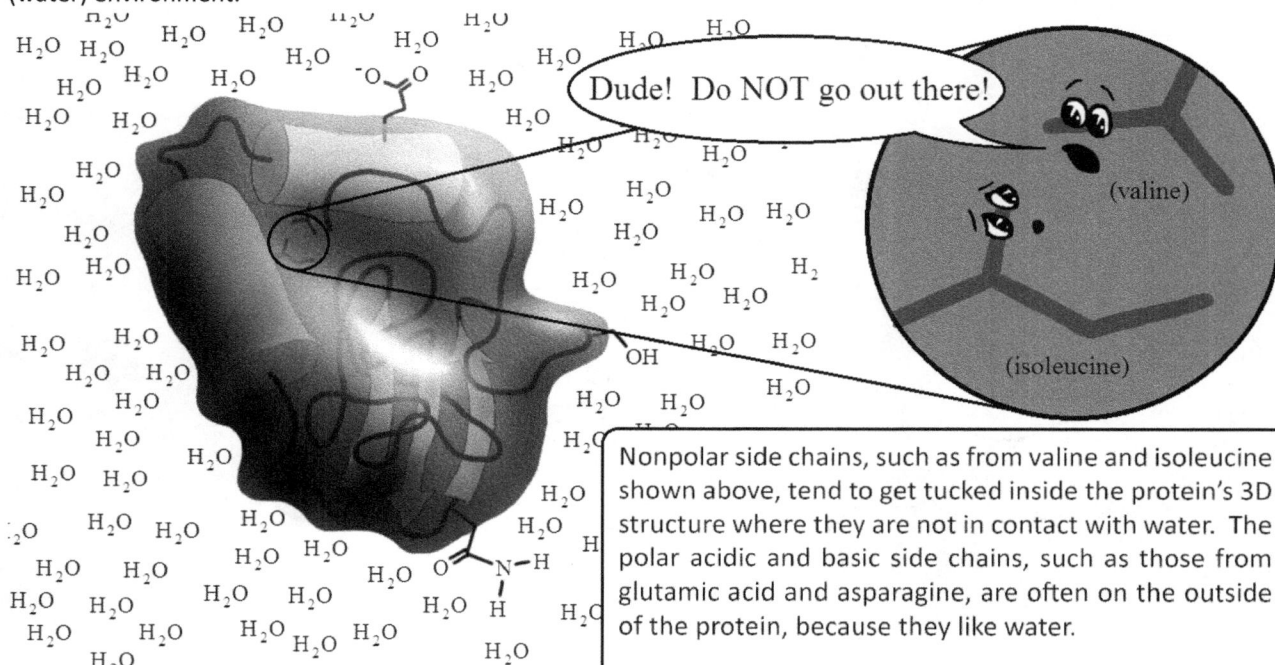

Dude! Do NOT go out there!

(valine)

(isoleucine)

Nonpolar side chains, such as from valine and isoleucine shown above, tend to get tucked inside the protein's 3D structure where they are not in contact with water. The polar acidic and basic side chains, such as those from glutamic acid and asparagine, are often on the outside of the protein, because they like water.

   To further stabilize this conformation, side groups that have sulfur in the them (cysteine and methionine) are clipped and adjusted so that a sulfur-sulfur (disulfide) bond forms between the two chains, like glue sticking a clump of hair together. The two cartoon alpha helices to the right are bonded in such a fashion.

   In addition to sulfur-sulfur bonds, we need to make note of the strange behavior of proline. It's shaped in a ring from the alpha-carbon to the nitrogen atom. This makes the backbone of proline bent. Proline, therefore, tends to bend the protein chain sharply wherever it appears. This makes it useful in positions where the protein backbone chain needs to bend at sharper angles in order to form a specific overall shape. To this end, proline is often found in sections that join helices and sheets together (see next page).

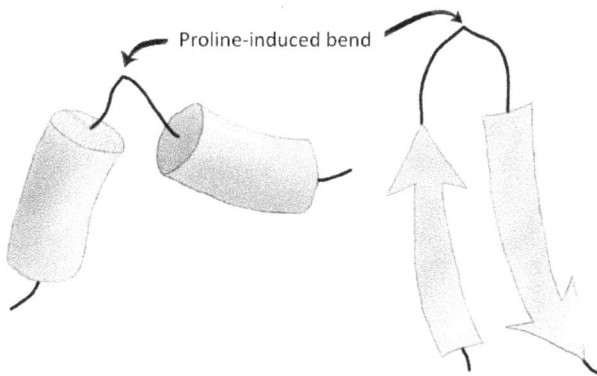
Proline-induced bend

Together with the sulfur-sulfur bonds, the proline bends, and the weird shape of the alpha-helices and beta-sheets, the protein chains look like blobs with tubes and flat arrows in them. Now, if multiple blobs clump together, (i.e. several individual protein chains), you arrive at the **quaternary structure** of proteins. Like the tertiary structure of proteins, the individual blobs are often held together by sulfur-sulfur bonds and connect in a way to minimize the number of nonpolar side groups that are exposed to water.

Here is an overview of protein structure:

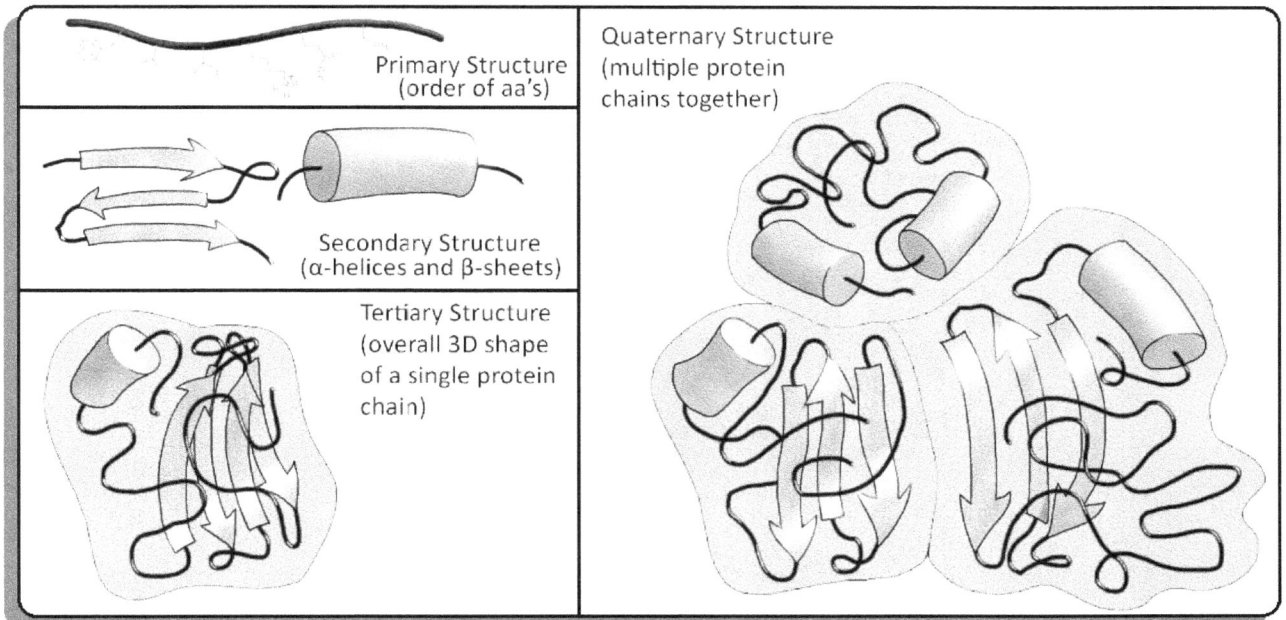

Primary Structure (order of aa's)

Secondary Structure (α-helices and β-sheets)

Tertiary Structure (overall 3D shape of a single protein chain)

Quaternary Structure (multiple protein chains together)

One final note on protein structure: domains. If you look at the overall quaternary structure of proteins, you will often notice discrete regions in the protein. We call these **domains** (see image on next page).

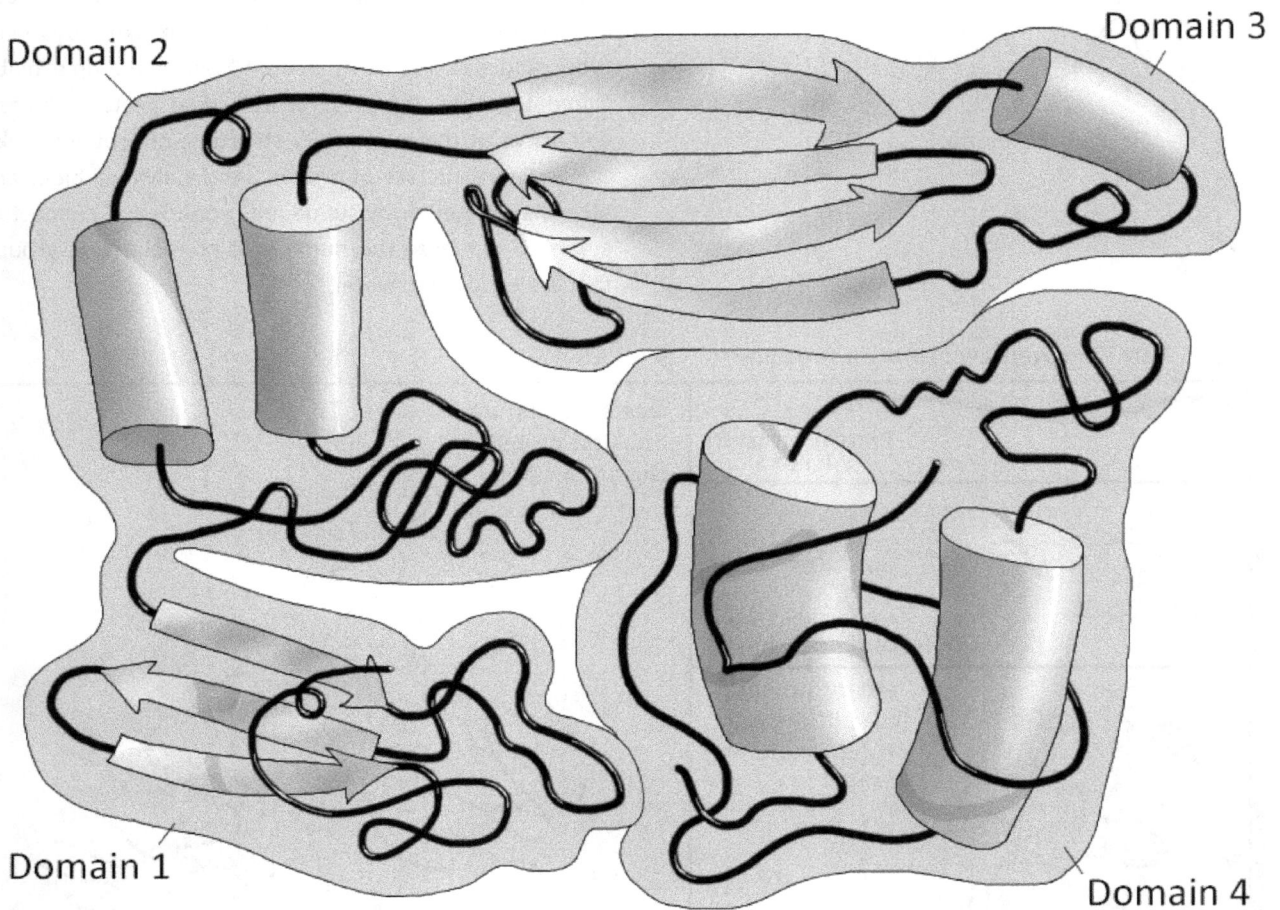

Domain 2

Domain 3

Domain 1

Domain 4

Domains can sometimes be made from different protein chains, but that isn't always the case. In this last image, three domains are made from one continuous protein chain. The fourth domain is made from a completely different protein chain.

Protein domains typically have their own chemical job they perform, and completely unrelated proteins can sometimes have very similar domains because parts of their chemical jobs are similar to one another. A full appreciation of the topic on protein domains requires understanding of genetics, evolution, and biochemistry, so we'll put off further discussion of domains for this book.

Protein Structure Determines its Characteristics

Proteins are complex three dimensional things, created from the combination of twenty different building blocks. The overall 3-D structure of the protein, in turn, often determines its uses in living systems. Let's look at a few of the roles that proteins play. Structural proteins typically fold up into long and thin blobs, such as those in hair or in cellular scaffolding. Sometimes they connect together like a chain, extending long distances. The strands they form not only provide support for the living system, but can also be used as a kind of road for other proteins to travel. The transport proteins, such as **dynein** and **kinesin**, often look like little blobs with long legs and feet. These "leg" domains will attach to the long structural proteins and "walk" along it, carrying stuff for up to the entire length of the organism.

Transport proteins carry cell materials along protein roads such as microtubules.

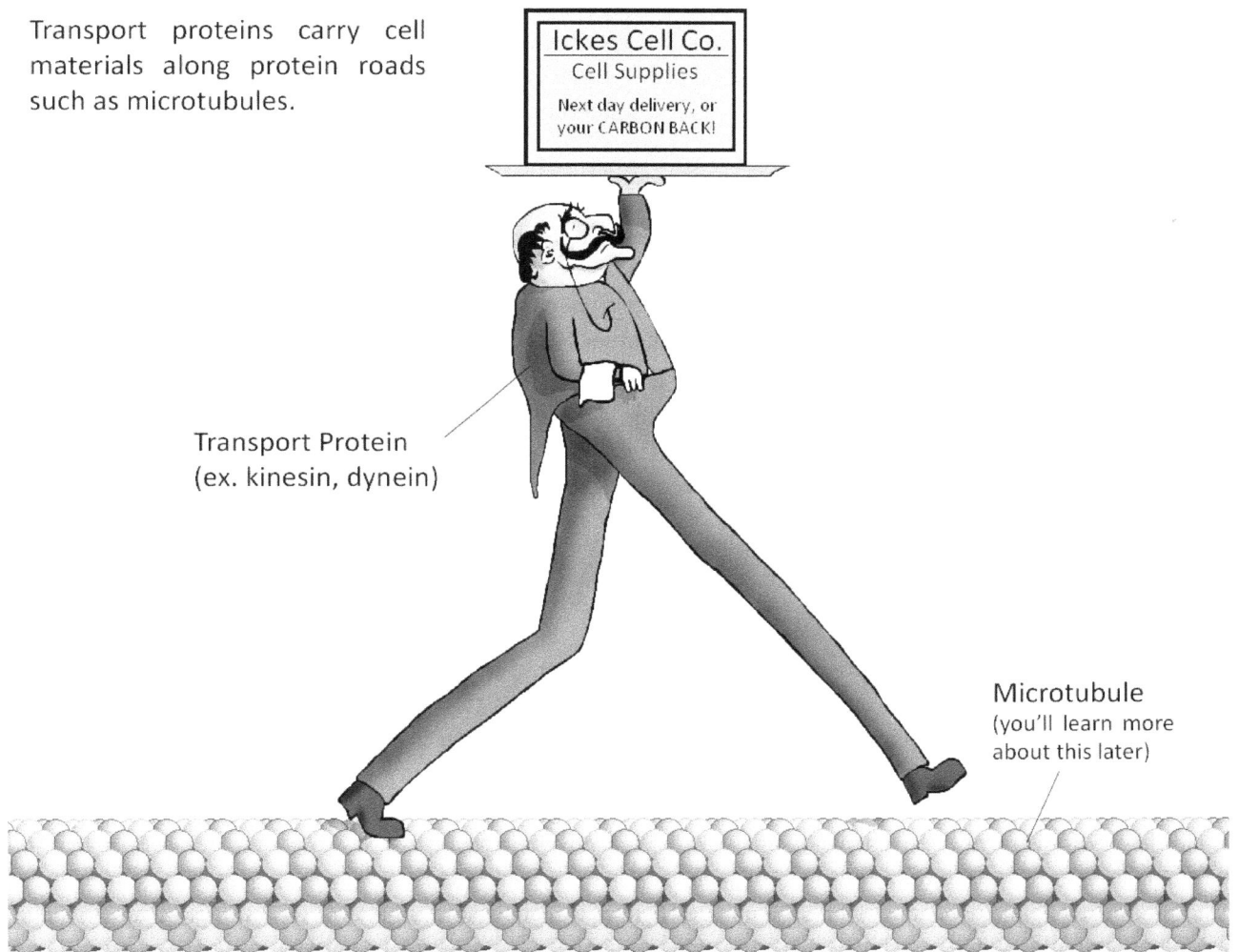

Ickes Cell Co.
Cell Supplies
Next day delivery, or
your CARBON BACK!

Transport Protein
(ex. kinesin, dynein)

Microtubule
(you'll learn more
about this later)

Proteins that act as enzymes, on the other hand, are characterized as blobs that fold so that they have little nooks and pockets into which very specific molecules fit. By creating a very special environment for these molecules, it speeds the chances of reaction between the two molecules that fit into the space. Not just any molecule can fit into these little crevices, though. These spaces are very specific. How? Two general mechanisms make this possible: 1) The shape tends to be the same as the specific molecule, and 2) the charge (and/or partial charges) tend to be optimal to attract the specific molecule. As such, protein enzymes recognize their target molecules with expert precision and make sure that reactions between the target molecules happen much faster than they would otherwise occur.

Not all the roles of proteins are necessarily characterized by highly recognizable structure. Other roles, such as immunity, are beyond the scope of the first semester in biology. Antibodies, for example, are a very diverse group, but the prototype structure looks like a capital "T", where the top ends of the T are shaped in a way that it specifically recognizes the surfaces of various bacteria, parasites, and viruses. Exactly how this works is usually the focus of second semester biology or later courses.

Protein Denaturation

The structure of the proteins determines its usefulness. This can be disrupted in a number of different ways, however. The first is to heat the protein. This causes the hydrogen bonds and the -S-S- bonds to snap, so that the protein becomes lax. The protein chains are still sticky, but when there is a lot of heat in the environment the chains move around too much to stick to themselves and each other. The second is to change the amount of salt or acid in the solution. This tends to affect the side chains that have carboxylic acids or amine groups in them, altering how they interact with the surrounding water. Lastly, you can hit the protein with ultraviolet light or other strong radiation. This causes the amino acids themselves to begin to degrade, so that the atoms within them begin rearranging in various chemical reactions.

When any of these things happen, we say that the protein has become **denatured**. A denatured protein loses its ability to perform its biological role. In fact, many serious diseases are a result of protein denaturing, or errors in the protein's production that leads to an incorrect structure. We'll be talking about that in the units ahead.

Immense Protein Diversity

Clearly, the primary structure of protein chains (the order of the amino acids) determines a great deal about how the overall protein will fold up in water and how the protein is to be used. The more combinations we can manage to put together, therefore, the more versatile an arsenal of protein chemicals that we can make. How many different proteins can we make? Let's see.

For the first amino acid, we could choose any of the twenty amino acids. For the next amino acid, we also could choose from any of the twenty aa's. So, after the second amino acid is added, we have a total of 20 x 20 = 400 different chemical combinations. For the third amino acid, we can again choose from any of the twenty available aa's. That pushes our total combinations to 8000 unique chemicals. Following this pattern, for a protein that is n amino acids long, the total number of chemical combinations would be 20^n. So, a simple amino acid chain sixteen aa's long can be made 6553600000000000000000 ways. Ultimately, this means that building proteins in such a modular fashion can result in a HUGE potential number of differently shaped chemical blobs, each having their own unique characteristics. That is a LARGE number of shapes and lots of "diners" where chemicals can meet up and interact. This is why proteins serve such a large number of roles in biology.

For the moment, however, we need to pause to consider an important question. Given all the possible amino acids, how do living systems know which amino acids to use and in what order? If we randomly put the amino acids together into a chain, we can certainly arrive at a great number of different structures, but living systems are not random. They seem to "know" what amino acid combinations make useful protein structures, but how? The answer to this lies in the order of genetic information that exists within the individual. The start of the details begins in the next unit.

Summary

In this section, we discussed proteins.

- Proteins fill many different roles in biological systems: enzymes, scaffolding / cellular structure, contraction and organism movement, cellular transportation, nutrition for young, some hormones, and immunity.
- Amino acids are characterized by having an N-C-C backbone, with a unique side group (usually labeled R). The N on the left is an amine group. The C on the right is a carboxyl group. The hydrogen of the carboxyl group is usually snapped off and attached to the amine group.
- There are four broad categories for amino acids: nonpolar, polar, acidic and basic. The nonpolar amino acids are usually tucked inside the protein. The other three types are usually on the outside, facing water.
- Amino acids are attached together using dehydration synthesis into long chains called proteins (or peptides). The order in which they are put together is called the primary structure. Since there are twenty amino acids, there is an astronomical number of possible primary structure combinations that a living system can make.
- Peptide chains are very sticky with lots of polar bonds on its N-C-C backbone. This causes them to take up many strange shapes.
- The secondary structure of a protein is characterized by alpha-helices and beta-sheets.
- The alpha helix is a structure in which the peptide chain spirals into a tube due to the interaction of polar bonds along the N-C-C backbone.
- The beta-sheet is a structure in which the peptide chain doubles over and lines up with itself to form a ruffled chip or wavy roof-like shape.
- The alpha helices, beta-sheets and remaining peptide chain forms into a blob known as the tertiary structure of the protein.
- Multiple peptide chains interacting with each other makes up the quaternary structure of the protein.
- Sulfur-sulfur bonds result when two sulfur containing amino acids (cysteine or methionine) are clipped and connected together via their sulfur atoms.
- Proline bends occur along the peptide chain where proline is. This is due to proline's unusual bent shape that occurs in its N-C-C backbone.
- Discreet areas in a protein are often called domains, and these have their own chemistry they perform. Different proteins can sometimes have very similar domains, because part of their overall function is similar.
- Cellular structural proteins tend to be long and thin, connecting into threads that span long distances.
- Transport proteins are characterized by having two leg-like domains, which "walk" along structural proteins to deliver various molecules to distant parts of the living system.

- Enzymes are shaped so they have nooks in them that only their target molecule(s) fit into. Enzymes speed the reactions of other chemicals.
- Protein structure denatures via heating, salt/acid changes and radiation. Protein denaturation can lead to illness.
- The incredible number of combinations of amino acid chains that living systems can make is stored in the organism's genetics.

In this section we cover:

1) The nucleotide building blocks

2) DNA structure

3) RNA structure

4) The Role of DNA as the information molecule

5) The Roles of RNA

6) The various roles of individual nucleotides

Unit 11

Nucleic Acid

You *cannot* get closer to the heart of biology than the study of nucleic acids. The major role of nucleic acid is to be the genetic material, within which all the information of life is stored. There are other roles that nucleic acids play, however, so we'll need to check those out as well. First, let's take a peek at the monomers of nucleic acids (the **nucleotides**) and how they come together to make the two major forms of nucleic acid: **DNA** and **RNA**.

Nucleotides

Nucleic acids are polymers made up of monomers called nucleotides. The primary structure of nucleotides is fairly complex, so let's first look at a very simplistic design:

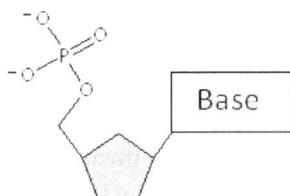

The gray bottom pentagon is a 5-carbon monosaccharide. In the upper left is a phosphate, and in the upper right is a confusing bunch of carbon rings called a "base." Let's slowly unpack the complexity, one part at a time. There are two different types of 5-carbon monosaccharides that can occur in place of the bottom pentagon: the sugar

73

deoxyribose and the sugar **ribose**. Deoxyribose has one less –OH than ribose. In a nucleotide, the five-carbon sugars would look like this:

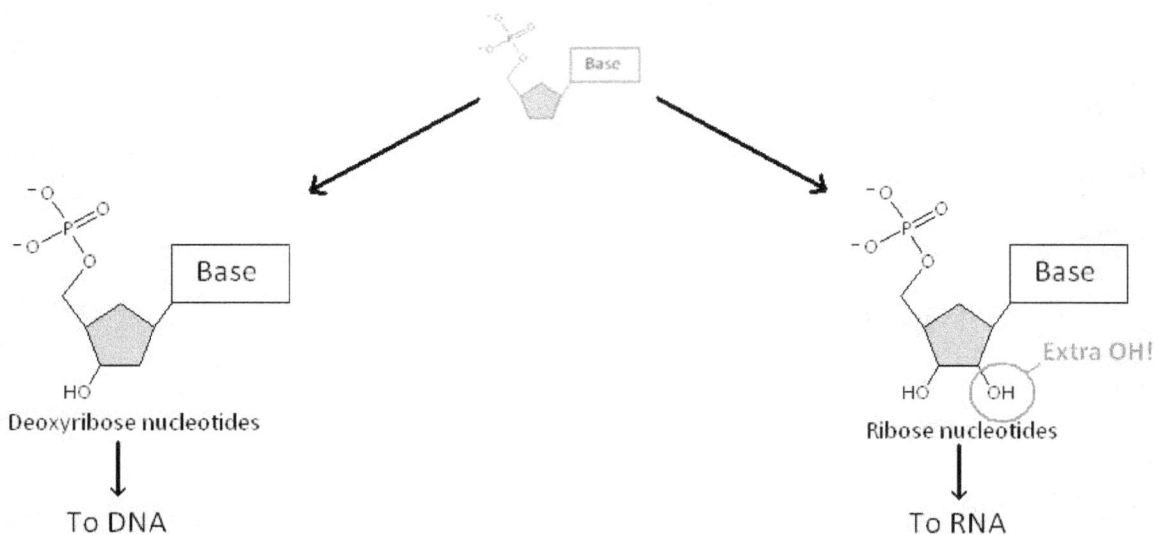

Deoxyribose nucleotides

To DNA

Ribose nucleotides

To RNA

The nucleotides that have deoxyribose (left) are used to make DNA (**deoxyribonucleic acid**). The nucleotides that have ribose (right) are used to make RNA (**ribonucleic acid**). Let's focus first on DNA.

Introduction to DNA Structure

There are four bases that can appear in DNA nucleotides:

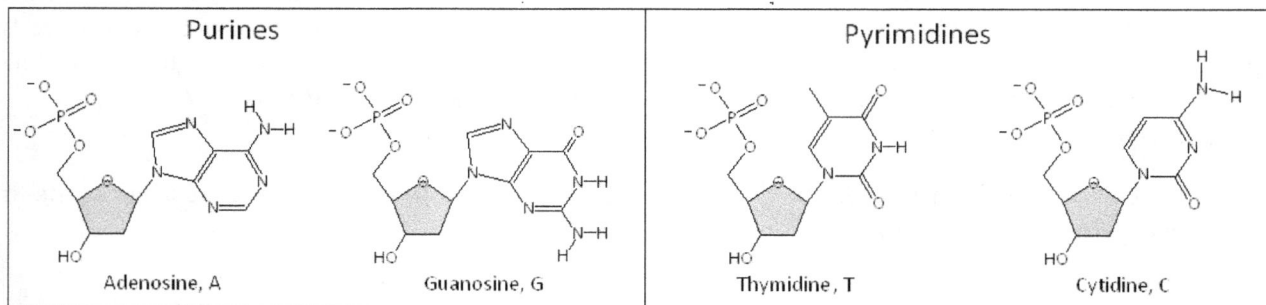

As you can see from this image, the bases are divided into two major groups. Those whose structures contain two rings are called **purines**. The bases with only one ring are called **pyrimidines**. The shorthand notation for these nucleotides is A, G, C and T, for **adenosine**, **guanosine**, **cytidine** and **thymidine**. These nucleotides are strung together to make DNA polymers by dehydration synthesis as shown here on the following page.[14]

---

[14] Technically, nucleotides start with three linked phosphates (nucleotide triphosphates) and not just one. Two phosphates are removed and one phosphate remains in the nucleic acid backbone. The end result is the same, but the essential pattern is harder to recognize. The important thing to understand is the recurring theme of dehydration synthesis. Two monomers are put together and water is released in the process.

Simplified representations

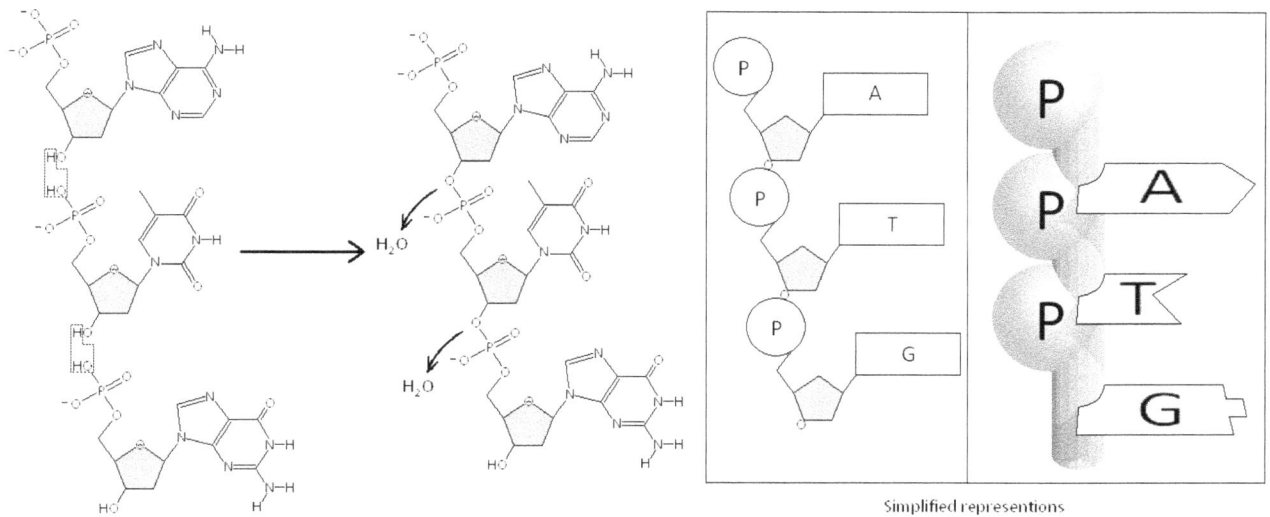

As you can see from the images on the far right, the resulting small polymer looks like half of a ladder, the backbone made from sugar-phosphate repeats. This particular half-ladder DNA chain is only three bases long. The DNA in living systems typically has on the order of billions of nucleotides! DNA bases have an interesting ability to "pair" with each other, and it is the pairing ability of the bases that makes nucleic acids unique, setting the stage for the inheritable nature of biological information. To clarify, let's zoom in on the bases themselves and see how they interact with each other.

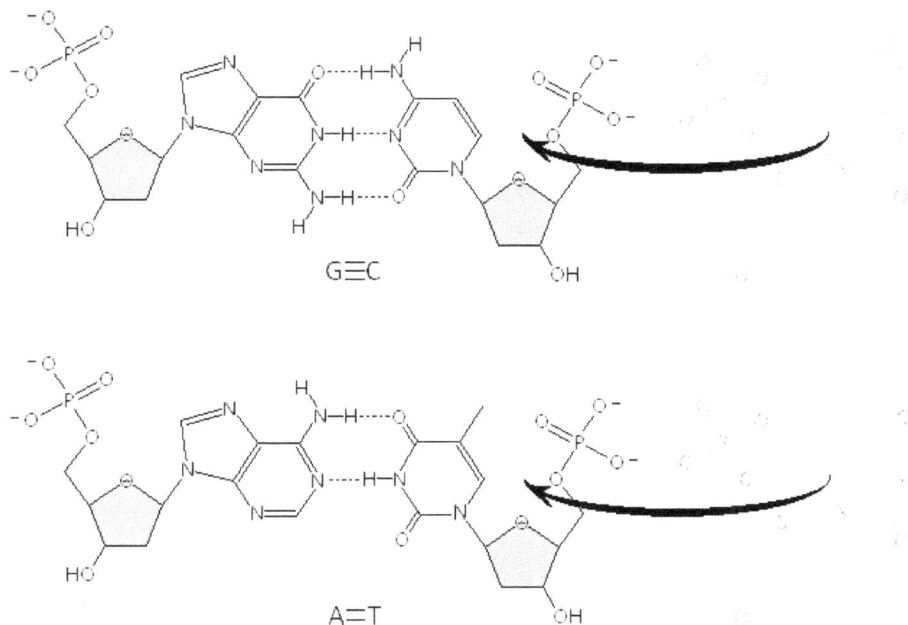

G≡C

A=T

In this last image, I've shown the base for guanosine interacting with cytidine (flipped to face left), and adenosine interacting with thymidine (also flipped to face left). The striking characteristic of these interactions is the hydrogen bonding between the bases (represented by the dotted lines). G has three hydrogen bonds that it shares with C, and A has two bonds it shares with T. This bonding pattern is very specific. Because of the specificity of interaction between these two bases, the two are referred to as a **complementary base pair.** You never see a G bonding with a T, or an A bonding with a C, for example. They simply don't fit that way. Due to this pairing nature and to the complexity of nucleotides, I will often draw them in a very cartoonish and stylized way.

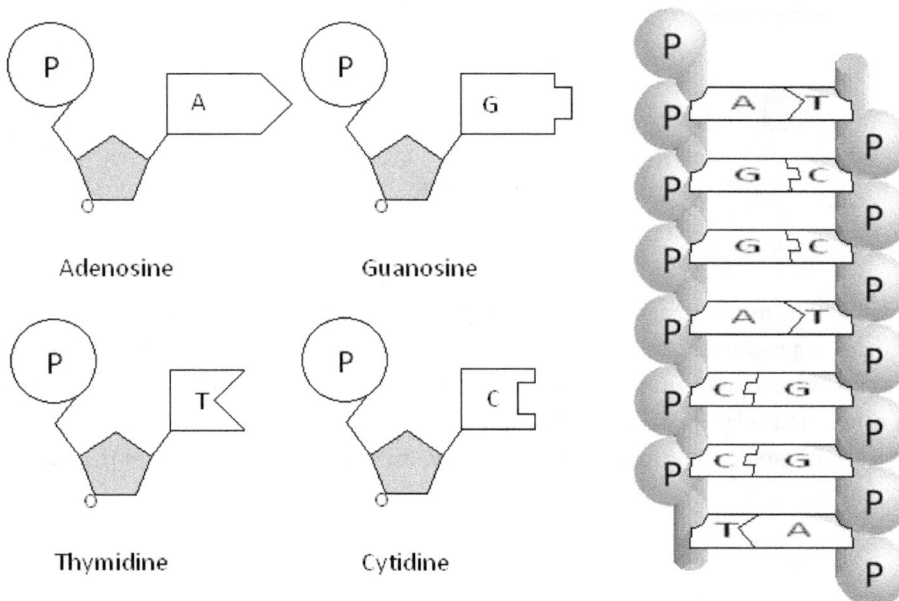

Adenosine   Guanosine

Thymidine   Cytidine

Based on previous images, you might think I've made a mistake with the order of phosphates in the backbones of the nucleotides on the far right. One sugar-phosphate backbone looks like it's going backwards! The sad truth, (and one of the most surprisingly difficult concepts in learning about DNA), is that when two nucleotide chains pair, their backbones end up going in opposite directions. Naturally, it's difficult to predict this outcome based on the simple models of the nucleotides we've looked at, but one must remember that the actual shape of a molecule is often grossly distorted by 2D drawings. In the case of nucleotides, the sugar ring, the phosphate and the base are not actually lying flat with each other. When the actual 3-D restraints are taken into account, the only way for the two chains to fit together is for their backbones to go reverse from each other.

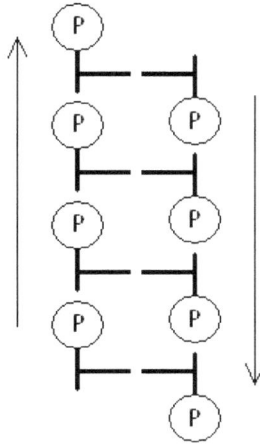

It gets even more complex than this, however, because the paired DNA chains (or DNA **duplex**) twist around each other as well, creating the famous **DNA double-helix**, shown here:

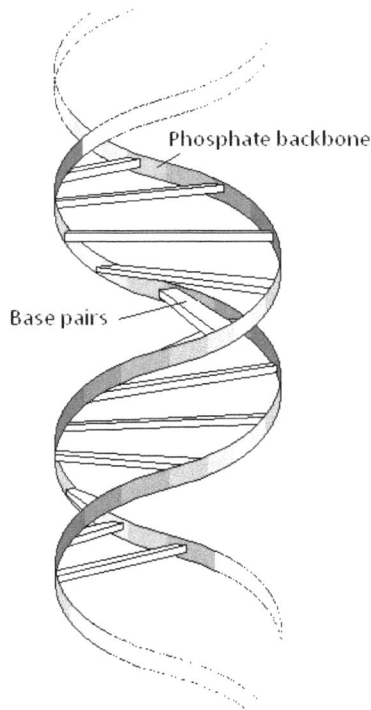

This is how DNA exists inside living systems such as you and me. It is an incredibly long and twisted helical pair of DNA strands that holds all the information for basic biological functions within living systems. In many of the future sections we will be expanding on the nature and function of this molecule and how its helical nature makes things tricky for living systems. For the moment, however, let's not forget about RNA, because they are equally important in the big picture.

RNA, A First Look

Now that we've had an initial glance at DNA structure, let's take a brief look at RNA. As we mentioned before, the major difference between the RNA nucleotides is the addition of the extra –OH on the sugar ring at the bottom-right of the nucleotide molecule. Also, the bases in RNA are almost identical to those in DNA, with one exception: the thymidine is replaced by **uridine**. If you look at the following image, however, you might notice that uridine is almost the same as thymidine, except for a missing methyl (CH_3) group. This nearly identical structure to thymidine means that it also pairs with adenosine (i.e., A = T for DNA, and A = U for RNA).

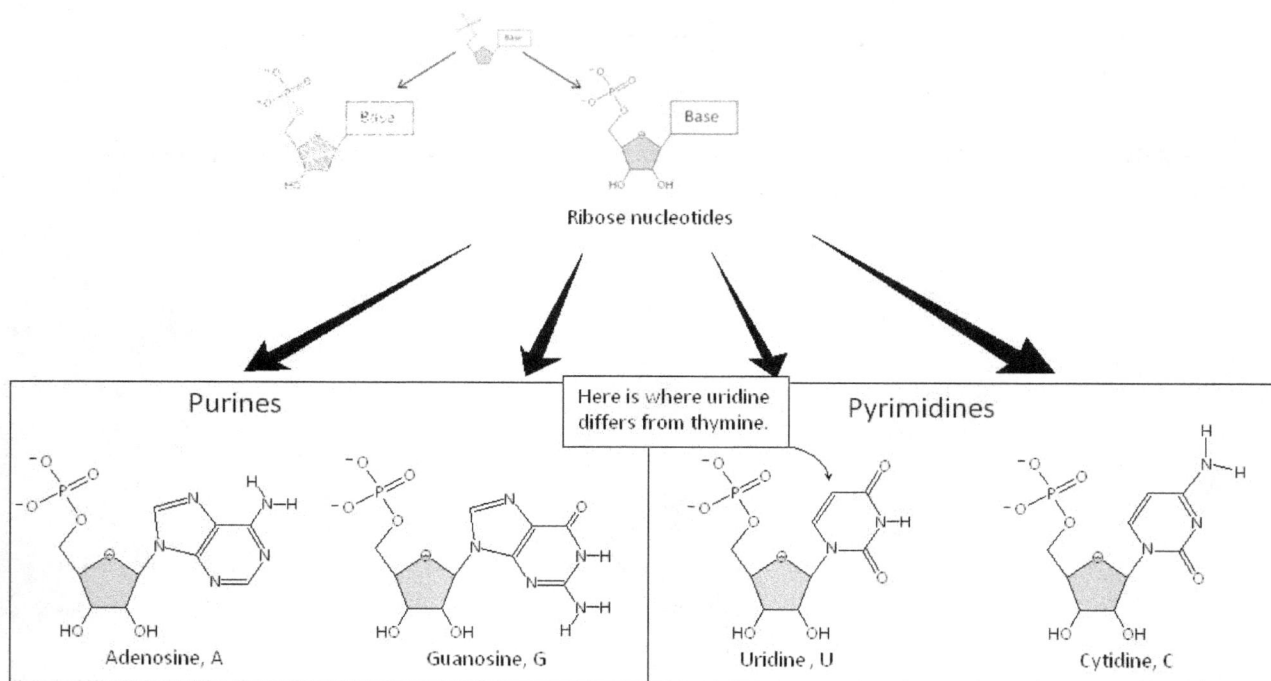

Ribose nucleotides

Purines

Here is where uridine differs from thymine.

Pyrimidines

Adenosine, A   Guanosine, G   Uridine, U   Cytidine, C

RNA nucleotides are put together into RNA polymers by dehydration synthesis, just like DNA and most other biological polymers. Although RNA can often twist into helices like DNA does, its twisting is not nearly as tight and compact as DNA. The reason is that the extra –OH tends to get in the way of the helix formation. In addition, RNA strands are not nearly as long as DNA strands, and each one tends to fold up into characteristic formations, with short stretches of helical parts.

As we shall see, the fact that RNA can clump into many weird shapes gives it a wide array of uses within living systems, kind of like what we discussed for proteins in the previous unit.

Now that we've seen the basic structure of nucleic acids, let's transition to get our first glimpse of their many roles in living systems.

The Many Roles of Nucleic Acids

DNA's main role: Genetic Information

Remember the argument about how twenty amino acid monomers could lead to an incredible number of different protein chains? A similar principle works here as well.

Let's calculate the number of ways we can put together DNA nucleotides into a chain, starting with the first one. Our first nucleotide can have one of four bases. Therefore, before we begin building the chain, we already start with four total possibilities. We now add the next nucleotide, which can also have any one of the four bases. That gives us $4 \times 4 = 16$ total possibilities. We add yet another, and once again we can choose among four. This leads us to $4 \times 4 \times 4 = 64$ possible combinations for a three nucleotide chain. As you can see, the pattern of 4^n soon emerges, where n is the number of monomers we've added to the chain. If we have 100 nucleotides, the total number would be 4^{100}, which is well over 1.6×10^{60}. To be perfectly clear how big that is:

100 nucleotides ---> 4^{100} combinations, which equals OVER:

1,600,000,000,000,000,000,000,000,000,000,000,000,000,000,000,000,000,000,000 combinations.

That's a LOT of combinations, and it's just from 100 bases! Still, it's nothing in comparison to the combinations in living systems. The number of DNA nucleotides in the **genome**[15] of any given organism could be in the *billions*. That would give us $4^{billions}$ combinations. I don't think this book is big enough to fit the number that comes out to be.

So what? It's just a bunch of mathemagic, right? Well, the fact that we can have such unimaginable combinations in the arrangement of nucleic acid chains leads to the ability to *store information*. For example, suppose we "read" the chain in segments of three nucleotides at a time. Every three nucleotides can be arranged in 64 ways. So, imagine assigning a letter or punctuation mark to each combination in the 64 combinations. You could, in theory, send a secret code to a friend by just sending the sequence of nucleic acids.

AGC CAT ATC GGC ACC ATG ATC TAT AGG CGT ACC GTC ATC CTT CAT. . . .

Could be the start of a message like:

"Dear Sam,

      I haven't the faintest clue what this idiot is talking about. I need a biology study guide to help me understand this study guide. . ."

Of course, DNA doesn't store messages in English (or any other human language). It stores a code that tells the rest of the living system how to behave on a biochemical level. How this code works and is interpreted is the focus of future sections. For the moment, it is only important to understand the significance of this system: Coded messages, stored as the sequence of base pairs in a long twisted molecule, are copied every time a living system has a child. The child, in time, copies the message as well, passing it along to the next generation, and so forth.

Copying the DNA is very easy, in principle. Since the nucleotides pair up in a very specific way, the living system can pull the DNA ladder slowly apart and build a new strand by matching up the bases on the old strand. In essence, the old DNA is used as a template for the new copy. More on that story later. . .

---

[15] Genome = complete collection of all hereditary genetic information within the nucleus

Primary roles of RNA

Protein production

RNA's primary role is in the production of proteins. Basically, RNA carries packets of information from the DNA to other parts of the living system, and these packets tell the living system the order in which to put amino acids together. If you'll recall from the unit on proteins, the order of the amino acids is the primary structure of proteins, and this structure determines the secondary, tertiary and quaternary structure as well. By determining the overall *structure*, the order of amino acids determines the overall *usefulness* of the protein. So, in this sense, RNA acts as a postal worker, carrying the blueprints from the DNA engineer to the factory where the blueprints are used to make the protein product.

Ribozymes - RNA Enzymes

Proteins are usually the first macromolecule that people think of when the word "enzyme" is mentioned, but some RNA molecules can also serve as enzymes. Whereas DNA is deliberately paired to have complementary strands, (and is thus limited to information storage), RNA is usually made without a complementary strand. They float around in water and fold up into weird shapes, primarily because many of the nucleotides are complementary to others within the *same* chain. As a result, the molecule assembles by folding and twisting, creating specific and elegant structures that we are currently referring to as "weird blobs". It shouldn't be too surprising then that they have little nooks and spaces that can serve as recognition and binding sites for small molecules to fit into. In this manner, some RNA molecules (called "**ribozymes**") act just like protein enzymes, speeding the rates of very specific reactions. Most reactions in living systems are catalyzed by protein enzymes, however; ribozymes are a definite minority as catalysts.

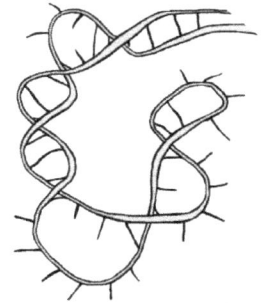

A generic but typical RNA 3-D structure

The Other Roles of Individual Nucleotides

Nucleotides have a small number of other roles besides becoming part of large nucleic acid chains. I'll list a few here.

Nucleotides as Energy Carriers

A few nucleotides are involved in energy management. In a way, you can think of a living system as an immensely complex network of chemical reactions that are all trading energy back and forth to each other. One of the most common ways in which reactions exchange energy is in the form of a molecule known as **ATP**, or adenosine triphosphate. ATP is, in a sense, the main energy "currency" of living systems, but it is not the only one used. A few other molecules, such as FAD (**flavin adenine dinucleotide**) and NAD^+ (**nicotinamide adenine dinucleotide**) are used as secondary energy currencies, but these are only exchanged between a limited number of reactions.

These particular molecules are all derived from adenosine, but the other nucleotides are used as well. **Guanosine triphosphate (GTP)**, for

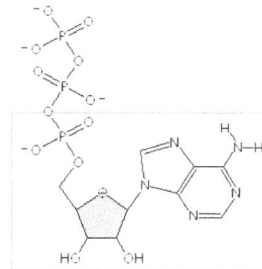

Adenosine triphosphate, ATP

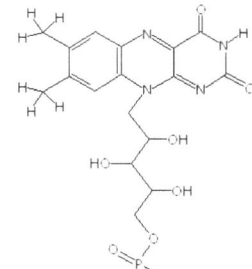

Flavin adenine dinucleotide, FAD

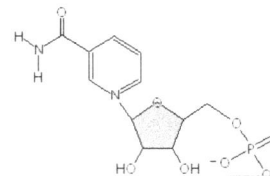

Nicotinamide adenine dinucleotide, NAD^+

example, is very similar in structure to ATP, but is used specifically as an energy source for protein production. Another example is **cytidine triphosphate (CTP)**, which is similar in structure to ATP and GTP. CTP is used primarily as an energy source for both phospholipid production and for attaching small sugar chains onto some proteins (a process called **glycosylation**). We'll be talking about the latter in a future unit.[16]

Nucleotides as Coenzymes

Some nucleotides are modified to serve as **coenzymes**. Coenzymes (meaning "with enzyme") are extra little molecules that proteins sometimes need to efficiently serve as enzymes. Think of them as assistants to proteins. Not all enzymes need them, but those that do generally cannot work well (or at all) without them. A good example of a coenzyme is Coenzyme A, which is derived from adenosine. It's shown on the next page.

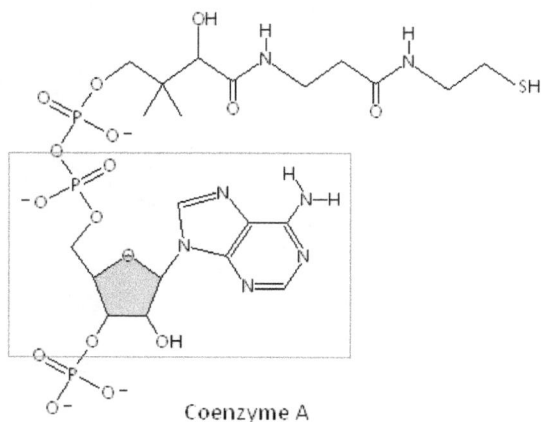

Coenzyme A fits into a couple different proteins that speed up the production or breakdown of fatty acids. It's also used in one of the final steps that prepares glucose to be fully broken down for energy in the cell. We'll be learning more about this later process in the units on glycolysis and the citric acid cycle (also called the Krebs Cycle or Tricarboxylic Acid Cycle). A detailed look at fatty acid production and breakdown is usually left for biochemistry courses.

There are lots of different coenzymes found in living systems. Not all are made from nucleotides, and not all nucleotides are converted into coenzymes.

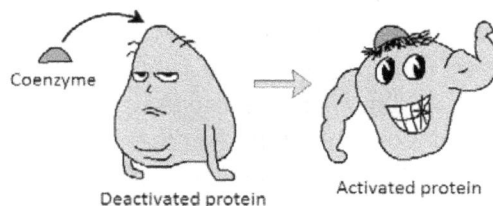

Coenzyme A

Nucleotides as Intracellular Messengers

A very small number of nucleotides are specifically made to serve as messengers within cells. They act to alert the cell to specific environmental changes, much like a scout might warn an outpost of an impending attack. These molecules usually hang out at the cell surface and start a chain reaction inside the cell when some kind of event happens outside the cell. The classic example of this type of molecule is **cyclic adenosine monophosphate**, or **cAMP**. Another example is **cyclic guanosine monophate**, or **cGMP**.

---

[16] The pattern would suggest the use of TTP, or thymidine triphosphate, as an energy source as well. Oddly enough, TTP is not used in this manner.

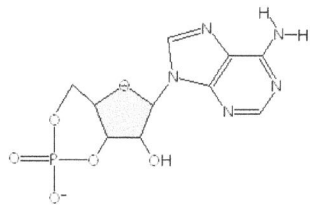

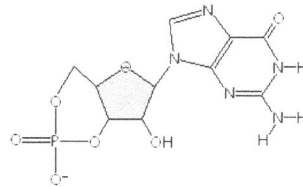

Cyclic AMP   Cyclic GMP

Summary

In this section, we looked at the basic structure of nucleic acids, starting from nucleotides and building up to the two main nucleic acid polymers: DNA and RNA. We also discussed the main roles of DNA, RNA, and a few of the individual nucleotides.

- Nucleotides are made from a 5-carbon sugar, a phosphate group and a complex carbon-ring structure known as a base.
- Nucleotides are divided into two broad categories depending on the type of 5-carbon sugar they have: deoxyribose nucleotides and ribose nucleotides. Ribose nucleotides differ from deoxyribose nucleotides by the extra –OH group attached to the 5-carbon sugar ring.
- Deoxyribose nucleotides are used to make DNA. They are typically called DNA nucleotides.
- Ribose nucleotides are used to make RNA. They are typically called RNA nucleotides.
- DNA nucleotides can have four possible bases. The molecules that result are: adenosine (A), guanosine (G), thymidine (T) and cytidine (C). Adenosine and guanosine are called purines. Thymidine and cytidine are called pyrimidines.
- Adenosine pairs with thymidine through two hydrogen bonds. Guanosine pairs with cytidine through three hydrogen bonds.
  - DNA:
    A pairs with T   G pairs with C
- RNA nucleotide bases are the same as DNA nucleotide bases, except for thymidine. RNA uses uridine in place of thymidine, but uridine looks just like thymidine except for a missing methyl group. Like thymidine, uridine pairs with adenosine via two hydrogen bonds.
  - RNA:
    A pairs with U   G pairs with C
- Nucleotides in DNA and RNA are connected together via dehydration synthesis to make DNA and RNA chains, respectively.
- Because nucleotides pair with each other, DNA chains form pairs as well. These paired chains twist around each other to make long DNA double helices.
- RNA chains are very short and many tend to curl up into blobs. This behavior is due to the nucleotides pairing to others within the same chain.

- DNA is used primarily for information storage. The sequence of nucleotides in the chain codes for the instructions on how the rest of the living system should behave biochemically.
- RNA has many roles, but its main role is to carry the information stored in DNA to the rest of the biochemical machinery of the living system to make proteins.
- RNA can also serve as enzymes called ribozymes. Like protein enzymes, these work to speed the rate of a very specific number of reactions. This is a very minor role for RNA.
- A few molecules, (most of them derived from adenosine), are used as energy currency within living systems. The most common is ATP.
- A few of the nucleotides, namely adenosine, can be converted into coenzymes, such as coenzyme A. A coenzyme is a small molecule that helps enzymes function properly.
- A few molecules derived from various nucleotides are used as messengers, informing cells of changes in chemical circumstances outside the cell. cAMP and cGMP are the most important examples.

In this section we cover:

1) Physical and chemical gradients

2) How gradients break down living systems

3) How living systems use gradients to get things done.

4) Osmosis and its dangers to cells

5) Isotonic, Hypertonic, and Hypotonic solutions

Unit 12

Diffusion and Osmosis

Until this point, we've talked about the basic types of molecules that are found in living systems. The next development in the discussion about the structures of life will be about cell membranes and then about the structures inside the cell. Before we can introduce that, however, it's important to understand the concepts of **diffusion** and **osmosis**. Once we have a firm grasp on these principles, it will allow us to develop a greater appreciation for the structure and importance of the cell membrane and other cell components.

Diffusion

The phenomenon of diffusion is simple, and a little experiment will help us outline it. Suppose you have a glass of water. Now, you put a drop of food coloring into the water. What happens? The food coloring slowly begins to spread out. Over time, it eventually spreads evenly throughout the water. This is diffusion.

Why does diffusion occur? Liquids are basically collections of molecules that are tumbling, rolling and shoving their way past each other in constant motion, like people in a dance club. The drop of food coloring is a huge clump of different molecules that are suddenly put into that highly energetic environment. That clump is said to be in high initial **concentration**,[17] meaning there is a lot of it squished into a small starting space. The water will push and shove these new molecules in random directions. Slowly over time the food coloring molecules are pushed farther and farther from each other, until finally they have all been moved at random throughout the entire liquid. They will

---

[17] Concentration is defined as the number of particles per given area of volume. A large number of particles in small volume gives a high concentration. A small number in a large volume gives a low concentration.

continue to be pushed and shoved, but there is no reason to suspect that they will all be pushed randomly back into the starting place. In fact, the odds are downright astronomically remote for that to *ever* happen, unless there is some specific mechanism or machine that is purposely doing that.

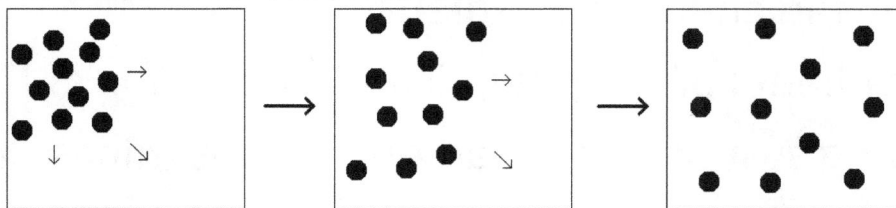

The end result of our little experiment is that the food coloring went from a high concentration (the space where it started) to low concentration (spread throughout the glass). Any time you have a difference in concentrations, there exists a **concentration gradient**. A gradient refers to a difference in intensity of some physical characteristic, like concentration, pressure, temperature, or energy. For example, here is what a concentration gradient might look like (high concentration at the left, low concentration at the right):

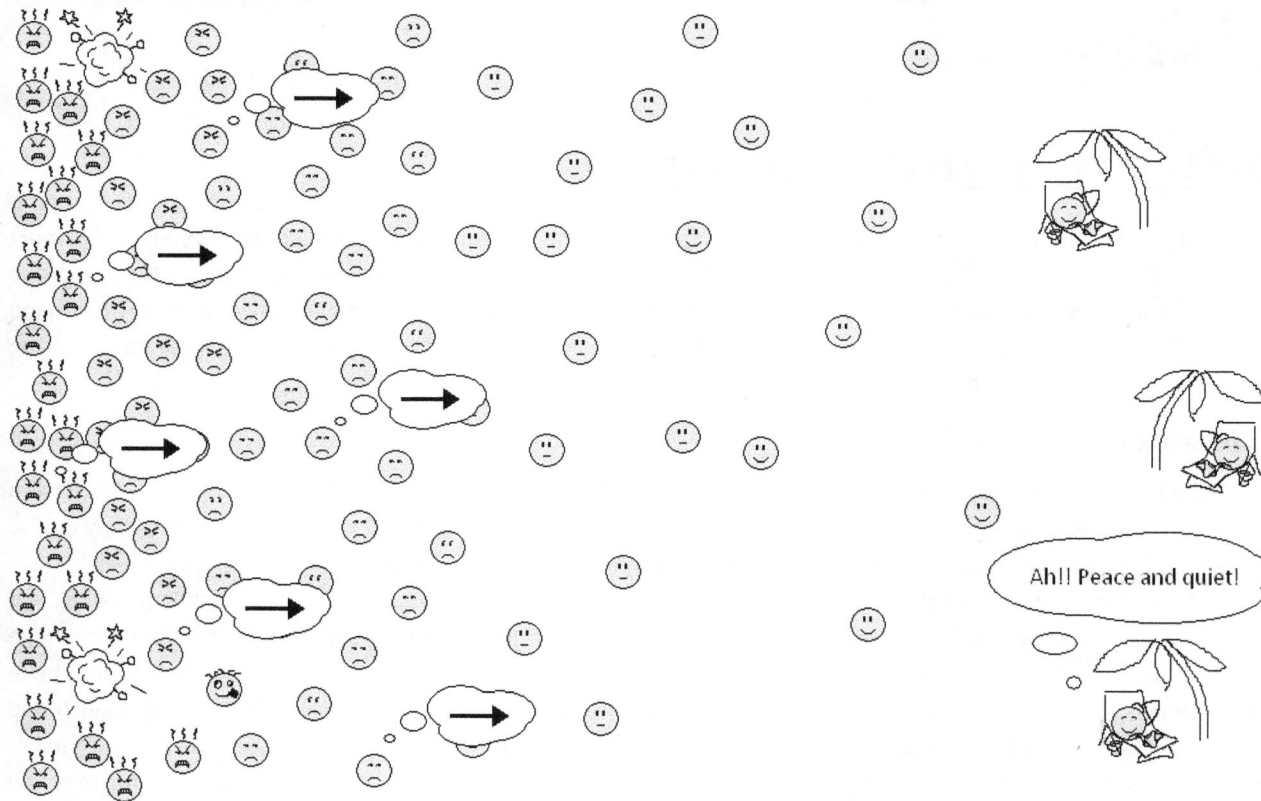

On the left, the faces are crowded and irritated. On the right, they are spread out and relaxed. As several of the little faces indicate, there is a general tendency for the chemicals to want to go to the right.[18] That is the direction of the gradient . . . "crowded" to "spread out," or high concentration to low concentration. It's always high to low.

If we expand upon this, a **temperature gradient** might look like the figure on the right. Suppose a hot stone is dropped into shallow water. The heat will spread from the hot region outward toward the cold region.

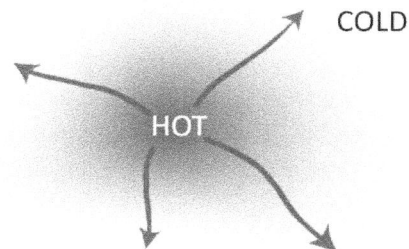

Once again, the gradient points from high to low, and eventually the heat will dissipate and the rock will no longer be hot. When gradients exist, physical systems tend to change in a way that eliminates the gradient. In that way, it behaves just like our food coloring did. This is the natural order of things, and unless we put work into reversing the process, all gradients will slowly degrade until the temperature, the concentration, the pressure and the energy are all spread out evenly. In that respect, pressure and energy gradients would look similar to our temperature gradient picture.

The idea of chemical and physical gradients is especially important for biological systems. Think of your body as a high concentration of all kinds of chemicals smooshed into a limited space. The natural tendency is for those chemicals to want to spread out. Tiny structures within your body are constantly breaking down and spreading out, and so without energy and materials, you quickly find yourself on the losing end of a timeless battle.

Pay energy fee, or no admission!

---

[18] Despite my drawings, try to resist thinking of chemicals as having human ambitions. Saying that chemicals "want" to move from high to low concentration is meant as an analogy that describes a property derived from the Laws of Thermodynamics.

Gradients aren't always bad, however, for on a microscopic level cells sometimes use gradients to get things done. For example, the little smiley faces we just encountered all want to move toward the right. Chemicals behave in a similar fashion, and cells often put barriers (such as membranes) in the way of that movement, letting the chemicals through only after they've donated a portion of their energy (see previous image). That energy is then used to get other things done.

So, while gradients are always working to break down the living system, the living system can use gradients to acquire energy to get things done. In the war against gradients, we all eventually lose, but in the short term we win little battles here and there, getting a few things done in the process. All of this is due, in essence, to the basic concept of diffusion. Let's expand upon this.

Osmosis

While cells can often acquire energy from chemical gradients using membrane barriers, they must be careful when doing so. There is a potential for a dangerous process called **osmosis** (i.e. "water movement"). Suppose I put a HUGE amount of Na^+ ions inside a cell, but I don't give them a way to escape. In other words, the cell's membrane around the perimeter is a barrier that doesn't let them leave very easily. Ideally, the Na^+ will want to diffuse down their concentration gradient, (i.e., leave the cell to spread out with as much water between them as possible). They're in a pickle of a situation, because they can't leave. What happens next is rather bizarre. Water begins diffusing through the membrane.[19] The water coming into the cell is successfully diluting the concentration of Na^+ inside the cell, but it also causes the cell to expand. If too much water comes in, the membrane breaks, causing the cell to pop and die.

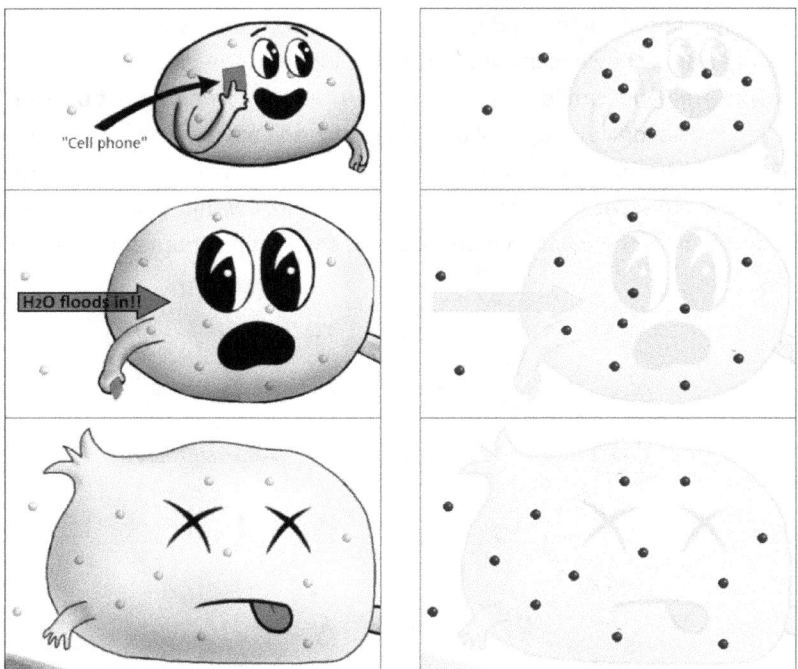

The second column in the image on the previous page focuses on the sodium ions themselves. I've grayed-out the cell drawings to concentrate on the real physical phenomenon. What we see is the classic chemical gradient behavior. The sodium spreads out, and in this case it bursts the cell in the process.

---

[19] As we will see in future sections, water can readily go through membranes.

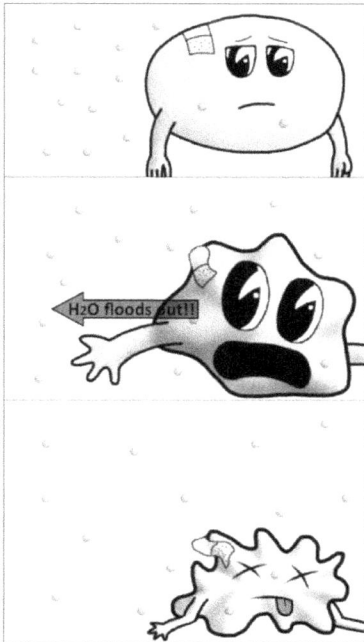

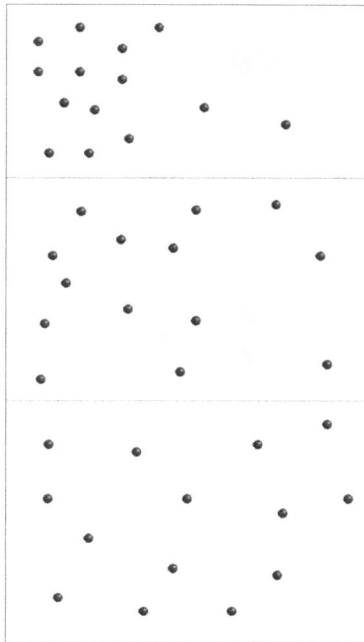

The direct opposite of this can happen as well (see image to the left). If there is a greater concentration of Na^+ outside the cell than inside, water will leave the cell, causing the cell to shrivel. Eventually, it loses too much water and dies.

Once again, the second column focuses on the sodium ions to point out the real physical phenomenon. Diffusion makes ions want to spread out, and if cells are not careful they will get crushed or broken in the process. We regularly use this phenomenon to our advantage by packaging food with high sugar or salt content to protect it from bacteria. When the bacteria attempt to invade the food, they encounter the incredibly high concentration of sugar or salt, which causes them to shrivel and die.

Wait. First I said that cells can use membrane barriers in order to acquire energy from concentration gradients. Now it seems that having such barriers causes the cell great harm. If differences in concentration across membrane barriers can be so disastrous, how and why do cells even do it?

To answer this, let's take a closer look at osmosis. Osmosis is the movement of water. In the above scenario of the exploding or shrinking cell, we only looked at the effect of the concentration of *one* ion type in the movement of water. In actuality, the water transfer in or out of a cell depends on the *total* concentration of *all* dissolved chemicals on both sides of the membrane. Individual concentrations may differ, but as long as the combined concentration is equal inside and outside the cell, the water will stay put. For example, if there is more Na^+ outside

the cell than inside, and more K^+ inside than outside, then the net effect is that water will not move. Think of it as Na^+ playing tug-of-war with K^+, each pulling on the water. The high outside Na^+ concentration wants to pull water out. The high inside K^+ concentration wants to pull water in. Their efforts cancel, and the water stays where it is. Neither chemical is able to spread out as it wants.

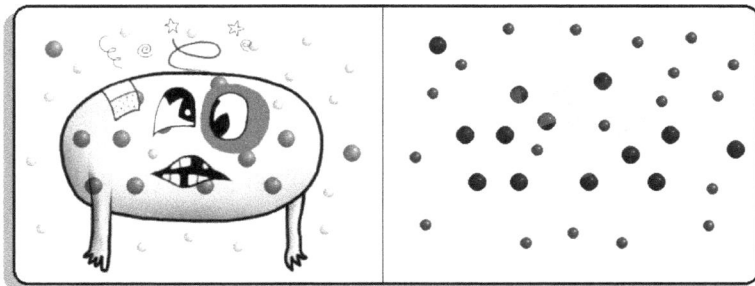

Another way that one might look at the phenomenon of osmosis is through the "osmotic strength" of a solution.[20] If one solution of chemicals has the same overall concentration as another solution of chemicals, the

---

[20] A solution is a liquid with other molecules mixed in. In the case of biology, we are usually talking about water with chemicals mixed in.

solutions are said to be **isotonic**, ("same strength"), to each other. If one solution has a greater combined concentration that another solution, it is said to be **hypertonic** ("greater strength") to that solution. The second, weaker solution is called **hypotonic** ("less strength").

Here is a picture to round out this part of the discussion:

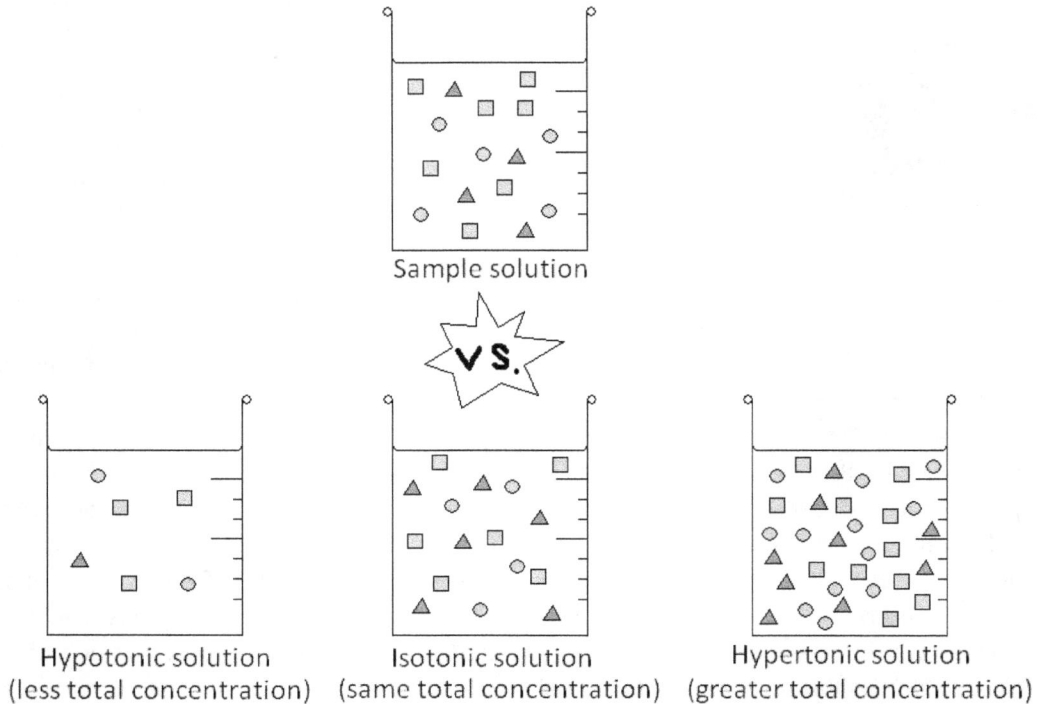

Sample solution

VS.

Hypotonic solution
(less total concentration)

Isotonic solution
(same total concentration)

Hypertonic solution
(greater total concentration)

So, let's summarize a cell's behavior in terms of isotonic, hypotonic and hypertonic solutions. In an isotonic solution, a cell will have the same osmotic strength as the solution, so no water moves; the cell does not shrink or expand. In a hypotonic solution, the cell has greater osmotic strength than the "lesser strength" solution, so the cell pulls water into it. The cell expands and pops. In a hypertonic solution, the cell has less osmotic strength than the "greater strength" solution; the cell shrinks as water is pulled from it.

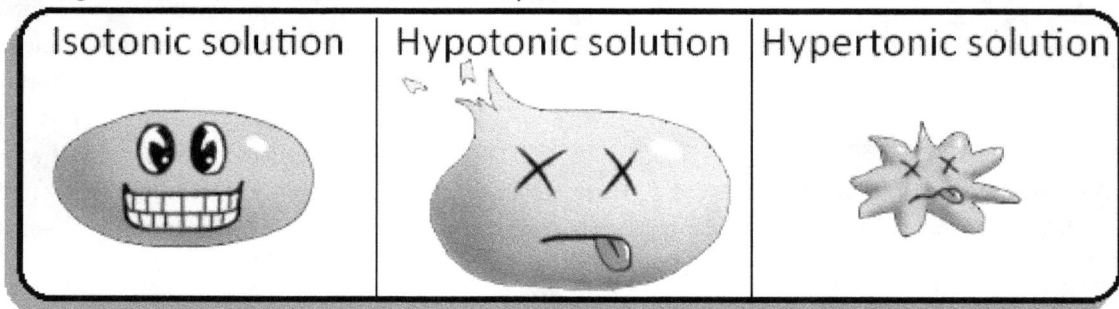

| Isotonic solution | Hypotonic solution | Hypertonic solution |

Summary

In this section, we discussed diffusion, concentration gradients and osmosis.

- A physical or chemical gradient is an intensity difference in some measurable quantity. A system containing a gradient will tend to change in a way that slowly reduces the gradient, unless some external force such as a machine counteracts it. So:
    - Chemicals are always trying to spread out, whenever possible.
    - Heat spreads out from high temperature to low temperature, eventually leveling out the temperature.
    - Differences in pressures tend to level out.
    - Differences in energy tend to level out.
- Living systems are constantly breaking down, in part because the chemicals in them always want to spread out.
- Living systems sometimes use membrane barriers between gradients, forcing the molecules in the high concentration area to use part of their energy to get to the low concentration area. This "taxed" energy can then be used to get other things done in other parts of the cell.
- Osmosis is the movement of water from an area of low ion/molecule concentration to an area of high ion/molecule concentration. The effect is due to the combined concentrations across a barrier.
- Isotonic solutions have the same osmotic strength, so no water will move between them.
- A hypertonic solution has greater osmotic strength than some test solution (or cell) in question. Hypertonic solutions will pull water from the other solution.
- A hypotonic solution has less osmotic strength than some test solution (or cell) in question. Hypotonic solutions will have water pulled from them by the other solution.
- Cells placed in isotonic solution do not expand or shrink. Cells placed in hypertonic solutions shrink. Cells placed in hypotonic solutions expand and burst.

Unit 13

The Cell Membrane, Part 1: Structure

The cell is the basic unit of life. As such, the **cell membrane** (or plasma membrane) is the cell's interface with the outside environment. The cell membrane is an incredibly complex structure and serves to play three key roles: 1) it keeps the cell's contents inside the cell and everything else out, 2) selectively exchanges material across it and 3) communicates with the world outside, including other cells.

Key Roles of Plasma Membrane:
1) Isolate the cell from the outside
2) Exchange materials selectively with the outside
3) Communicate with the outside

The main molecule type in membranes is the phospholipid. Phospholipids line up one by one with their nonpolar tails snuggling up against each other in an attempt to avoid water. Their polar heads all poke into the watery surroundings. In this way, they form a sheet of phospholipids, like a crowd of people jammed into a stadium, standing room only. The nonpolar tails of the phospholipids touch the nonpolar tails of yet another layer of phospholipids whose heads point in the opposite direction, much like the same crowd standing on one massive mirror. Altogether, this arrangement is called the **lipid bilayer** (or **phospholipid bilayer**).

This lipid bilayer acts to isolate the contents of the cell, called the **cytoplasm**, from the water outside the cell, called the **extracellular fluid**. Ions and polar compounds are hydrophilic and usually have a hard time passing through the lipid bilayer, because it is a nonpolar barrier. The smaller the polar molecule, however, the less perfect its

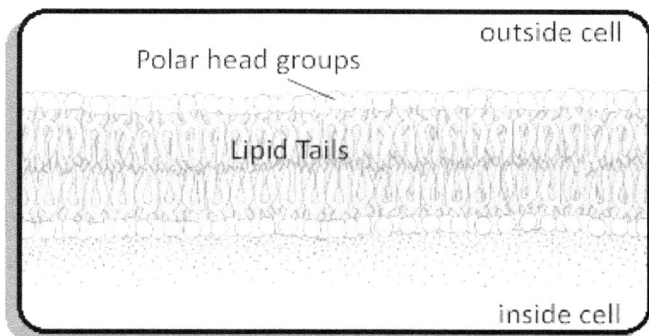

exclusion from the cell may be. Water, as we have seen in the case of osmosis, can pass a little at a time through the membrane, leading to potential complications for the cell.

The cell membrane has more than just phospholipids. There are also embedded proteins called **membrane proteins** that dot the cell's surface like little islands in a sea of lipids. Many membrane proteins are attached to fibrous proteins that run just under the inside surface of the cell membrane, forming a rigid skeleton that not only keeps the membrane proteins from moving, but also gives the cell its shape. There are membrane proteins that are NOT attached to the underlying network as well. These proteins drift around like icebergs or ships on a membrane sea. This vision (i.e. membrane proteins moving about on the cell surface as if floating in a fluid) is referred to as the **fluid mosaic model**.

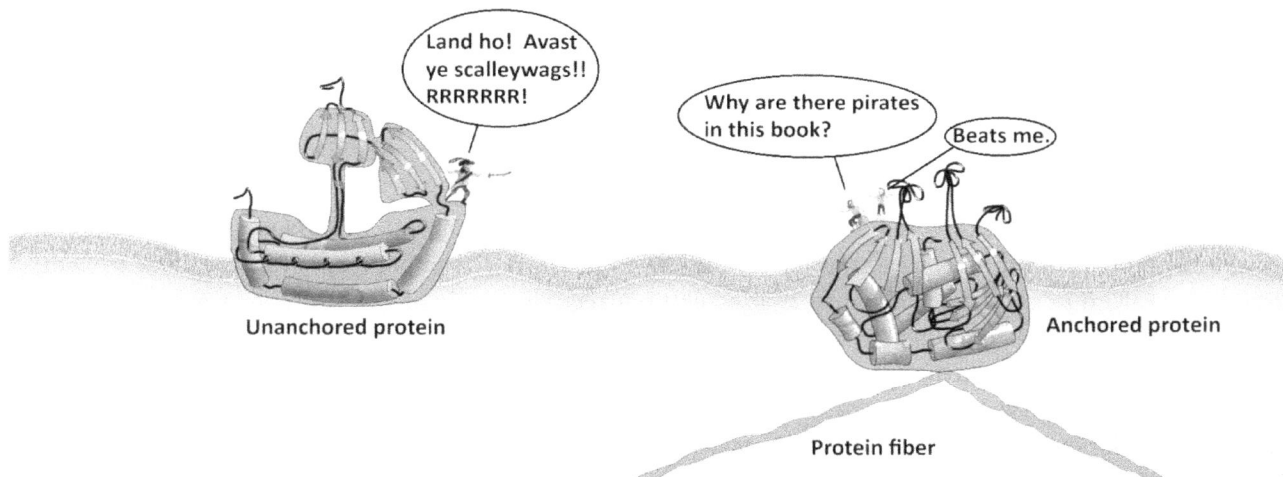

Many membrane proteins have carbohydrate chains attached to them, like the palm trees on the island. The proteins that have these carbohydrates are called **glycoproteins**. These are added in the process called glycosylation, which we referred to briefly in the nucleic acids section, Unit 11. There are a number of purposes for glycoproteins. First, their carbohydrates have a lot of hydrogen bonding capacity so that they tend to gather and hold a lot of water around them. That makes cell membranes rather mucousy on the outside. Second, glycoproteins serve the purpose of cell recognition, helping other biological molecules know what kind of cell it is, (liver

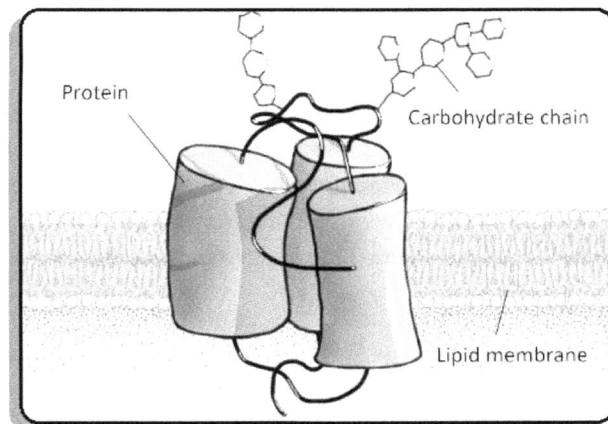

93

cell vs. heart cell vs. brain cell, etc.).

Finally, the lipid bilayer also contains numerous cholesterols (steroid molecules). The more cholesterol that is contained in the membrane, the less stiff the membrane will tend to be. Cholesterols also make the membrane less permeable to little molecules, such as water.

Membrane Proteins

So we know that membranes have lots of little island-like proteins in them, some moving, some not. Let's focus a little more on these things. There are three major types of membrane proteins: receptor proteins, recognition proteins and transport proteins. **Receptor proteins** bind to very specific molecules that float around in the extracellular fluid, and when these molecules attach, receptor proteins set off a series of chemical reactions within the cell.

The chemical reactions that they set off generally start with the production of cAMP or cGMP, two nucleotide-derived chemicals we talked about in Unit 11. Think of receptor proteins as little switches on the outside of the cell that only certain chemicals can turn on, and when they do the cell changes behavior in some way.

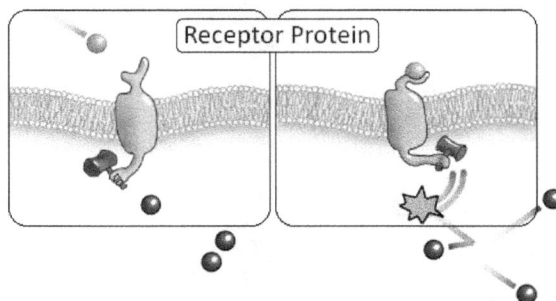
Receptor Protein

Recognition proteins serve to mark cells, like tags (i.e. lung, intestine, liver, etc.). For multicellular organisms with immune systems, immune cells can cruise around and check for the recognition proteins on a cell to tell what type of cell it is (friend or foe). Recognition proteins can bind to recognition proteins from other cells and inhibit the cell from growing or replicating. This is known as **contact inhibition**. Failure of contact inhibition to stop a cell from reproducing is believed to be one possible cause for cancer.[21] As mentioned above, most recognition proteins are glycoproteins, meaning that they have polymers of carbohydrates sticking out into the extracellular fluid.

Transport proteins serve to move ions and polar molecules across the membrane, which have a hard time passing through the lipid bilayer due to its nonpolar nature. As such, transport proteins act as specific gates through which some chemicals can pass. Transport proteins are very specific, however, and there are a number of different types. The first type is the **channel protein**. Channel proteins are transport proteins that are basically just hollow tubes through the membrane. The size of the opening usually matches the size of very simple ions, such as K^+ or Na^+, and so these proteins allow only one type of ion to enter or leave the cell.

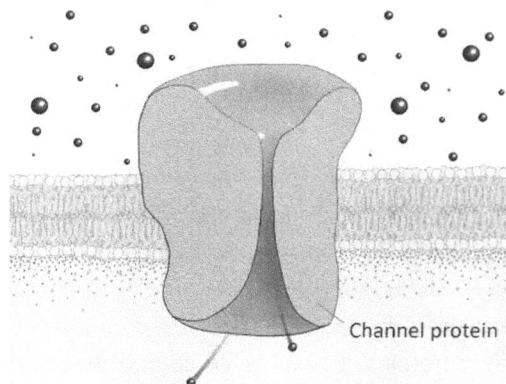
Channel protein

---

[21] A cancer is a group of cells that grow and multiply inside the body without regulation, causing problems to the surrounding organs because they take up greater and greater space. After a certain point, some of these cancer cells detach from the original growth and move to other areas of the body. This is called **metastasis**, and is generally the point at which cancer becomes much more difficult to treat medically.

A second type of transport protein is a **carrier protein**. Carrier proteins bind to specific molecules, and when they do, they change shape and eject the chemical to the other side of the membrane. Carrier proteins usually require energy to complete their cycle, and they are often used to force molecules to move backwards up their concentration gradient. The best example of such a carrier protein is the **sodium-potassium pump**. This particular protein grabs 3 Na^+ and throws them out of the cell, and then pulls 2 K^+ ions into the cell from outside. For many multicellular animals, the sodium concentration is higher outside the cell, so sodium would rather be inside the cell. Likewise, the potassium is higher inside the cell, so it would prefer to leave. The sodium-potassium pump thus works to force each type of ion to go in a direction it doesn't want to go. In other words, the Na^+-K^+ pump helps to maintain the sodium-potassium gradients, which would otherwise diminish (as gradients always try to do). Without the Na^+-K^+ gradients, the osmotic strength from *other* chemicals would cause water to rush into the cell. This kind of fate can be observed in cells that have malfunctioning or poisoned[22] Na^+-K^+ pumps; they swell and pop. In addition, cells use the sodium gradient to help pull in sugar, which otherwise would not want to enter the cell. Further, some cells, like nerve and heart cells, use the sodium-potassium gradients as part of their electrical conduction capabilities. That is a topic that is typically studied in the second semester of biology and beyond, so we'll make a graceful exit from that discussion.

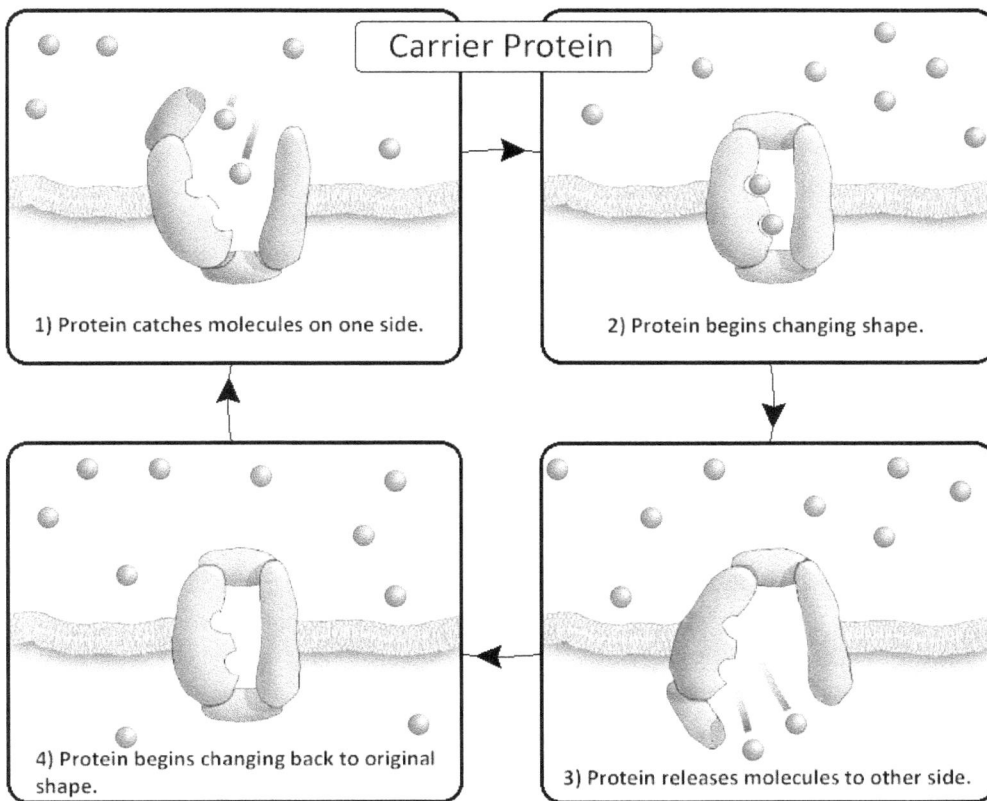

Carrier Protein

1) Protein catches molecules on one side.

2) Protein begins changing shape.

3) Protein releases molecules to other side.

4) Protein begins changing back to original shape.

---

[22] Ouabain, an African arrow poison, will bind to the Na^+-K^+ pump and cause it to stop working. This will make the cell expand and burst.

One final note on the topic of membranes should be relayed. Cellular membranes aren't only found at the perimeter of the cell. Membranes also surround many of the smaller pieces that are found inside the cell, and these latter membranes typically have the same overall structure and functions. We'll be seeing more about that in later units. More immediately, we'll be taking a closer look at the concept of membrane transportation. It's a complex topic involving more than just transport proteins, so we'll be devoting the next unit to it.

Summary

In this section, we discussed the structure and function of the cell membrane.

- Cell membranes serve three major roles:
    1) They keep the cell components inside and keep the external world out of the cell.
    2) They selectively exchange materials in and out of the cell.
    3) They communicate with the outside world, notifying the cell if conditions have changed.
- Cell membranes are made of two layers of phospholipids. Molecules of each layer line up so that their nonpolar tails are resting against each other to minimize their exposure to water. The two layers touch each other at the tips of their tails.
- Within the phospholipid membrane are proteins. Some are anchored to a fibrous protein network just inside the cell. These proteins do not move. Other proteins are not anchored and float around freely in the membrane. This is the basis for the fluid mosaic model.
- The phospholipid membranes are also called lipid bilayers. They are very nonpolar. Ions and most polar molecules have a hard time passing through the membranes.
- There are three categories of proteins found in the plasma membrane: receptor proteins, recognition proteins and transport proteins.
- Receptor proteins bind to molecules from the outside and start a chain reaction inside the cell. The chain reaction alerts the cell about the conditions outside the cell.
- Recognition proteins are used to identify the type of cell that it is. Cellular recognition usually leads to contact inhibition, which stops cells from overgrowing their neighbors to become cancerous. Recognition proteins are also used by immune cells to distinguish friend from foe at the cellular level.
- Transport proteins are used to move chemicals from one side of the membrane to the other. There are two broad categories for transport proteins: channel proteins and carrier proteins. Channel proteins are like tubes that allow specific ions/molecules to pass. Carrier proteins bind to specific ions/molecules, change shape and then release those entities across the membrane.
- The sodium-potassium pump is an important carrier protein. Without it, the sodium-potassium gradients cannot be maintained, and cells pop.
- Membranes can also be found surrounding certain smaller pieces inside the cell. These membranes look and behave in many of the same ways as the membrane surrounding the entire cell.

Unit 14

The Cell Membrane, Part 2: Transportation across the Membrane

As we saw in the last unit, the cell membrane is a complex barrier made of a phospholipid bilayer and embedded proteins. Most molecules have a hard time getting through this barrier, so the cell has to have a number of ways to help the chemicals. We talked briefly about transport proteins. In this unit, we expand on the topic of how things get across the membrane, as the full topic is much more complex than simple transport proteins.

There are two main types of membrane transport: **passive transport** and **active transport**. Let's look at them one at a time.

Passive Transport

Passive transport occurs when molecules simply diffuse down their gradients across the membrane. There are two types. The first type is **simple diffusion**. This happens when molecules diffuse across the membrane with absolutely no assistance from the cell; the cell does not need to do anything to get it to happen. We touched upon an example when we talked about osmosis. Water simply diffuses passively across the cell membrane. Other examples include many gases and lipid soluble compounds, (such as cholesterol-based hormones). Gases are usually small, nonpolar molecules, and lipids are nonpolar, so that neither chemical type has difficulty crossing the lipid bilayer. Simply put: simple diffusion is for molecules that do not have trouble getting across the membrane on their own. They diffuse easily and get into the cell faster than any other molecule, no matter the means.

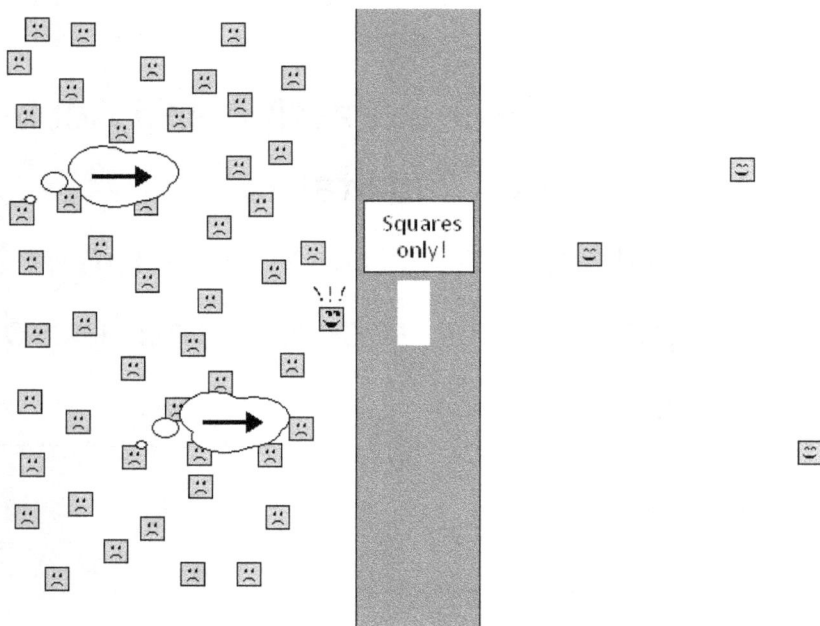

The second type of passive transport is **facilitated diffusion**. This happens when molecules want to diffuse across a membrane, but need special help in order to do so. This is often the case for most polar molecules and ions. Imagine, for example, a lot of square polar molecules on one side of a membrane, and very little on the other side, as depicted in the drawing to the left. Their concentration gradient is across the membrane, but they can't cross it because the membrane is nonpolar. Now imagine that a door has just been installed in the membrane with a sign that says "squares only." Any square that is floating around can go through the door and down its gradient, so we say its diffusion is "facilitated." In the same way, a cell can have a barrier membrane that permits only specific molecules.

Another form of facilitated diffusion occurs when a molecule diffuses into a cell *against* its concentration gradient because it is paired with a different molecule that is diffusing *down* its concentration gradient. That sounds a bit complicated, so let's take an example. Suppose we reconsider the situation of the many squares outside the cell that REALLY want to get in. Now suppose there are triangles inside the cell that want to leave. The cell doesn't want

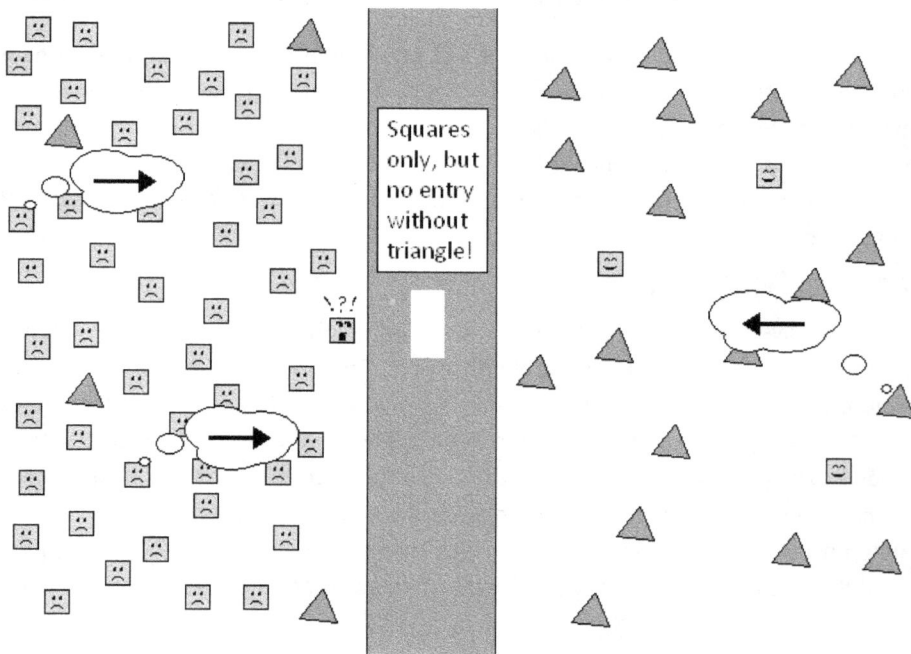

98

the triangles to leave. In fact, it wants more triangles. This would be the case if the triangle represents a molecule that the cell needs for energy or as a component of the cell. So, the cell alters the door to say: "Squares only, but no entry without a triangle." In this way, the cell uses the tendency of one molecule to float in (the square), to force another molecule (the triangle) to come in against its will, so to speak. The diffusion of the second molecule, here the triangle, has been "facilitated" by pairing it with a square.

A real chemical example of this is Na^+ and glucose. Sodium is like the squares; it wants in. Glucose is like the triangles; it wants out, but the cell doesn't want it to leave. To get more glucose to come in, the cell forces sodium to bring glucose with it, or it doesn't get admission into the cell.

Active Transport

Passive transport uses the energy locked in the chemical gradients. Active transport, on the other hand, relies on the cell's energy to move chemicals across the membrane. There are three broad types: **protein active transport**, **endocytosis** and **exocytosis**.

We saw an example of protein active transport when we looked at the sodium-potassium pump in the previous unit. The protein goes through a cycle in which it 1) takes up the chemical(s) in question, 2) changes shape, 3) deposits the chemical on the opposite side of the membrane and then 4) changes back into its original shape for another cycle. This cycle takes chemicals and moves them backwards up their gradients. Since this goes against the gradient, it requires energy to do so. To power the cycle, cells use high energy molecules (usually ATP, see Unit 11). These high energy molecules bind temporarily to the protein pump, give it the energy it needs, and then break off, exhausted and spent like a dead battery. In the case of ATP, it breaks into ADP and phosphate during the process of charging the protein.

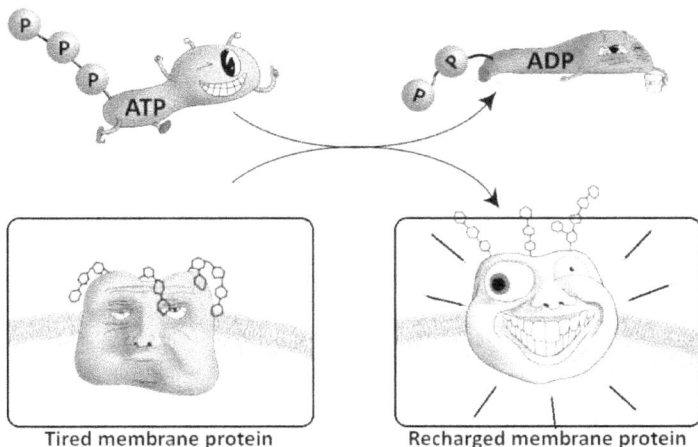

Tired membrane protein   Recharged membrane protein

As this image shows, ATP gives its energy to the protein, and the protein is recharged to do its job. A charged sodium potassium pump will eject 3 Na^+ from the cell and pull 2 K^+ into the cell in one cycle, then it needs to be recharged again by another ATP.

Whereas protein active transport involves small molecules pumped a few at a time across a membrane by a simple protein, endocytosis and exocytosis involve the mass transport of BIG things into and out of the cell, respectively. By "BIG things," I refer to objects such as another cell or a piece of a cell that is too large for protein transport to handle and certainly too large for any other means of diffusion. There are a few different types of endocytosis, depending on what is being taken in and the mode of delivery. Those types are **phagocytosis**, **pinocytosis** and **receptor-mediated endocytosis**. Phagocytosis is akin to the cell "eating" a large thing floating by in the extracellular fluid. The cell senses the object through various protein interactions at its surface (the details being beyond the scope of this book) and slowly wraps its membrane around the object, much like a mouth eating a piece of food. When the membrane wraps around the object entirely, it joins together, creating a membrane pouch (or

vesicle) around the object, which is now inside the cell. Once in this state, the cell employs a variety of means to break down the object. The pieces that come off of the object leave the pouch by simple diffusion, facilitated diffusion or protein activated transport.

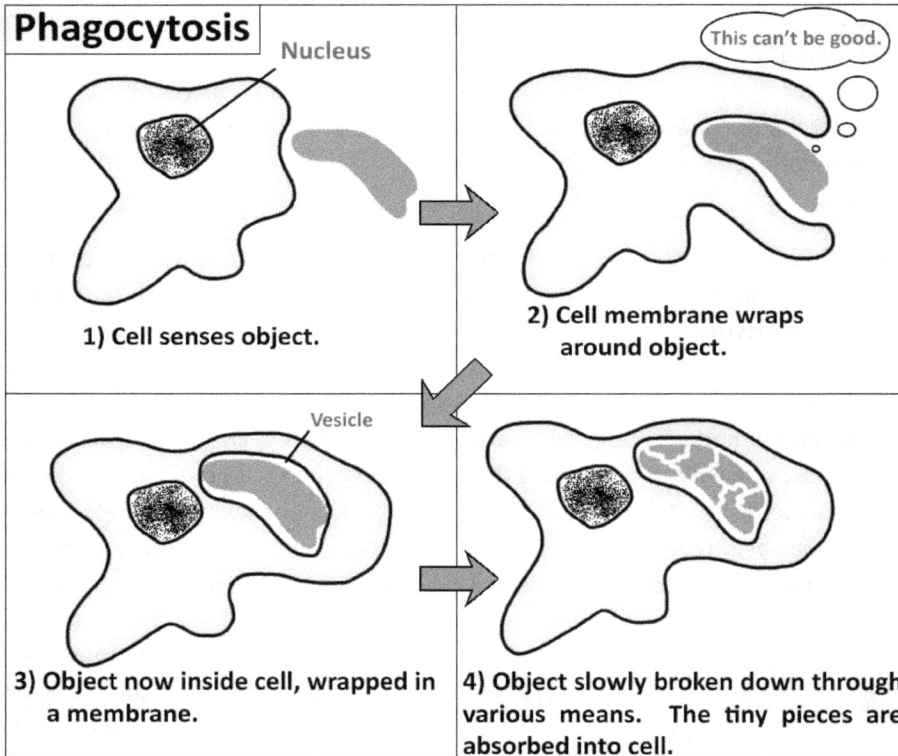

Phagocytosis

Nucleus

This can't be good.

1) Cell senses object.

2) Cell membrane wraps around object.

Vesicle

3) Object now inside cell, wrapped in a membrane.

4) Object slowly broken down through various means. The tiny pieces are absorbed into cell.

Pinocytosis is a similar process, but instead of the membrane wrapping around an object, it wraps around a big section of extracellular fluid, like biting water. As with phagocytosis, this creates a membrane-bound vesicle inside the cell. Inside the vesicle are extracellular fluid and whatever chemicals were in it. The materials in the pouch then diffuse through the membrane into the cell, (like it does in phagocytosis).

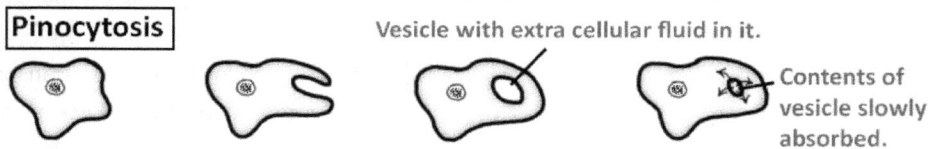

Pinocytosis

Vesicle with extra cellular fluid in it.

Contents of vesicle slowly absorbed.

Pinocytosis is considered non-specific. The cell is simply gulping down a volume of extracellular fluid. Receptor-mediated endocytosis is a similar process, but unlike pinocytosis it is initiated by specific molecules binding to various receptor proteins that sit in the cell membrane. The binding of these molecules causes a protein network (usually proteins called **clathrin**) to pinch a section of membrane inward. This pulls a vesicle full of the extracellular fluid into the cell.

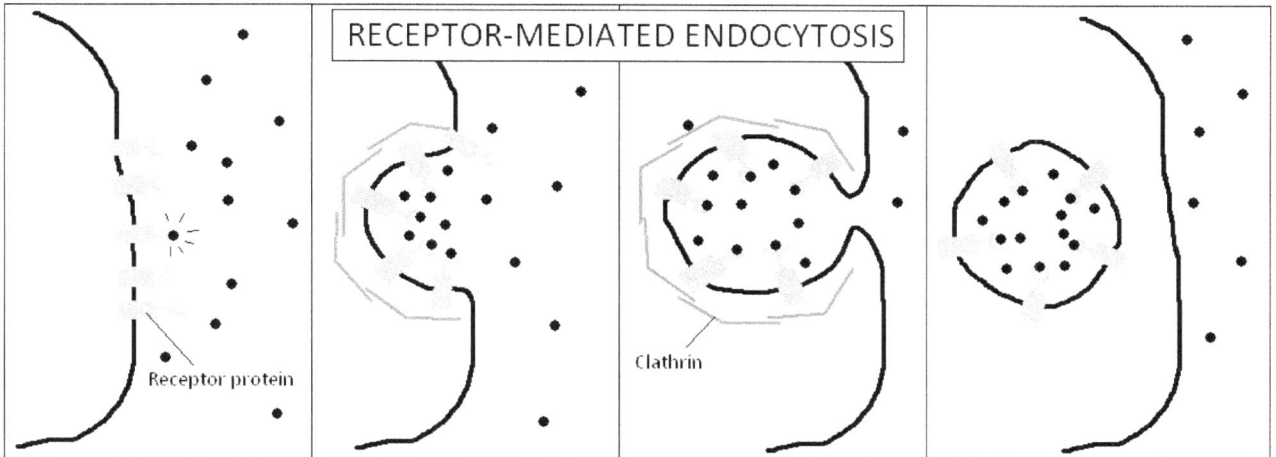

RECEPTOR-MEDIATED ENDOCYTOSIS

Receptor protein

Clathrin

Exocytosis is the opposite of endocytosis. In exocytosis, a vesicle combines with the cell membrane from the inside, like a bubble coming to the surface of water. When such a bubble pops, the air inside the bubble is released to the atmosphere. Similarly, when the vesicle combines with the membrane and "pops," the material inside the vesicle is released to the outside of the cell.

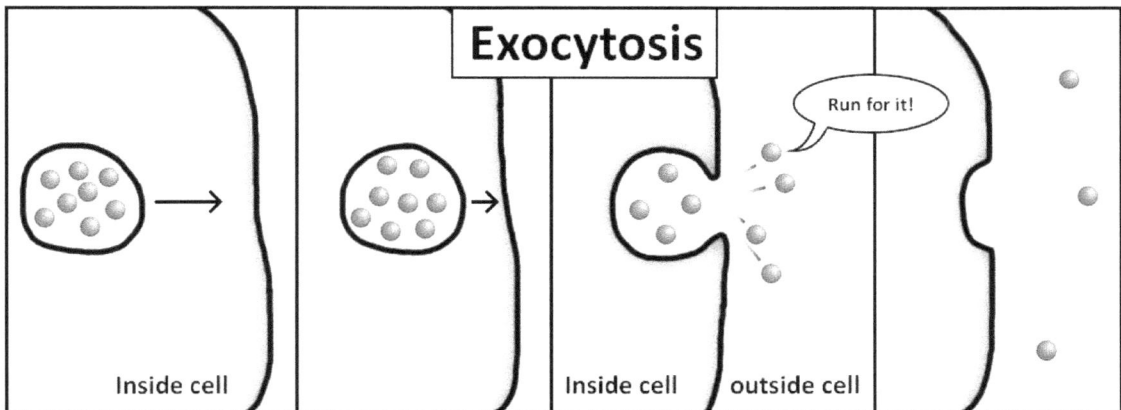

Exocytosis

Run for it!

Inside cell

Inside cell | outside cell

With that, we bring to close our discussion of the basic methods of transporting materials across the cell membrane. In the next section, we'll conclude our discussion of the cell membrane by taking a look at plant cell walls, bacterial walls and a few other additional structures found near the membranes.

Summary

In this section, we talked about the basics of transportation across cell membranes.

- There are two broad categories for membrane transportation: passive transport and active transport.

- Passive transport uses the energy in chemical gradients to help chemicals cross membranes. There are two types of passive transport.
    - <u>Simple diffusion</u> – molecules that can cross the membrane on their own, like water and gases. This is the fastest means that chemicals enter or leave cells. The most important example is osmosis, (the movement of water).
    - <u>Facilitated diffusion</u> – protein channels help chemicals diffuse down their gradients. Sometimes one molecule diffuses down its gradient, dragging another molecule up its gradient, like Na^+ and glucose into the cell.
- Active transportation requires the cell's energy to get activated. There are several types.
    - <u>Protein active transport</u> – uses proteins to pump molecules up their gradients. These pumps require high energy molecules like ATP to keep charged.
    - <u>Endocytosis</u> – large scale movement of materials across the membrane via vesicles that are made when the membrane surrounds the materials.
    - <u>Phagocytosis</u> – the cell folds its membrane around a large object, like another cell, and pulls it in.
    - <u>Pinocytosis</u> – the cell gulps a volume of extracellular fluid by wrapping the membrane around it and then pulling it in.
    - <u>Receptor-mediated endocytosis</u> – specific chemicals bind to receptors in the cell's membrane, activating a process in which the membrane invaginates inward to make a vesicle. Clathrin is a typical protein that helps to make the vesicle form.
    - <u>Exocytosis</u> – this is the opposite of endocytosis. A vesicle inside the cell combines with the membrane and then pops to release its contents outside the cell.

Unit 15

The Cell Membrane, Part 3:
The Cell Wall and Cell Coat

Cell Walls

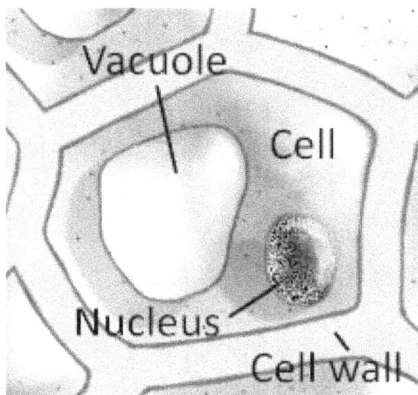

In addition to the cell membrane, some cells have a tough **cell wall**. These typically occur in plants, fungi and bacteria. The cell wall creates a strong barrier that protects the cell against a variety of physical forces, including the threat of popping when too much water flows in by osmosis. When this happens, the water coming in causes the cell to swell up and push against the inside of the cell wall, building what is called **turgor pressure**. The turgor pressure stops more water from coming in, allowing organisms that use cell walls to have much more resistance to environments that are hypotonic. Further, when the turgor pressure builds up, the entire cell (with cell wall) becomes stiff, giving the organism greater structural strength. This is important in plants.

The cell wall also provides the basis for structural support for plants and many macroscopic fungi. By surrounding the cells in a tough casing, the entire organism can withstand the effect of gravity. In this sense, the cell walls of plant cells act like a rigid skeleton that keeps plants and fungi from crumbling into a massive pile of goo.

In this last image, I've drawn a little circle inside the cell called a **vacuole**. Vacuoles are little sacs inside plants cells that act (in part) as waste dumps for the cell. Instead of having a sewage system that takes cell garbage away, the plant cell just puts its garbage in the middle. There is a secondary purpose to the vacuole, however, for it swells up and helps to establish the turgor pressure that contributes to the overall rigid structure of plants.

Plant Cell Walls

The structure of cell walls depends on the type of organism. Plant cell walls are made of primarily cellulose, the long-stranded carbohydrate we saw when we discussed carbohydrates in Unit 8. The sugar chains in cellulose are lined up next to each other, making a sheet. Then another sheet is laid on top of it in such a way that the strands face a different angle (usually 60 to 90 degrees from the previous layer), so that the result is a netlike structure. It's similar to thatched baskets, except that the carbohydrates are not interwoven.

Holes and gaps are frequently left in these sheets, so that water, air, and other small molecules can freely pass through. Plant cell walls also have other carbohydrate compounds in them, such as **pectin** and **hemicelluloses** (show below). These occur in lesser amounts and are used to hold the cellulose strands together.

When plant cells make their walls, they first lay down a **primary cell wall**. This particular wall is somewhat elastic and capable of expanding as the cell grows. When the cell stops growing, it produces a **secondary wall** that is thicker and more rigid. This secondary wall is deposited inside the primary wall, and unlike the primary wall, it contains a significant amount of lignin (a complex molecule that does not readily fall into the four primary macromolecule types we discussed). Where two plant cells make contact, there is a connecting space called the **middle lamella** that has a lot of pectin in it. When the middle lamella breaks down, such as when fruit rots, the cells break free of each other, making the entire collection of cells mushy.

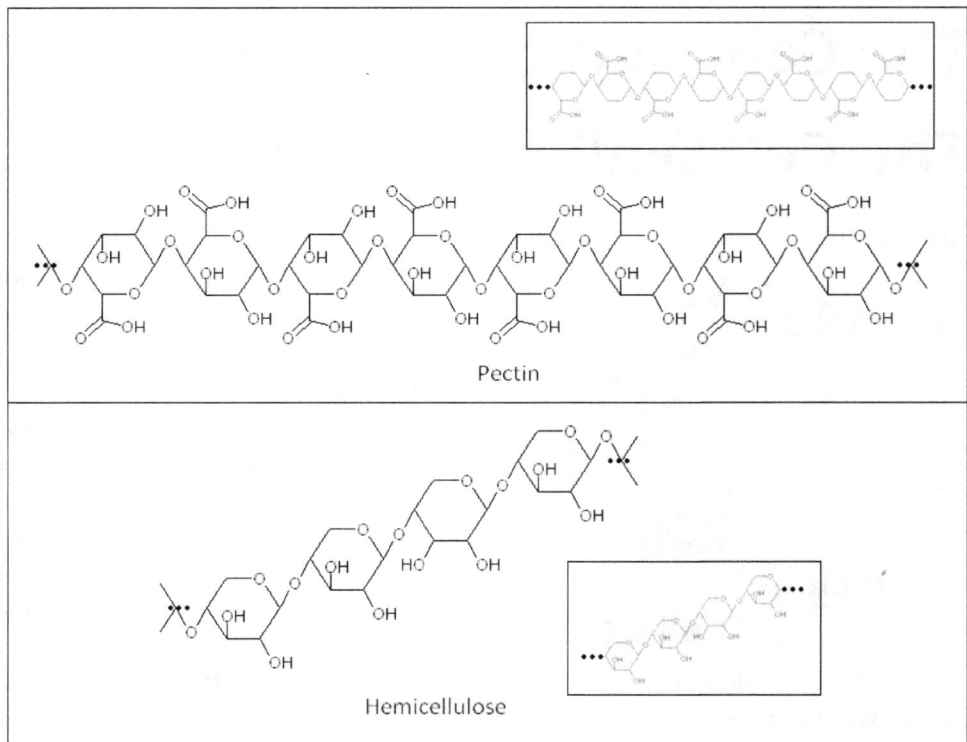

Pectin

Hemicellulose

The air spaces form long tubes that help to deliver various atmospheric gases throughout the plant.

104

Plant cell and wall layers

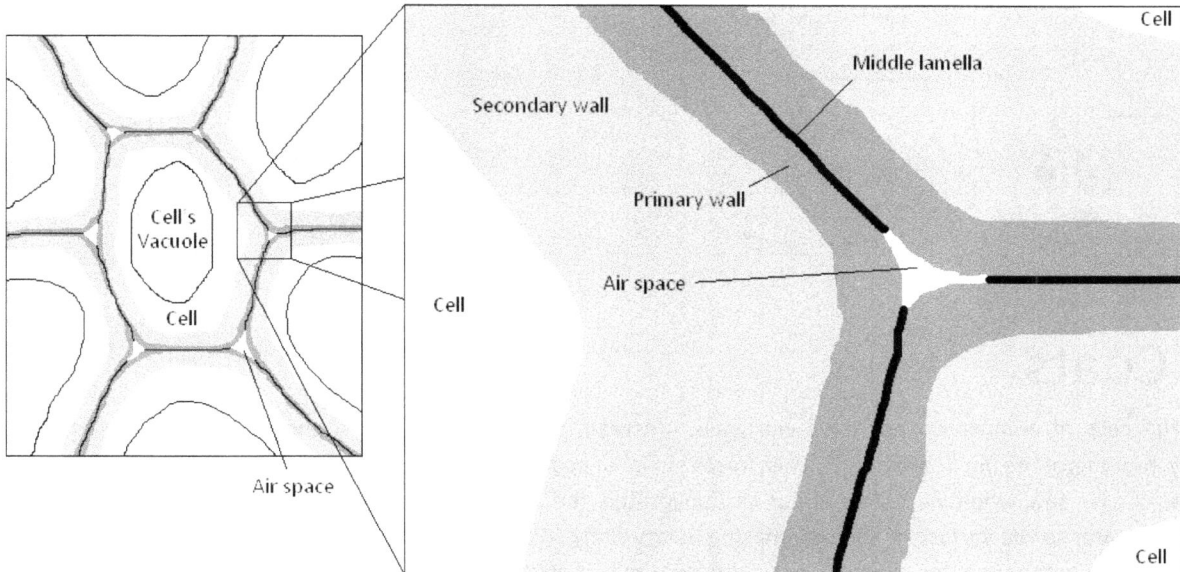

Fungal Cell Walls

Fungal cell walls are made from **chitin**, the polymer created from monomers of glucosamine we looked at in Unit 8. Chitin is the same material that is used in the shells of insects and crustaceans (crabs, shrimp, lobsters, etc).

Bacterial Cell Walls

Bacterial cell walls are made from **peptidoglycans**, which are basically long chains of polysaccharides cross-linked by short proteins. **Penicillin** kills bacteria by inhibiting the production of the bacterial cell wall, making the bacteria susceptible to osmotic changes. The result is that they pop and die. Broadly speaking, there are two types of bacterial cell walls: **Gram positive** and **Gram negative**, named after a staining technique used to see bacteria under microscopes.

Gram positive bacteria have thick cell walls containing a great deal of teichoic acids in them. Teichoic acids are large polymers of variable structure that commonly have a lot of glycerol and phosphate repeats in them. They attract a lot of metal ions and provide the wall with increased toughness. Gram negative bacteria have thin cell walls with no teichoic acid. They also have a second lipid bilayer membrane surrounding them. The Gram staining process is rather complex, but it works essentially like this: A blue pigment gets trapped within the cell wall and cell of the Gram positive bacteria. The Gram negative bacteria cell walls are too thin to trap the blue pigment, which then washes away. After the blue pigment step is completed, a red pigment is added. This makes the blue Gram positive bacteria turn purple and the colorless Gram negatives turn red. A few other mechanisms also occur to cause reds and blues to appear, but they are outside the scope of this book.[23]

---

[23] For example, some Gram positive cells are too fragile for this process and break, causing them to lose their blue pigment and appear red at the end. Altogether, the process can get complicated and is best left for courses on microbiology.

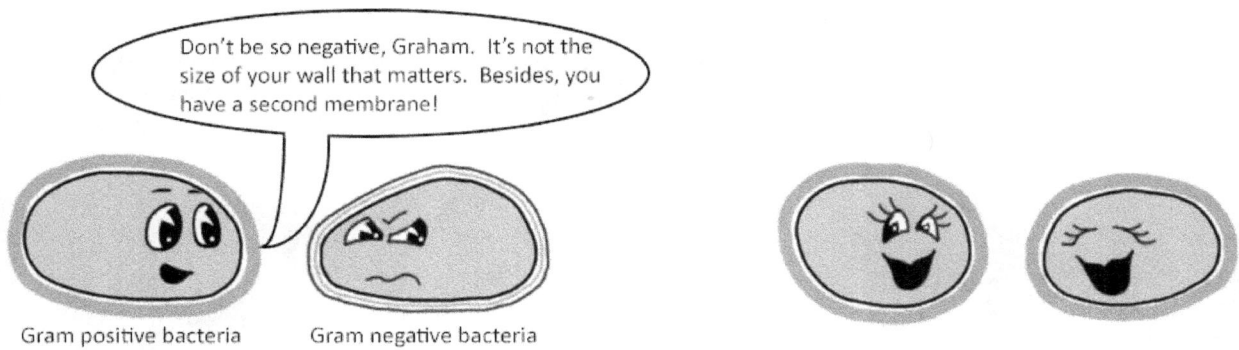

Don't be so negative, Graham. It's not the size of your wall that matters. Besides, you have a second membrane!

Gram positive bacteria   Gram negative bacteria

Cell Coats

The cells of animals do not have cell walls. Instead, they have a layer of short carbohydrates that are covalently bound to the lipid bilayer.[24] This forms a hairy coat-like covering around the cell that is called the **glycocalyx**. As we saw when we talked about carbohydrates, the sugars are very polar, so the glycocalyx attracts a great deal of water to the surface of the cell, making it very slimy. The primary purpose of the glycocalyx is to provide recognition. Each organ type, for example, has a different mix of oligosaccharides in its coat, so heart cells, and kidney cells, and intestinal cells will each have different glycocalyx coats. Secondarily, the glycocalyx serves as a form of contact inhibition. This means that nearby cells sense the glycocalyx of another cell and stop growing if there is not enough space. If they lacked this preventative measure, the cells would continue to grow and divide out of control, forming a cancer.

---

[24] Until now these short carbohydrates have only been drawn attached to random membrane proteins (i.e. glycoproteins).

Glycocalyx

Glycocalyx

Outside cell
(extracellular fluid)

Glycolipid

Inside cell

Glycoprotein

Glycocalyx

Typical Animal Cell

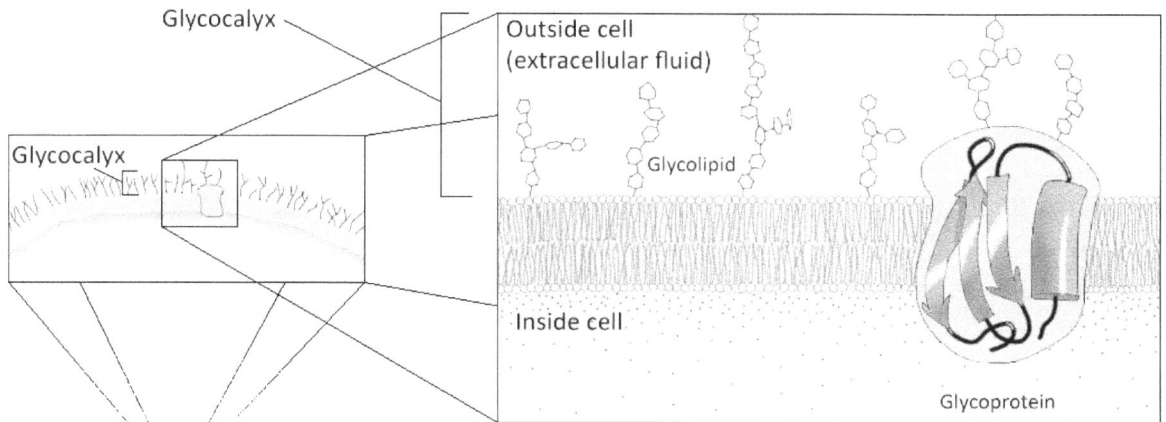

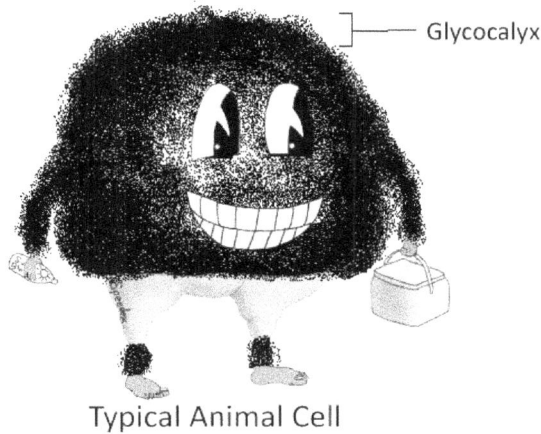

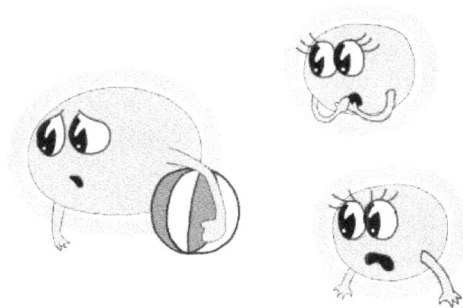

This brings us one set closer to closing our discussion about the outside of cells. We've got one more topic to cover about cell membranes, then we'll transition into talking about the insides of cells and how they work.

Summary

In this section, we discussed cell walls and the glycocalyx.

- Cell walls are tough outer coverings that protect cells from physical stresses as well as osmotic popping from hypotonic environments.
- Water that flows into cells will make them swell up against their cell wall. This creates turgor pressure that prevents further swelling and strengthens the overall structure.
- Plants have two types of cell walls: the primary cell wall (the outermost) laid down first, and the secondary cell wall (the innermost) laid down afterwards. The primary cell wall is more elastic and is thinner. The

secondary cell wall is much thicker and contains a lot of lignin. It is very rigid and helps to give plants their rigid stems and leaves.

- Plants store their cellular garbage in sacks called vacuoles. The vacuoles help to further strengthen the rigid framework of the plant.
- Fungi have cell walls made of chitin. It is the same material used by bugs and crustaceans when they make their shells.
- Bacteria use peptidoglycans to build their walls. Peptidoglycans are long polysaccharide chains with short proteins connecting them together into a network.
- Gram positive bacteria have thick cell walls and lots of teichoic acid in them.
- Gram negative bacteria have thin cell walls and no teichoic acid. They also have a second cell membrane on the outside of the wall.
- Gram negative bacteria are so named because the gram stain does not stick well to the cells. The turn out red, whereas the gram positive cells turn out bluish purple.
- Many antibiotics, like penicillin, work by disrupting the cell wall, causing the bacteria to burst.
- The glycocalyx is the furry polysaccharide coat found on many types of animal cells. It gives the cell a slimy covering, provides it with a distinctive coat that tells other cells what kind of cell it is, and helps to give the cell contact inhibition so that it does not overgrow its neighbor cells.

In this section we cover:

1) Spot Desmosomes

2) Belt Desmosomes

3) Tight Junctions

4) Gap Junctions

5) Plasmodesmata

Unit 16

The Cell Membrane, Part 4:

Cell Membrane Structures

In the previous unit, we saw that neighboring cells can detect each other by sensing their oligosaccharide glycocalyx coat on the outside of cells. There are a few other structures that cells use to communicate with neighboring cells. These include the **spot desmosome**, the **belt desmosome**, the **tight junction**, the **gap junction** and (in plants) the **plasmodesmata**. These structures are shown in the large image on the next page. The image is meant to be a generic cell; not all cells have these structures.

Spot Desmosome

Spot desmosomes are coin-like structures that sit on the inside of the plasma membrane opposite similar structures of neighboring cells. They have proteinaceous threads (made of a membrane protein called **cadherin**) that attach them to the opposing desmosome, running through the plasma membranes of the cell and its neighbor. On the cytoplasmic side of these disks are other protein threads made of **keratin**. The keratin is part of the internal network of fibrous threads called the **cytoskeleton**, which we will discuss later. Spot desmosomes are commonly described as little rivets that help hold neighboring cells together.

Plate 1 - Tight Junction

Plate 3 - Spot Desmosome

Plaque
(plakoglobin,
desmoplakins)

Cadherin molecules
(Desmogleins, Desmocollins)

Tonofilaments
(keratin,
    intermediate filaments)

Plate 4 - Gap Junctions

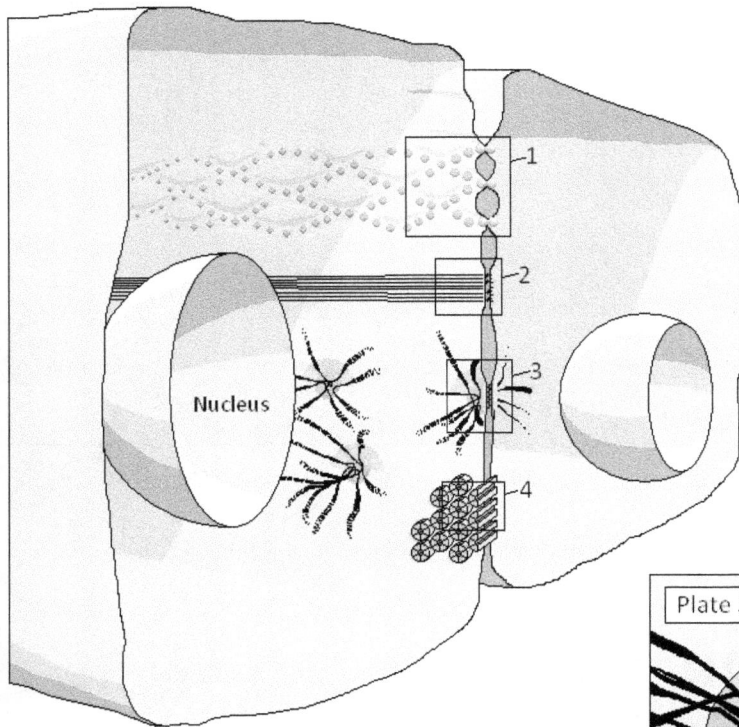

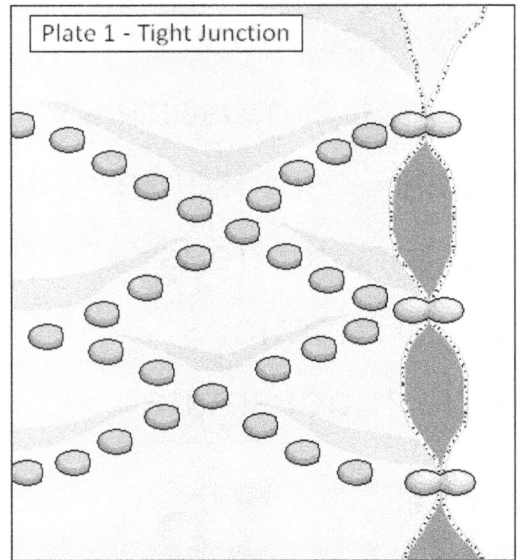

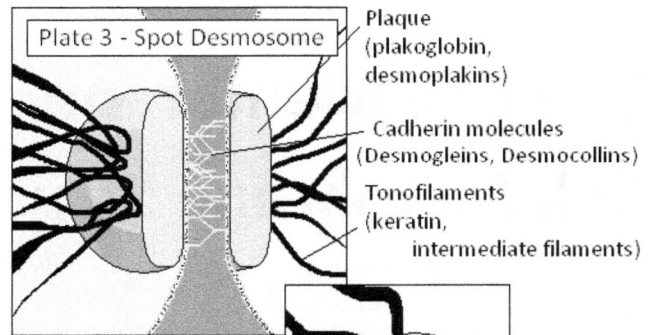

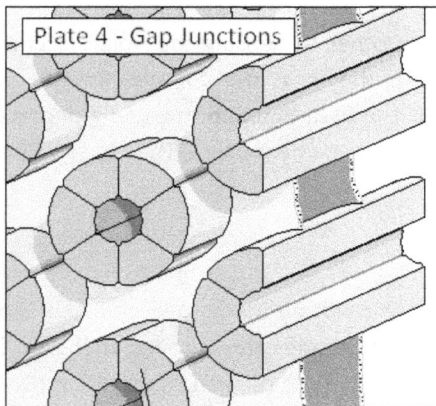

Connexion monomer

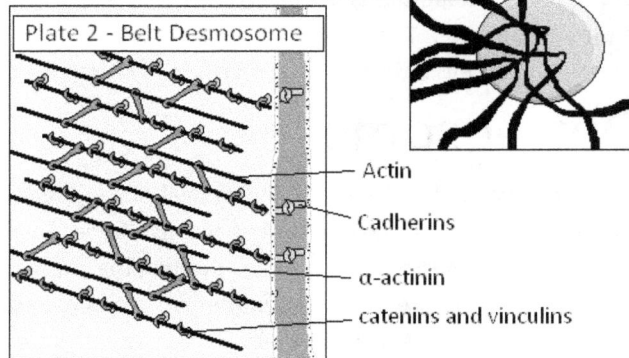

Plate 2 - Belt Desmosome

Actin

Cadherins

α-actinin

catenins and vinculins

Nucleus

Belt Desmosome

Belt desmosomes are lines of proteins that sit in the plasma membrane and form a ring around the circumference of the cell. These proteins are connected with a network of **actin** proteins on the inside of the cell.

110

Actin, as we will see, is a common protein involved in contraction (such as in muscles). The exact roles of belt desmosomes are still debated, but they are believed to serve as an internal support for the cell, like a regular array of hooks from which you could string ropes across a room, say.

Tight Junction

Tight junctions are lines of transmembrane[25] proteins that connect with similar proteins in neighboring cells. These lines are interwoven and pull the plasma membranes of neighboring cells tightly together. Tight junctions are very typical of intestinal cells, where they prevent things from flowing between the cells.

Gap Junction

Gap junctions are proteinaceous tubes that span the plasma membranes of neighboring cells, forming a means of direct communication between the cytoplasms of the two cells, like underwater tunnels.

Plasmodesmata

In plants, tough cell walls provide the majority of cell support. Communication is achieved through breaks in the cell walls between adjacent cells; these breaks are called plasmodesmata. Water and nutrients can flow directly through these breaks. Sometimes, the cell membranes of the adjacent cells line the walls of the plasmodesmata and meet each other at the center. Sometimes the plasma membrane melds directly with the plasma membrane of the next cell, allowing the cytoplasm to flow directly from one cell to another.

---

[25] Transmembrane = a membrane that spans across the membrane. This word describes most of the membrane proteins.

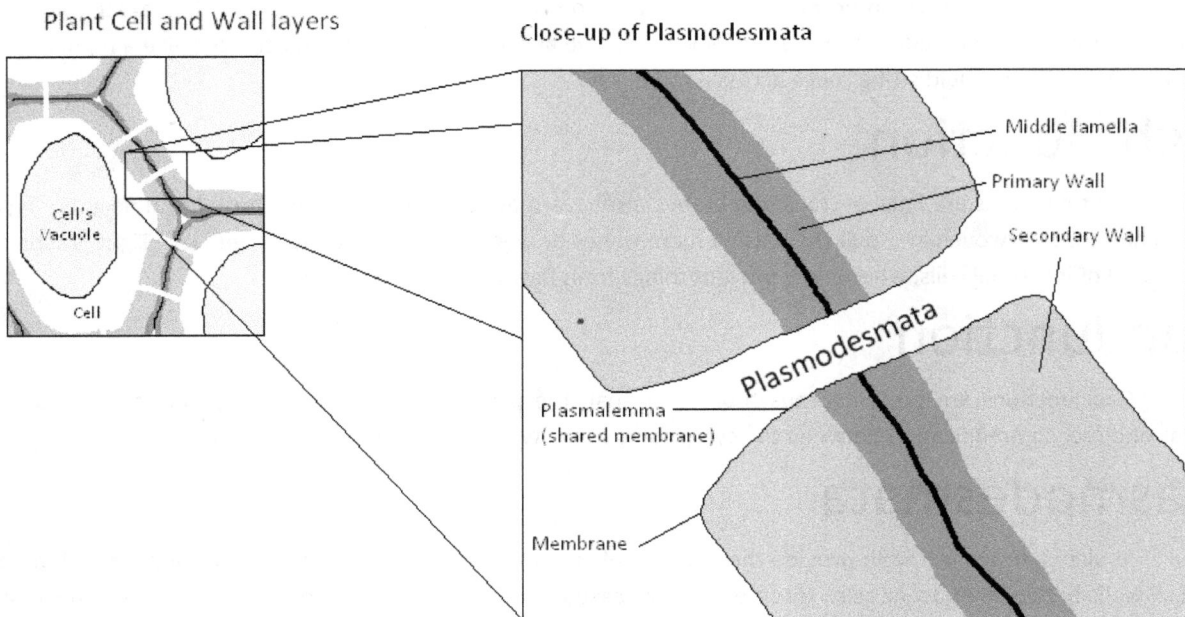

Plant Cell and Wall layers

Close-up of Plasmodesmata

Middle lamella

Primary Wall

Secondary Wall

Plasmodesmata

Plasmalemma (shared membrane)

Membrane

Cell's Vacuole

Cell

Summary

In this section, we looked at several of the basic cell membrane and cell wall structures.

- Spot desmosomes are like discs welded to the side of the cell, connected to a similar spot desmosome in the next cell. They act to keep the cell firmly attached to the next cell. They also act as an anchor for intermediate protein threads that run throughout the cell.

- Belt desmosomes are lines of actin proteins that circle around the cell. Their exact role is still debated, but they are thought to act as an anchoring place for other protein threads that run into the cell.

- Tight junctions are interwoven lines of proteins that tightly bind the membrane of one cell with the membrane of another. They prevent material from seeping between cells and are very important in hollow organs like the intestines.

- Gap junctions are little protein tunnels that directly connect the cytoplasm of one cell with another. This allows materials to transfer rapidly between cells.

- Plasmodesmata are openings in the cell wall of plants that allow the plant cells to communicate with each other. The plasma membranes that run through these openings can either be fused so that the cytoplasm flows directly, or they may lie against each other forming a double-membrane barrier midway across the plasmodesmata.

Unit 17

Cell Anatomy

At this point, we've talked about the cell membrane at great length. We discussed how it transports molecules and materials into and out of the cell, and conversed over the kinds of structural things you might find at the cell surface.

We will now discuss the many things you can "see" inside the cell. First, "see" is a bit of a tricky word to use. The structures we are going to talk about have no color at all. If you plop a cell under a microscope in an attempt to marvel at its many innards, you'd be lucky enough just to see a vague circle, or oval, or hotdog shaped thing that represented the cell as a whole. The only way to actually see the stuff inside the cell is to stain the cell with some kind of molecule that has color. The stain you use depends on the structure you want to see. We use a whole slew of different colored molecules, each having some tendency to attach to a particular type of structure inside the cell. Over the years, we've created newer ways to see inside the cell, including fluorescence and electron microscopy. As these more advanced imaging techniques emerged, we've been able to look at these structures in new ways, increasing our impressive understanding of the cellular workings of life.

When you look at a cell under a microscope, you won't always see all the structures that we will be discussing in this unit. There are many reasons for this, the first one we already mentioned. If you don't include the proper dye, you won't see a particular structure because it has no inherent color. Also, cells are 3D structures. Microscopes only focus at one depth at a time. So, if you are focused on one particular depth, say, just below the cell's surface, you will not see the structures that are positioned much farther down. Finally, there are a wondrous variety of different cells out there in nature, and not all cells have all the structures in their arsenal of pieces-parts. Many of the structures that we discuss here are fairly universal though. So, without further delay, let's begin our discussion of the anatomy of the cell.

The Organelles

Just as you have organs that make up your body (i.e. the heart, the intestines, the kidneys, and for most of us the brain), the cell has individual parts as well. These we call the organelles. We talked about two already: the cell membrane and the cell wall. Just as the cell has a membrane around it, some of the organelles inside the cell have their own membranes around them, as well. The most obvious of these, (and typically the first organelle to learn about), is the **nucleus**. The nucleus is the headquarters of the cell, where the DNA is stored. Surprisingly, not all cells have a nucleus. The presence or absence of a nucleus allows scientists to divide all organisms into two broad categories. The first category is the **eukaryotes**, which have nuclei. All plants, animals (including humans), fungi, and some microscopic organisms are eukaryotes.[26] The second category is the **prokaryotes**, which do *not* have nuclei. Their DNA is mixed in with the rest of their cytoplasm. Prokaryotes are typically believed to be more primitive forms of life, even more so than my roommate. Unlike my roommate, however, nearly all prokaryotes are microscopic organisms. Bacteria are an example of prokaryotes.

On the next page is a list of the major organelles of the cell, as well as their general function. Refer to the image that follows for an idea of what they look like. It also shows the visual differences between eukaryotes and prokaryotes.

---

[26] Technically, many organisms, such as animals, have specialized cells inside them that have had their nuclei removed, such as the case with red blood cells. Premature red blood cells have nuclei, but the nuclei and DNA are eventually kicked out the cell when the blood cell matures. Even though the blood cell has no nucleus, it is still considered a eukaryotic cell, since it started with a nucleus.

Organelle	Major Function(s)	Eukaryotic/Prokaryotic
Cell Membrane	Barrier to retain cytoplasm Cell recognition Material Transportation	Both
Cell Wall	Strong structural support Prevents osmotic bursting	Eukaryotes (plants, fungi, algae), Prokaryotes
Chromosome	A large polymer of DNA used to store genetic information Becomes discreet and observable in eukaryotes when the cell divides	Both
Chloroplast	Membrane bound organelle in plants where light is captured and turned into chemical energy Has its own unique set of genetic material	Eukaryotes (plants and protists)
Flagellum (plural flagella)	Protein whip-like structure that protrudes from the cell and serves as a propeller, pushing the cell through the water	Both
Golgi Apparatus	Membrane bound organelle responsible for the processing of material destined to be secreted from the cell Makes lysosomes	Eukaryotes
Intermediate filaments	Small molecular threads used for structural stability.	Eukaryotes
Lysosome	Involved in the breakdown cell waste, engulfed objects, worn-out proteins, carbohydrates and lipids Involved in the transport of the cell waste to the cell surface for removal	Eukaryotes (mainly animals)
Microfilaments	Small molecular threads used for structural stability	Eukaryotes
Microtubules	Small molecular threads used for structural stability, cell material transport, cell movement, and cell division	Eukaryotes
Mitochondrion (plural mitochondria)	Membrane-bound organelle that serves as the energy harvesting structure of the cell Has its own set of unique genetic material	Eukaryotes

Table continued. . .

Organelle	Major Function(s)	Eukaryotic/Prokaryotic
Nucleolus	Circular shape within the nucleus seen when the cell is not dividing Contains RNA, active DNA, and various proteins used to create protein blueprints called messenger RNA, or mRNA	Eukaryotes
Nucleus	Contains chromosomes (the genetic material, DNA) Surrounded by two membranes that separate it from the cytoplasm	Eukaryotes. The presence of a nucleus is the defining feature of Eukaryotes
Plasmid	Small circular DNA molecules found in many bacteria. This is in addition to their large circular chromosome.	Prokaryotes, though they can also be found in chloroplasts and mitochondria
Peroxisome	Vesicle that metabolizes long chain fatty acids, less commonly found amino acids, and some sugar molecules. Uses peroxide in its reactions	Eukaryotes
Ribosomes	Large protein/RNA structures within the cytoplasm that are responsible for the manufacture of proteins	Both
Rough endoplasmic reticulum	Same as smooth endoplasmic reticulum, but with ribosomes attached to it Involved in the initial stages of protein secretion (transportation of proteins out of cell)	Eukaryotes
Smooth endoplasmic reticulum	Tubular system that is continuous with the nuclear double membrane Stores and sorts materials for transit throughout the cell or directs materials to be secreted from the cell	Eukaryotes

Eukaryotes

Typical Animal Cell

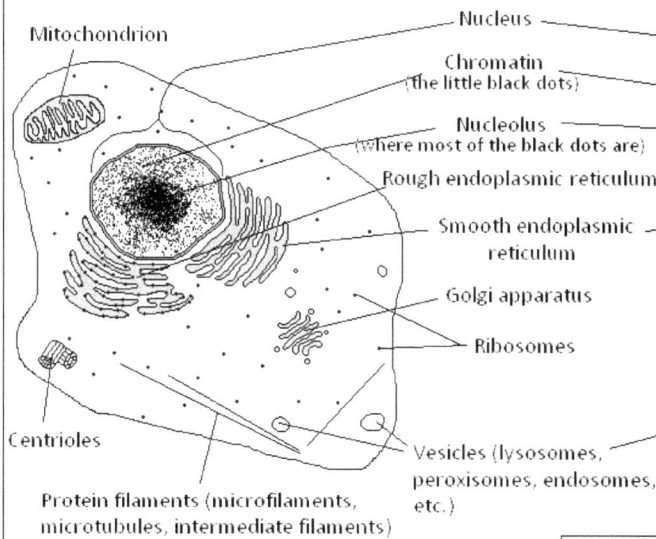

Mitochondrion

Nucleus

Chromatin
(the little black dots)

Nucleolus
(where most of the black dots are)

Rough endoplasmic reticulum

Smooth endoplasmic reticulum

Golgi apparatus

Ribosomes

Centrioles

Vesicles (lysosomes, peroxisomes, endosomes, etc.)

Protein filaments (microfilaments, microtubules, intermediate filaments)

Typical Plant Cell

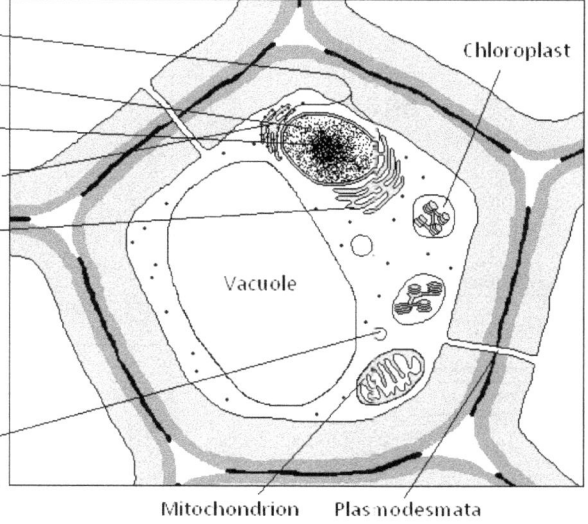

Chloroplast

Vacuole

Mitochondrion   Plasmodesmata

Eukaryotic nucleus as cell approaches division

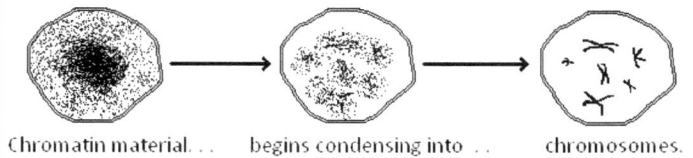

Chromatin material. . .   begins condensing into . . .   chromosomes.

Prokaryotes

Typical Bacterial Cell

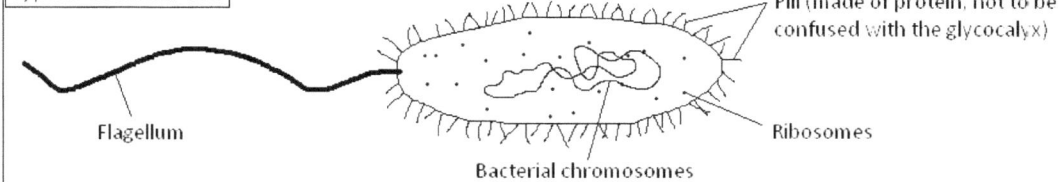

Pili (made of protein, not to be confused with the glycocalyx)

Flagellum

Bacterial chromosomes

Ribosomes

The Nucleus, Chromosomes, and Nucleolus

The headquarters of the cell, the nucleus stores the DNA and is separated from the cytoplasm by a double-membrane called the **nuclear membrane** or **nuclear envelope**. Each polymer of DNA is wound tightly upon lots of protein complexes called **nucleosomes.**[27] The nucleosomes look like yo-yos with the DNA wrapped twice around each one. These protein yo-yos are then attached to each other to create a compact spiraling structure that is tightly

---

[27] Nucleosome actually refers to the protein complex (made of 5 smaller proteins called histones) as well as the DNA wound around it.

wound. This spiral is, in turn, gathered into a densely packed arrangement to form a **chromosome**. A great many proteins are required to complete these structures, but they are outside the scope of this book. Here is a simplified view of how chromosomes are organized.

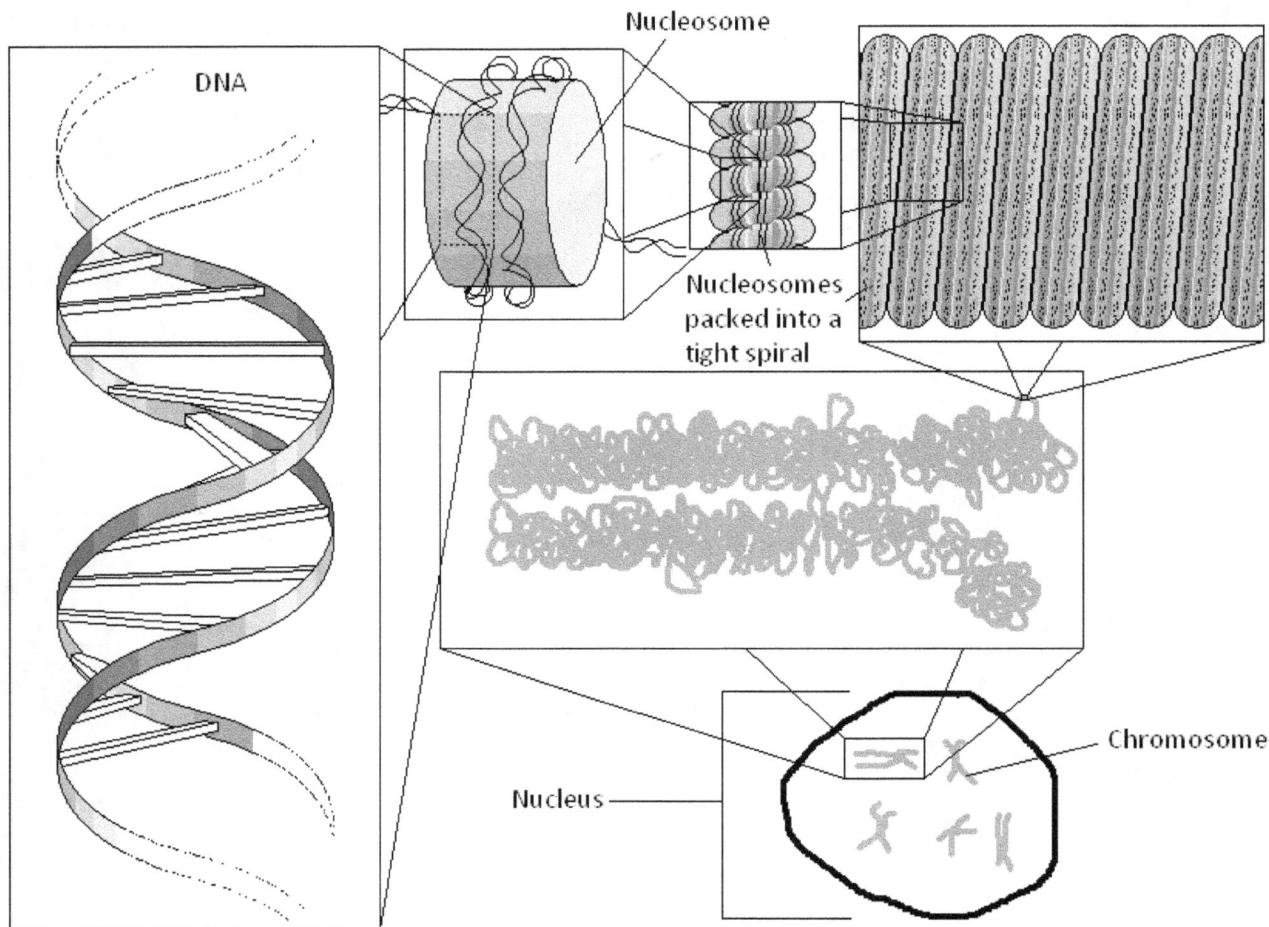

When the cell is dividing, the chromosomes look similar to how they are shown in the bottom right of this latest image. With the proper dye, you can see discreet, individual chromosomes, and if the focus is lucky, you might be able to see them all at once. The total number of chromosomes that a cell has, however, depends on the species. Humans, for example, have 46 total chromosomes.[28]

When the cell is *not* dividing, the chromosomes are less tightly packed, and under the microscope they can appear as cloudy material within the nucleus. This cloudy material is called **chromatin**. The darkest part of this material, usually in the center of the nucleus, is called the **nucleolus**. The nucleolus is a collection of active DNA, RNA and various proteins that are interacting during the preparatory stages of protein production, which we will discuss in

---

[28] More specifically, we have 23 pairs of chromosomes. We'll learn more about this later.

future sections. In short, "active" DNA is DNA that is being used to make the blueprints for protein. All the molecules involved in this process are gathered in one place, making the nucleolus look darker and more crowded than the rest of the nucleus. The blueprints that are made are shipped out of the nucleus; the actual protein production takes place in the cytoplasm.

Cell Nucleus

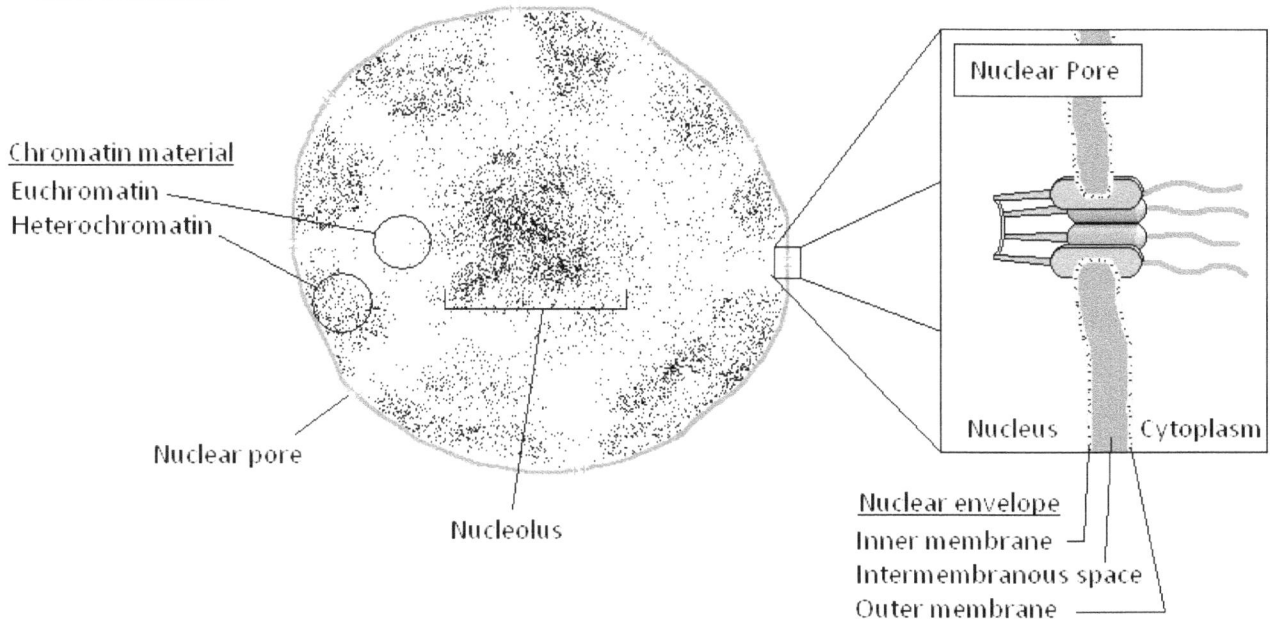

As this latest image shows, the chromatin outside the nucleolus is divided into two categories: **euchromatin** and **heterochromatin**. Euchromatin refers to the areas where the DNA is less dense. It forms little canals between the denser heterochromatin areas. The heterochromatin that is pushed up against the inside of the nuclear membrane is the DNA that is not currently being used.

Also in this image, notice the small openings in the nuclear membrane. Within these openings are protein structures called **nuclear pores**. These protein complexes help various large molecules get into and out of the nucleus.

Unlike eukaryotes, prokaryotic chromosomes are not as tightly wound and are typically circular in structure, as if the ends of the chromosomes had been sown together. Furthermore, prokaryotes typically have additional small circular DNA molecules called **plasmids**. These are basically small packets of DNA that carry information for extra proteins that confer additional abilities to the bacteria. We'll encounter them again much later.

Ribosomes

Ribosomes are structures that float around in the cytoplasm of the cell. They are large compared to most proteins, but still small enough that they are little more than dots inside the cell. Ribosomes are divided into a large subunit and a small subunit. Together they look like little gloves interlocking each other. Each subunit is made from both protein and RNA. This RNA is called **rRNA**, which stands for **ribosomal RNA**.

The purpose of ribosomes is to synthesize protein, a process we will look at more in depth later. For the moment, a quick overview is as follows: RNA polymers called **messenger RNA**, or **mRNA**, are made in the nucleolus using information in the DNA. These are the "protein blueprints" referred to previously. The mRNA finds its way out into the cytoplasm where it encounters a ribosome. It then travels through a channel between the subunits of the ribosome, interacting with various cellular molecules inside the ribosome to make protein. The new protein slowly appears from the other end of the tunnel that pierces the large subunit. Here is an image to help clarify the structure and concept:

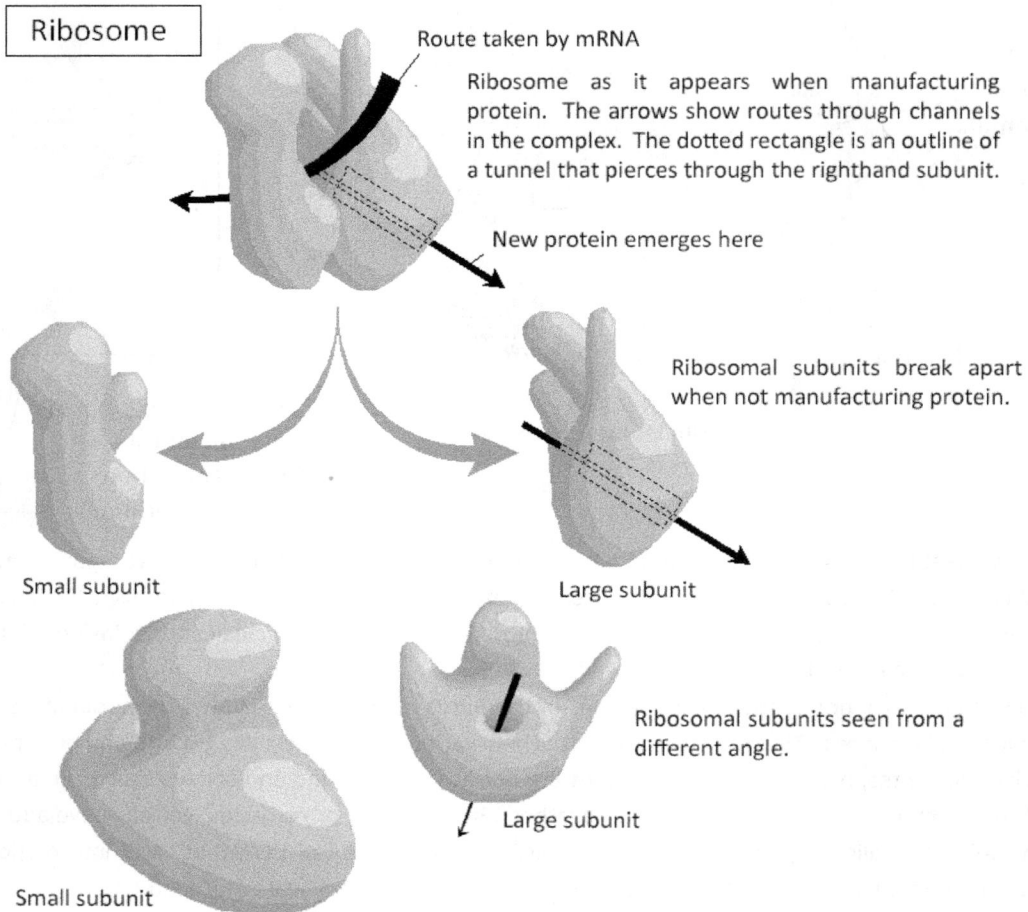

Ribosome

Route taken by mRNA

Ribosome as it appears when manufacturing protein. The arrows show routes through channels in the complex. The dotted rectangle is an outline of a tunnel that pierces through the righthand subunit.

New protein emerges here

Ribosomal subunits break apart when not manufacturing protein.

Small subunit

Large subunit

Ribosomal subunits seen from a different angle.

Large subunit

Small subunit

As we will see shortly, ribosomes can be found in two different places. First, they can be found floating around in the cytoplasm. The proteins they make there will be released into the cytoplasm. Second, they can also be found attached to the endoplasmic reticulum (discussed below). The protein they make there get put inside the endoplasmic reticulum and routed to various vesicles. It may then eventually be exported from the cell.

The Endomembrane System

The **endomembrane system** is the name given to the collection of organelles that include the smooth and rough endoplasmic reticulum, Golgi apparatus and the various vesicles that transit about the cell. The nuclear membrane is technically also a member of the endomembrane system, though a discussion of it was more appropriate in relation to the nucleus (above). At any rate, the endomembrane system works to transfer membrane phospholipids and various proteins throughout the numerous membranes found in the cell. The endomembrane system is also involved in the secretion of proteins and other molecules out of the cell. Let's look at the individual parts of the endomembrane system presently.

Endoplasmic Reticulum, Smooth and Rough

The **endoplasmic reticulum** is a series of membrane tubules that connect directly with the nuclear membrane. There are two varieties, the **smooth endoplasmic reticulum (SER)** and the **rough endoplasmic reticulum (RER)**. The phospholipids that are in the lipid bilayer membranes of the cell are made in the SER. As the phospholipids are produced, they begin to form into little bubbles called **vesicles** that bulge from the SER. Then, these little vesicles depart from the smooth endoplasmic reticulum and are eventually taken to the cell's membrane. At that point, the vesicle merges with the cell membrane, so that the phospholipids in the vesicle are now part of the cell membrane. Since the lipids can roam around freely in the membrane (i.e. the fluid mosaic model), these new lipids will diffuse across the membrane and all around the cell.

The SER also stores calcium, which is important for processes such as flexing muscles, a topic usually covered in the second semester.

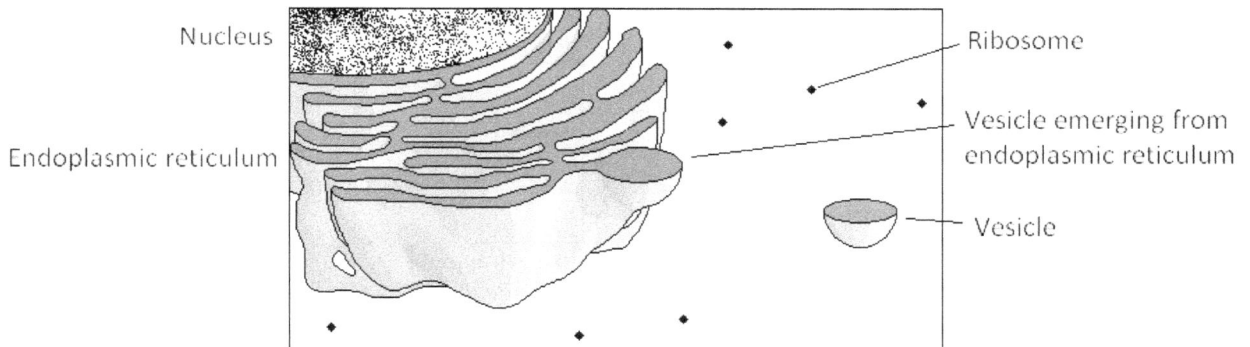

The rough endoplasmic reticulum is like the SER, but with ribosomes attached to its surface. These ribosomes make proteins and thread them into the hollow inside (i.e. the **lumen**) of the endoplasmic reticulum. Alternately, the ribosomes attached to the SER embed the new proteins directly into the membrane of SER.

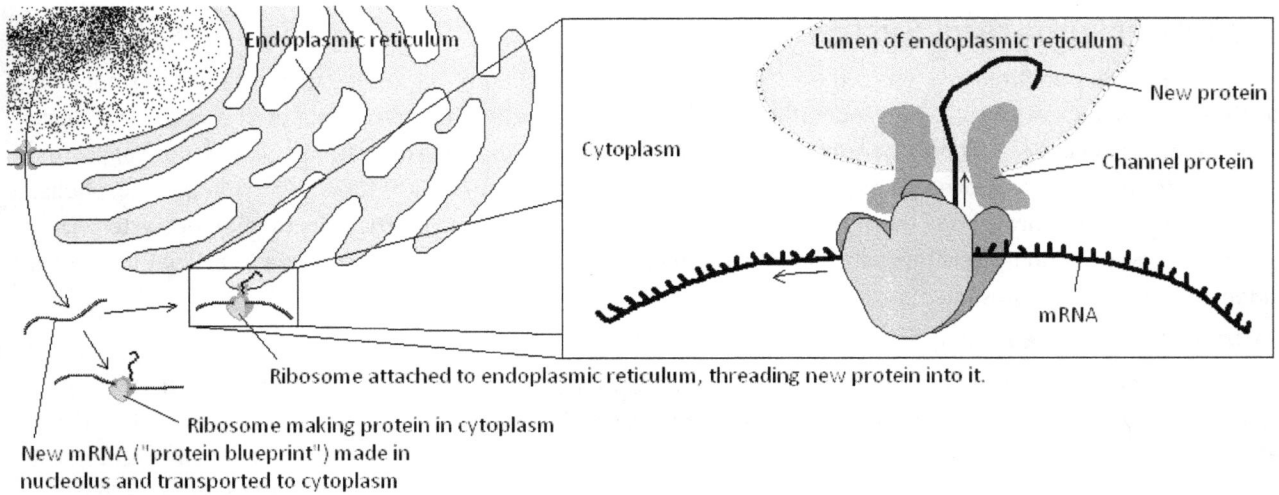
Endoplasmic reticulum

Lumen of endoplasmic reticulum

New protein

Cytoplasm

Channel protein

mRNA

Ribosome attached to endoplasmic reticulum, threading new protein into it.

Ribosome making protein in cytoplasm

New mRNA ("protein blueprint") made in nucleolus and transported to cytoplasm

When a vesicle forms from the endoplasmic reticulum, the proteins that are floating in the lumen get trapped inside the vesicle. Likewise, the proteins that were embedded in the membrane of endoplasmic reticulum become part of the membrane of the vesicle. When the vesicle departs for the cell membrane (or other destination), the materials inside it and the proteins in its membrane tag along for the ride. As the vesicle combines with the cell's membrane, the proteins inside the vesicle get released to the outside and the proteins embedded in the vesicle's membrane become part of the cell's membrane proteins.

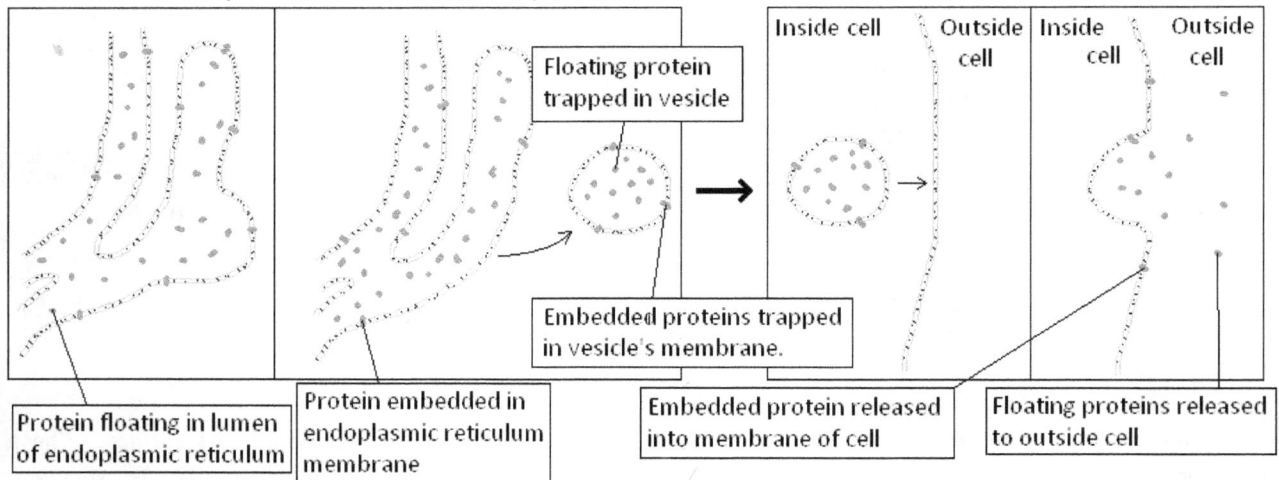
Floating protein trapped in vesicle

Inside cell | Outside cell | Inside cell | Outside cell

Embedded proteins trapped in vesicle's membrane.

Protein floating in lumen of endoplasmic reticulum

Protein embedded in endoplasmic reticulum membrane

Embedded protein released into membrane of cell

Floating proteins released to outside cell

The vesicle's trip is actually slightly more complex though. Before it reaches the membrane, it first goes through the **Golgi Apparatus**, so let's look at that.

Golgi Apparatus

The **Golgi apparatus** looks like a stack of hollow curled pancakes surrounded by bubbles. Vesicles that depart from the endoplasmic reticulum fuse with the Golgi apparatus, delivering proteins and phospholipids to it. Once in the Golgi apparatus, the proteins and lipids are modified by the addition of sugars and phosphates (processes called

122

glycosylation and **phosphorylation**, respectively). From here, the Golgi apparatus packages these molecules into new vesicles and sends them off to the cell membrane to be secreted or sends them into the cytoplasm to become part of the cell's population of vesicles, such as **lysosomes** (see below).

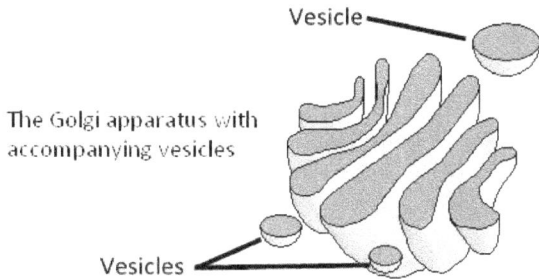

Vesicle

The Golgi apparatus with accompanying vesicles

Vesicles

In the following image, we see the route that a molecule such as a protein would take to be either secreted from the cell or packaged into a vesicle like a lysosome. The route that a protein takes depends on the primary structure of the protein in question. Proteins destined for secretion typically have characteristic amino acid sequences that cause the Golgi to package it into secretion vesicles. Likewise, lysosome-bound proteins have characteristic amino acid sequences that the Golgi recognizes and uses to direct it into newly forming lysosomes. Vesicles, such as lysosomes, depart the Golgi apparatus just like vesicles leave the endoplasmic reticulum: they bubble-up and detach.

A protein's path through the Golgi apparatus

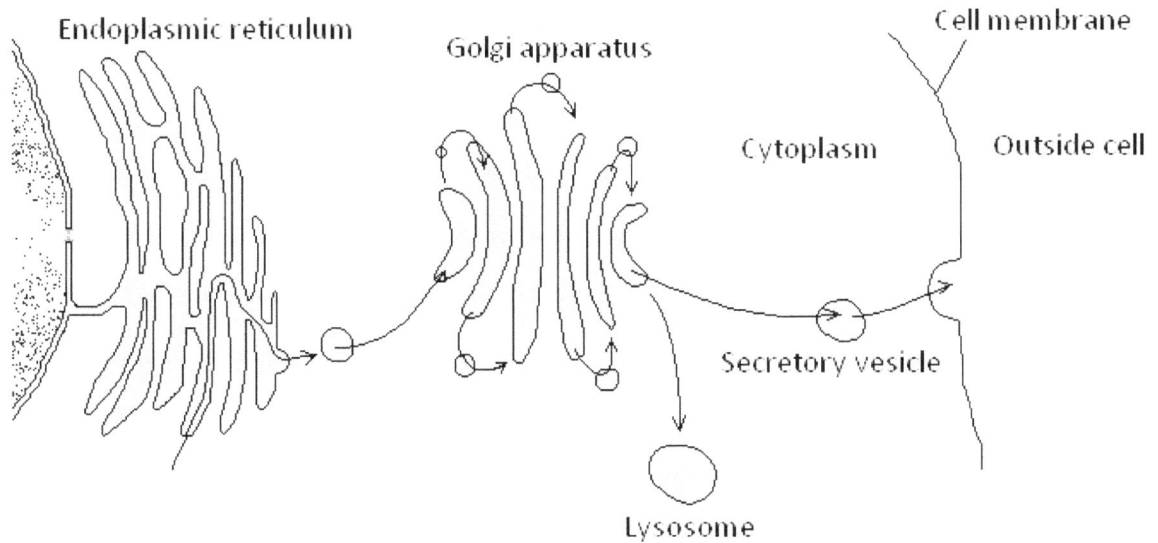

Endoplasmic reticulum

Golgi apparatus

Cell membrane

Cytoplasm

Outside cell

Secretory vesicle

Lysosome

Lysosomes

Like the many vesicles you would see in a cell, **lysosomes** look like little bubbles. They are made by the Golgi Apparatus, and their job is to be the waste control specialists. Lysosomes contain enzymes that break down other proteins and chemicals, old cellular debris, or bacteria and viruses that have been phagocytized. The lumen of a lysosome is very acidic (i.e. low pH/high H^+ concentration). This acidic environment helps the enzymes inside the lysosome do their job to break down materials.

Whenever the lysosome is going to break something down, it fuses its membrane with the membrane of the target, usually another vesicle or worn-out mitochondrion or chloroplast. Once the membranes fuse, the contents of the lysosome mix with the contents of the target. The target is slowly digested. The basic chemicals that result from the breakdown of the target are usually transported out of the lysosome and into the cytoplasm for the cell to reuse, or are ejected from the cell when the lysosome fuses with the cell's membrane. The lysosome's enzymes are neutralized before this happens, however.

Another important use for lysosomes is in a process called **apoptosis**, or cell suicide. Basically, the lysosome is directed at the cell itself, digesting its critical structures and causing the cell to die. This may seem sad and tragic, but the health of the overall body depends on some cells removing themselves from the population. If the number of cells in the body cannot be controlled, cancer can result.

Lysosomes are generally only present in animals. Plants and fungi use their vacuoles to take care of waste products. Bacteria do not have lysosomes.

Peroxisome

Peroxisomes are vesicles that are found in nearly all eukaryotes. They play an important role in the metabolism of some long fatty acids, less common amino acids, and a few carbohydrates. The telltale feature of peroxisomes is that they use peroxide during the course of their activity.

Endosome

One final vesicle we'll cover in the endomembrane system is the **endosome**. An endosome, whose name means roughly "in-body," is a little membrane bubble that results when something is engulfed during endocytosis. If you'll recall, endocytosis is the process of the membrane folding inward to make a bubble that is full of whatever was outside the cell. That bubble is called the endosome. Endosomes are typically found near the surface of cells that undergo endocytosis or phagocytosis. They contain extracellular debris, and they are eventually joined with lysosomes, whose enzymes and acid digest the material inside the endosome.

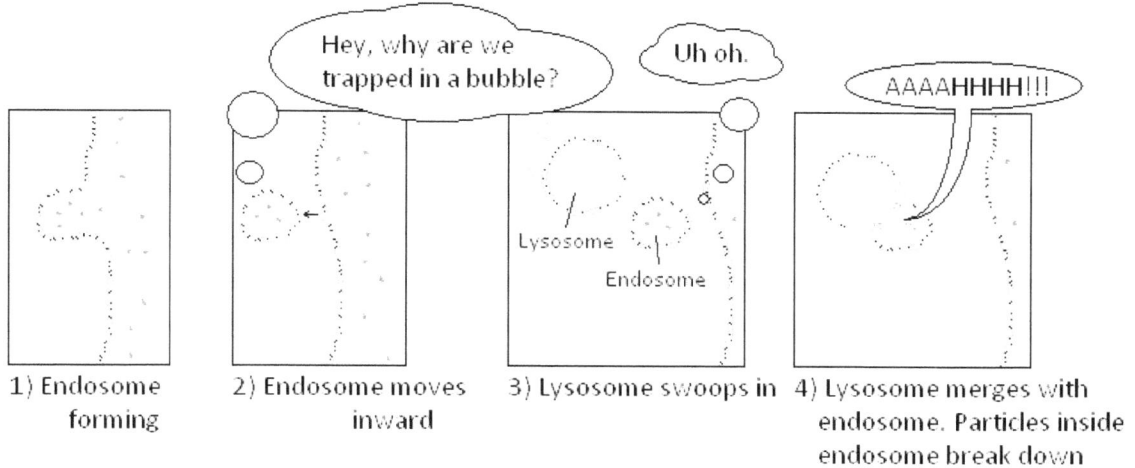

1) Endosome forming 2) Endosome moves inward 3) Lysosome swoops in 4) Lysosome merges with endosome. Particles inside endosome break down

The Symbiotic Organelles

The word **symbiotic** means that two organisms live in harmony, helping each other to thrive. There are two organelles within the cell (the **mitochondrion** (plural mitochondria) and the **chloroplast**) that are believed to be symbiotic to eukaryotic cells. The idea is that long ago these organelles were bacteria-like organisms that originally preyed upon eukaryotic cells, perhaps invading eukaryotic cells from time to time to take advantage of the resources inside.   Over great lengths of time

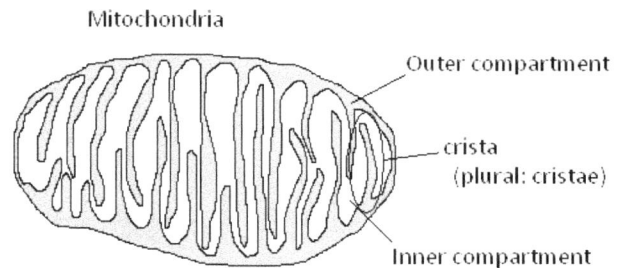

(millions and millions of years), this relationship gradually shifted so that both the cell and the invading organism benefited from the interaction. The mitochondrion and chloroplast are both involved in energy management, and so the eukaryotes that harbored them gained the benefit of more efficient energy handling. By offering their services of energy management, the ancient mitochondria and chloroplasts no longer had to live the rough and rowdy life of a cellular pirate.

The reason this theory was generated is that both the mitochondrion and the chloroplast have their own DNA. In order for living organisms to survive and reproduce on their own, they must have DNA, suggesting that once upon a time mitochondria and chloroplasts lived on their own. If so, their DNA is mostly out-of-practice now, however, as it would have been millions and millions of years since they lived on their own, but the DNA they still have still serves as the blueprint for many (though not all) of the proteins found inside them.

That said, let's look at these two organelles more in depth.

Mitochondria

Mitochondria are membrane-bound organelles where a great deal of energy is harvested from several specific basic organic chemicals. They are, in a sense, power plants for the cell. Within their membrane is a second membrane, like a balloon within a balloon, only this inside membrane is ruffled, like a rug that is scrunched up.

Within this inner ruffled membrane are the important proteins that function in the reactions that generate the energy for the cell. These particular reactions are called the **Krebs Cycle** (also called the **citric acid cycle**, or the **tricarboxylic acid cycle**), which we will discuss in detail much later. For the moment, a brief description of that cycle is as follows: glucose (a sugar) is broken down into smaller pieces, and these pieces are eventually transported into the mitochondria where most of the energy contained in the molecules is extracted and pumped back into the cell. In addition to this, the mitochondria also play a role in the metabolism of fats.

New mitochondria are made when other mitochondria divide, much like a bacteria does.

Chloroplasts

Chloroplasts are membrane-bound organelles found in plant cells and look like bags that contain connected stacks of tightly-packed hollow pancakes (called **granums**). Chloroplasts can be thought of as a plant's solar panels. They are responsible for capturing energy from the sun in a process called **photosynthesis** (which we will discuss in an upcoming unit). Chloroplasts are full of green pigments. They are the reason leaves are green. The details of their anatomy are shown here:

Like mitochondria, chloroplasts have their own DNA, but many of the genes responsible for the chloroplast's upkeep are now in the cell's nucleus. Chloroplasts divide like bacteria do as well.

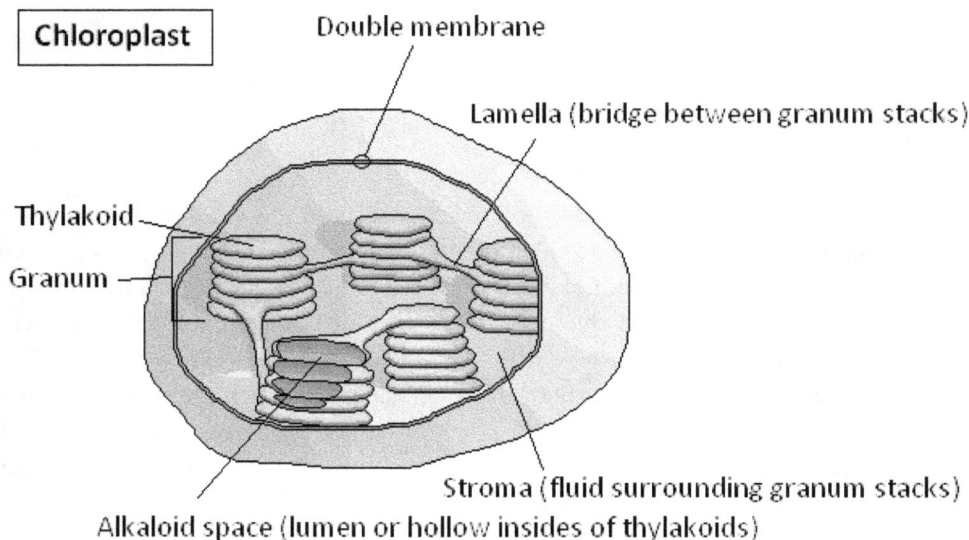

Chloroplast

Double membrane

Lamella (bridge between granum stacks)

Thylakoid

Granum

Stroma (fluid surrounding granum stacks)

Alkaloid space (lumen or hollow insides of thylakoids)

The Cytoskeleton

The **cytoskeleton** is a system of long fibrous proteins that crisscross the cell. Their general role is to provide structure and support for the cell and its many other organelles, although some fibers serve to assist in cellular movement and material transportation as well. There are three broad categories of protein fibers in the cytoskeleton: the microfilaments, the intermediate filaments and the microtubules. Refer to the large image near the end of this section for a visual description. We'll look at each individually.

Microfilaments

The **microfilaments** are the smallest of the three categories of protein fibers in the cytoskeleton. The best example of a microfilament is actin, which we encountered before in our study of belt desmosomes. These tiny filaments are made of monomers of actin (called G-actin) that come together to make a long polyprotein (meaning "many protein") polymer. Curiously enough, the actin monomers tend to add to the fiber from one end, while they leave the fiber from the other end.

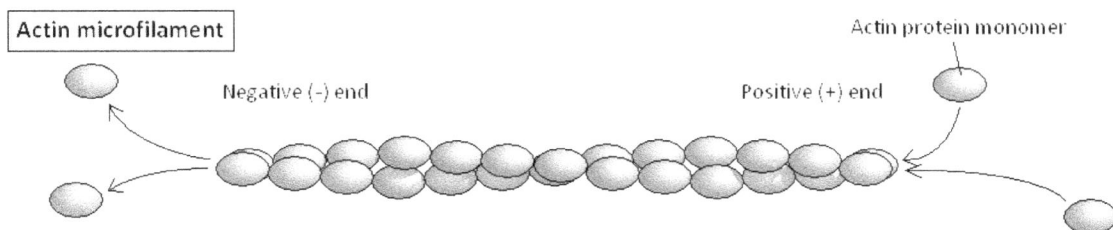

Actin microfilament
Negative (-) end
Actin protein monomer
Positive (+) end

The peculiar nature of how the monomers combine to make actin gives these fibers a polarity of sorts, (although not in the chemical sense that we talked about earlier in the book). This polarity allows other proteins that interact with the actin to "know" direction. It's like finding a road in the wilderness that has arrows painted on it, directing you to the nearest sign of civilization. Similarly, various transportation proteins (usually from a family of proteins known as **myosins**) find the actin fiber and orientate toward or away from the positive end of the actin. These proteins, in a sense, walk along the actin in one way or the other, carrying materials across the cell to wherever they need to go. One type of protein might attach and always travel towards the positive end, while another type of protein might attach and always travel towards the negative end.

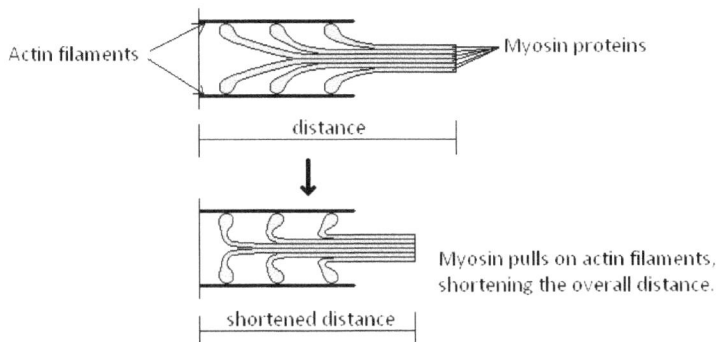

Actin filaments
Myosin proteins
distance
Myosin pulls on actin filaments, shortening the overall distance.
shortened distance

Actin is also very important in muscles. Basically, myosin proteins are lined up with actin filaments, and when the signal is given all the myosin proteins pull on the actin filaments to cause the cell to shorten. It's like a team of people playing tug of war with a wall – they just pull themselves toward the wall. To the left is a very simplified version of what happens.[29]

---

[29] For those interested, the image shown is only half of a structure called a sarcomere, which is covered more in depth during the second semester of biology or in courses on anatomy and physiology.

Depending on the type of cell, actin is found in highest density around the inside of the cell's membrane. There it sometimes helps to move the cell membrane (by a mechanism similar in principle as muscle) to assist in endocytosis (i.e. engulfing material).

Intermediate Filaments

Intermediate filaments are larger in size than microfilaments and have a more complex structure. The basic unit of intermediate filaments contains four short protein threads that come together to make a pair of dimers (dimer = two monomers). Since there is a *pair* of dimers, we call the basic unit a tetramer (tetramer = four monomers). These tetramers are lined up back-to-back-to-back in very long threads.

The primary purpose of intermediate filaments is structure and support. Their design tends to make them rather elastic in nature, which affords them the ability to survive deformations. There are lots of different kinds of intermediate filaments, and they are found in a variety of places. A common example is a family of filaments called **lamins**. These form a network just on the inside of the cell's inner nuclear membrane. This network helps to give the nucleus its round shape. Without this scaffolding, the nucleus would just be a flimsy squishy bag flapping around in the cytoplasm.

Another common family of intermediate filaments is the keratins. We saw these when we looked at spot desmosomes. Keratin is very common in hair, nails, and animal horns.

Microtubules

Microtubules are the largest of the protein fibers in the cytoskeleton. Their basic unit is a pair of protein blobs called tubulin. One is alpha tubulin and the other is beta tubulin. This dimer of protein blobs lines up back-to-back-to-back with other alpha-beta tubulin pairs to make long threads (called **protofilaments**). Thirteen of these protofilaments line up next to each other to make a tube. Like actin, microtubules also have directionality to the way they fit together. The negative end is usually anchored somewhere near the center of the cell, while the positive end continues to grow outward by the addition of more alpha-beta tubulin dimers. Also like actin, this directionality gives microtubules an important role in transportation of cell material. For example, when a protein called **dynein** binds to the microtubule, it will begin to "walk" toward the negative end, and in doing so it carries materials with it toward the cell's center. Another protein, called **kinesin**, will bind to microtubules and walk toward the positive end, carrying cell material *away* from the cell's center.

Microtubules play important roles in cell division, as we will see in future sections.

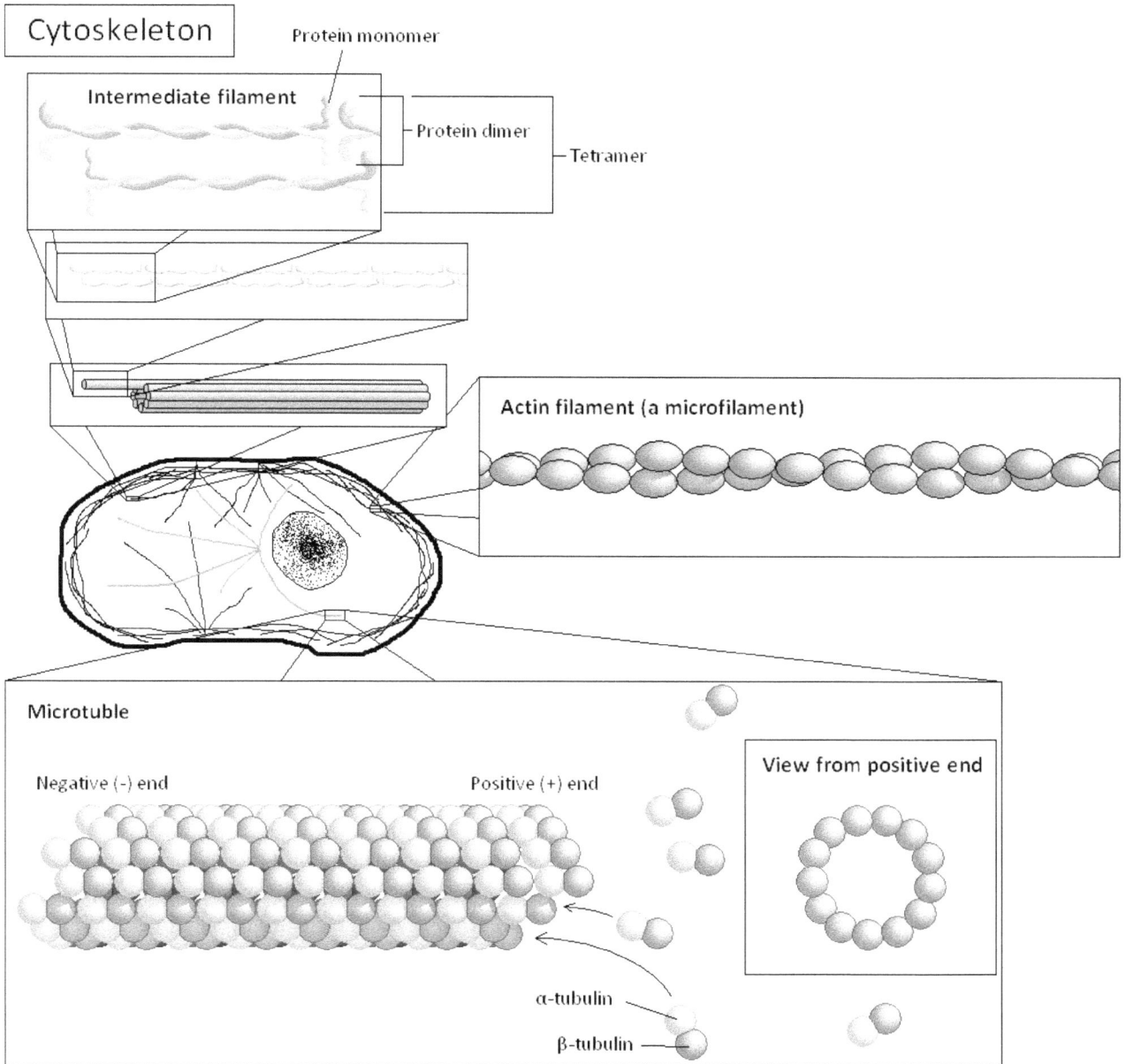

Cytoskeleton

Protein monomer

Intermediate filament

Protein dimer

Tetramer

Actin filament (a microfilament)

Microtuble

Negative (-) end

Positive (+) end

View from positive end

α-tubulin

β-tubulin

Centrioles

The **centriole** is a barrel-shaped structure that is very large compared to microfilaments, intermediate filaments and microtubules. Centrioles are usually found in pairs positioned at roughly 90 degrees from each other, an arrangement called a **centrosome**. Centrioles are made of lots of microtubules. The general design is a set of nine bundles of microtubule **triplets** arranged in a star-like fashion (see image, next page).

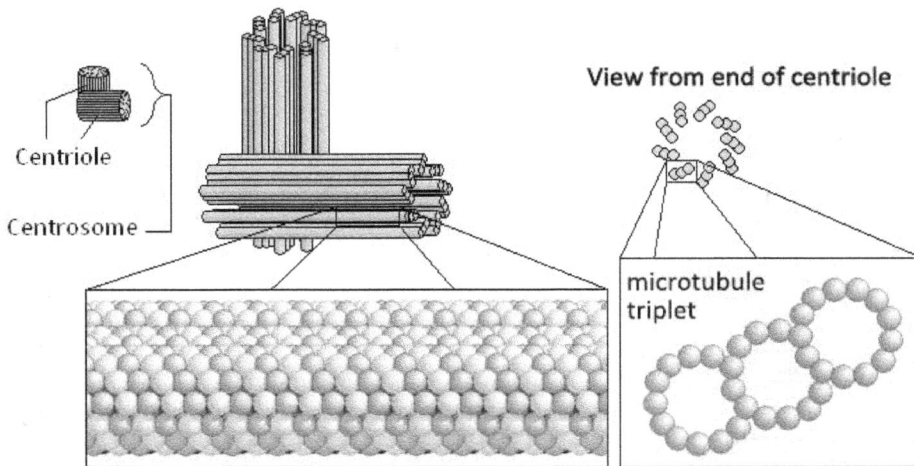

Two Centrioles Arranged into a Centrosome

Centriole

Centrosome

View from end of centriole

microtubule triplet

Theories concerning the purpose of centrioles have changed over time, but they are presently believed to be important for determining the position of the nucleus. Also, they seem to have a clear role in the positioning and functioning of cilia and flagella (see below). Although centrioles are not always drawn in simple diagrams of the cell, when they are, they are usually depicted near the nucleus when the cell is preparing to divide.

Cilia and Flagella

Cilia are small protein-containing hairs that cover the surface of some kinds of eukaryotic cells. These hairs wave back and forth to cause one of two different effects. 1) If the cell is stuck in place, the cilia force water to flow in one direction, which is important in many organ systems such as the lungs and female mammalian reproductive tract. 2) If the cell is mobile, the cilia help to move the cell around, like oars on a ship.

Cilia have a centriole-like microtubule system called an **axoneme** in the center of their column. The axoneme is shaped like a rim of microtubule-fused pairs (**doublets**) around the outside, with two independent microtubules on the inside (see top-right part of the following image). Between each of the doublets are dynein arms. If you recall from the discussion about microtubules above, dynein tends to grab hold of microtubules and walk in one direction. The dynein in cilia sprout from one doublet to grab the next. As they do this, the dynein try to "walk" up the nearby microtubule, causing the entire structure to bend. All of the dynein arms in the cilia are designed to work in an orchestrated fashion with each other so cilia bend this way and that, waving back and forth.

Since axonemes have nine fused microtubule pairs around the rim and two independent microtubules on the inside, the structure is typically dubbed the **9 + 2** structure. At the very base of the cilia's axoneme, just inside the cell's membrane, is a **basal body**, which is a structure that is nearly identical to (and derived from) the centriole.

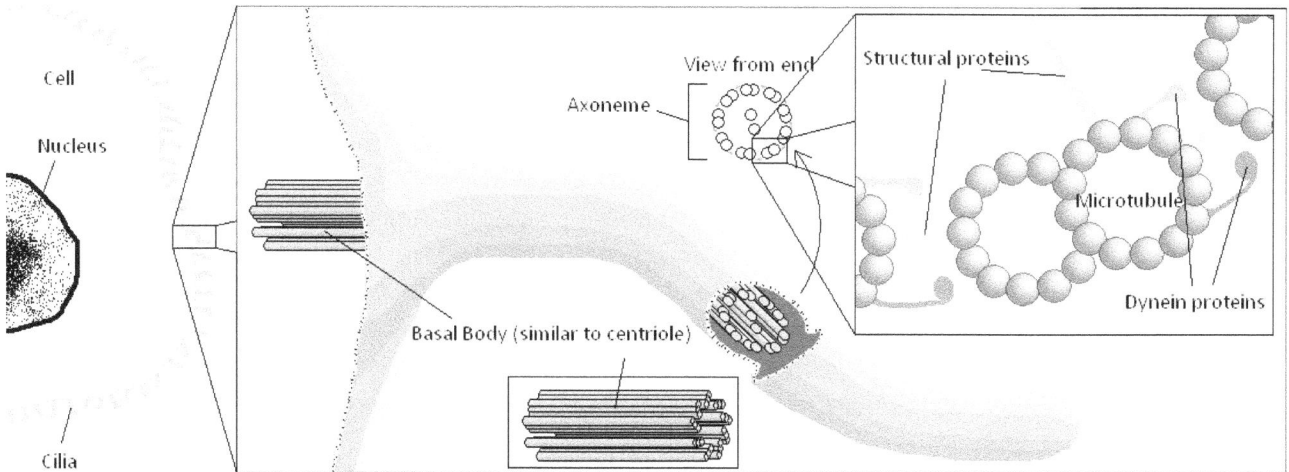

Flagella are longer than cilia and are used primarily for locomotion. The flagella found in eukaryotes, such as in sperm, have a structure that is essentially identical to cilia, only longer. The flagella in prokaryotes, however, are structurally complex and very different. We will not go in depth into that topic as it is outside the scope of this book, suffice it to say that prokaryotic flagella work by spinning around their axis (like a propeller pushing a boat through water), while eukaryotic flagella wave back and forth (like an oar or a fish's tail).

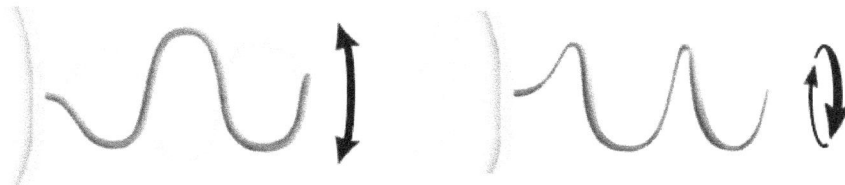

Eukrayotic flagella motion:
wavelike, back-and-forth

Prokaryotic flagella motion:
rotating

One final note should be made about the cilia and flagella. In previous sections, we talked about the glycocalyx that many cells have. The glycocalyx, as you'll recall, is made from small carbohydrate molecules attached to proteins and lipids in the membrane. This creates a very tiny fuzzy covering attached to the membrane. Do not confuse this with the hair-like cilia or flagella. Cilia and flagella are MUCH larger than the glycocalyx. You would have a hard time seeing anything like the glycocalyx with any equipment short of an electron microscope, whereas structures like cilia and flagella can be seen with light microscopes (and the appropriate dye). Cilia are still small enough, however, that you would probably have a hard time seeing them in the light microscope as well.

131

Summary

In this section, we talked about the anatomy of the cell.

- We divide cells into two broad categories: eukaryotes and prokaryotes. Eukaryotic cells have nuclei, while prokaryotic cells do not.

- Cells are made of organelles. These include: the nucleus, nucleolus, chromosomes, ribosomes, endoplasmic reticulum (rough and smooth), Golgi apparatus, lysosomes, peroxisomes, mitochondria, chloroplasts, microfilaments, intermediate filaments, microtubules, centrioles, cilia and flagella.

- The nucleus is the headquarters of the cell. It houses the DNA.

- A chromosome is a complete polymer of DNA coiled up around protein complexes called nucleosomes. These are, in turn, tightly wound together in a very compact manner. Chromosomes are most compactly wound during the process of cell division.

- The nucleolus is the part of the nucleus that contains unwound DNA, proteins and RNA which are involved in the initial stages of the protein production. This is where the "blueprints" for proteins are made.

- The nucleus is surrounded by a double membrane. This double membrane is continuous with the endoplasmic reticulum. The nuclear membrane has numerous openings that are guarded by nuclear pore protein complexes. These complexes facilitate and regulate the traffic between the cytoplasm and nucleus.

- Ribosomes are large protein/RNA complexes. The RNA in ribosomes is called ribosomal RNA (or rRNA). Ribosomes have a large and small subunit that fit together like two gloves.

- Ribosomes make protein by receiving blueprints in the form of RNA (called messenger RNA, or mRNA). This RNA threads through a channel in the ribosome. The new protein emerges from a small tunnel in the backside of the ribosome.

- The endoplasmic reticulum (ER) is a series of flattened tubes directly attached to the nuclear membrane, and is part of the endomembrane system of the cell. There are two kinds of endoplasmic reticulum: the smooth endoplasmic reticulum and rough endoplasmic reticulum. The rough endoplasmic reticulum has ribosomes attached to it. These particular ribosomes thread the proteins they are making through a small channel in the endoplasmic reticulum, so that the proteins ultimately end up in the space inside the endoplasmic reticulum. This space is called the lumen, a word that means roughly "hollow space within." Alternately, ribosomes may embed new proteins into the membrane of the ER.

- One of the most important roles of the ER is to manufacture triglycerides. These triglycerides, as well as any proteins embedded in the membrane, get packaged into small vesicles that depart from the ER toward the Golgi apparatus.

- The Golgi apparatus looks like a stack of hollow pancakes with bubbles around it. Vesicles merge with the Golgi from the ER. Once in the Golgi, the transported material gets chemically modified. When complete, the Golgi releases a vesicle (with chemicals inside) that goes to either the cell membrane or toward the system of lysosomes in the cytoplasm. The Golgi also makes lysosomes in this fashion.

- Vesicles that merge with the cell membrane release their contents to the outside of the cell.

- Lysosomes are vesicles that contain acid and enzymes that breakdown cellular debris and engulfed microorganisms.

- Peroxisomes are vesicles that metabolize long chain fatty acids, uncommonly found amino acids and some sugars. They use peroxide to metabolize many of these chemicals.

- Mitochondria harvest energy from various basic biological molecules. Mitochondria are believed to have once been free-living organisms. They have their own DNA.

- Chloroplasts are found in plants. They contain chlorophyll (which is what makes plants green). The chlorophyll captures light and helps the chloroplast turn the light into chemical energy. This chemical energy is used by the rest of the plant.

- The cytoskeleton is a series of structures made from microfilaments, intermediate filaments, and microtubules. Centrioles, cilia and flagella are included in this system.

- Microfilaments are the smallest elements of the cytoskeleton. Actin is the most commonly cited example. Actin is made of globular actin protein that connects together to form a double spiraling filament. Actin is used for structure, cellular transportation and cellular movement. Actin is found in muscle, where it takes an active role in muscle contraction.

- Intermediate filaments are intermediate in size to microfilaments and microtubules. They are made from threadlike proteins that pair up. These pairs combine with other pairs to make tetramers. The tetramers line up to form very long threads. Intermediate filaments tend to be elastic in nature. They are commonly found in hair, nails and horns.

- Microtubules are the largest of the cytoskeletal filaments. They are made of pairs of alpha and beta tubulin that line up into protofilaments. Thirteen protofilaments arrange in a circular tube fashion. Microtubules serve structural roles as well as roles in cell material transportation and movement.

- Centrioles are large barrel-shaped structures (usually found as pairs of barrels positioned at roughly 90 degrees from each other). Centrioles are believed to help determine the position of the cell nucleus and are important in the functioning of cilia and flagella.

- Cilia and Flagella are hair-like protein structures found in the cell's membrane. Cilia are smaller. They have two roles: 1) they make water flow over stationary cells, and 2) they move non-stationary cells around. Flagella are much larger and are used to move the cell around. Eukaryotic flagella have the same structure as cilia but are longer. Prokaryotic flagella have a different (more complex) structure than eukaryotic flagella.

Unit 18

Energy and Photosynthesis

In the previous sections we learned about how particles come together to make atoms, how atoms come together to make molecules, and how molecules come together in the reactions of life to make larger structures within the cell, such as the cell membrane, the organelles, and the various macromolecules like DNA, polysaccharides and the fatty acids. In this section, we begin our discussion of energy.

Energy

The precise definition of energy requires a great deal of confusing philosophy, but the definition has to do with the ability of one system to do "work" on another. For most novice students, this jargon brings up images of time cards and annoying bosses ordering people around, so let's instead try a makeshift definition to get our feet wet. By approximation, *energy is the ability of one thing to change another thing.*[30] If one object has no energy, it will be unable to alter another object. Similarly, if the first object has a lot of energy, it has the potential to cause great change to another object. The Universe is a complex thing, however, and the ways in which one object can change another depends on the circumstances and the types of energy present. Scientists recognize many forms of energy, such as thermal energy (heat), light, sound, chemical, nuclear, etc. Let's look at a few examples to get some practice with the concept.

Consider heat energy. If one object has more heat energy than another object, it can change the second object by heating it up, *changing* its temperature. The change isn't necessarily permanent. If you remove the hot object, the second object will cool down again as the heat dissipates into the air. The change isn't always obvious either. If you put a battery (chemical energy) into an electronic device, the battery will begin to cause microscopic

---

[30] For example, one energetic object might change another object's speed, temperature, structure, appearance, etc.

changes by forcing electrons to move around inside the device.[31] Naturally, you would not notice this form of change unless the device had some kind of output or observable behavior, such as making sound, light, heat, etc. For example, the sound energy made by the device briefly *changes* the positions of little hairs in your ear. This starts a chain reaction that alerts your brain to the sound. As another example, the light energy from the electronic device briefly *changes* tiny molecules in your eyes. Once again, this starts a chain reaction that ultimately alerts your brain of the light energy. To summarize, the energy from the battery gets transferred to the electronic device, which then causes changes inside your body.

Battery-powered devices are not the only things changing your body, however, and you can't always tell when it happens. In fact, energy that brings change that is hard to detect can be very dangerous. For example, in the beginning of this book we talked briefly about what radioactivity is. Radioactivity results from nuclear energy. Nuclear energy causes changes in the nuclei of atoms, which then eject radioactivity into the environment. This radioactivity can cause tiny changes inside your body, such as in your DNA. If the change occurs at the wrong place, it can lead to cancer. Nuclear energy isn't the only form of energy changing your body though. As we saw with the electronic device, other forms of energy are constantly changing your insides . . . heat, vibrations, sound, electricity, chemical. . . They are attacking you all the time. Sometimes it's subtle. At other times it's obvious and even violent. With all the changes happening in your body, you are faced with the constant threat of illness and death. The only solution, however, is to fight fire with fire, as they say. We must use energy to reverse the changes. We require energy to thrive. If we do not have it, we cannot correct the changes taking place within us or around us. Our insides and outsides will gradually become less suitable for life, and then we'll die.

Though we are surrounded by energy, we can't use just any ol' type. Our bodies require specific forms, primarily that of food and drink. We must eat to both minimize harmful change and encourage healthy change. It is a fact of life. It's like the old saying: there are only four constants in life: taxes, death, our need to eat and loud neighbors. Energy powers it all.

Energy and Sun

Let's take a step back to get a grasp on where all this starts. The energy that biological systems use begins with the sun. The sun is a big ball of nuclear energy. That energy is constantly rearranging the sun's atomic nuclei, and this emits massive amounts of radioactivity. Some of that radioactivity is called **electromagnetic radiation**, or EM radiation, which is just a fancy name for a whole range of particles of light (**photons**) that have different energy values. The details are vast and complex, but basically EM radiation travels as waves and comes in a very large spectrum, from the extremely energetic **gamma rays** to the relatively weak **radio waves**. Between the two, within an intermediate range of energy, is *visible* light. Consider the analogy of rain. Some drops are large, while others are small. EM radiation is similar. Some photon particles have incredibly high amounts of energy, while others have less. All of these particles of light are racing past us all the time, but more so during the day. Although it is difficult to show in grayscale, the picture on the next page should give you an idea of the spectrum of electromagnetic radiation.

---

[31] The movement of electrons is basically what electricity is.

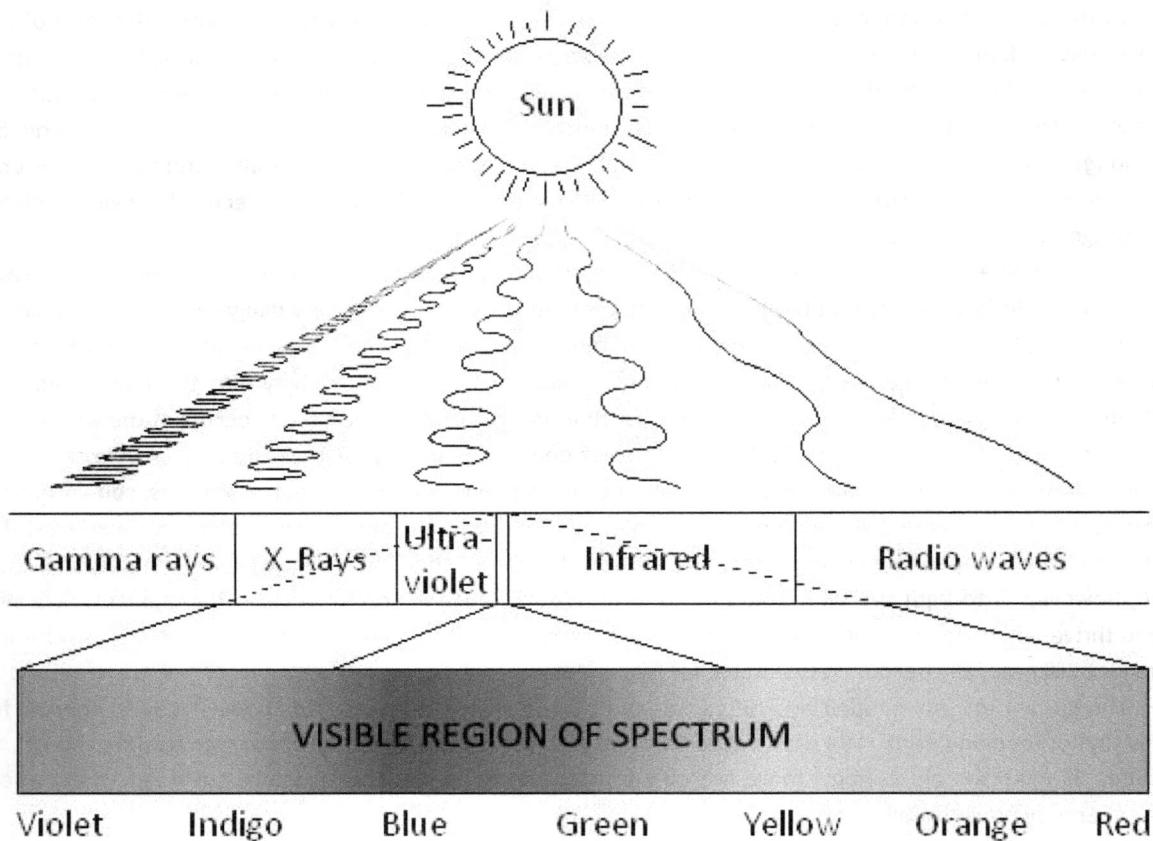

Gamma rays | X-Rays | Ultra-violet | Infrared | Radio waves

VISIBLE REGION OF SPECTRUM

Violet   Indigo   Blue   Green   Yellow   Orange   Red

Collectively, EM radiation represents a *lot* of energy hitting the Earth at any given instant. A long time ago, living systems found a way to use that energy to power the changes they needed to survive: they developed the ability to absorb the energy that was in light. Up until then, living systems had to scavenge around for whatever energy they could find in their inorganic environment. With the advent of harvesting energy from the sun, things began to change radically. These light-absorbing organisms began storing the energy they caught from the sun's rays as highly energetic molecules. The scheme worked very well. The harvest was so bountiful, in fact, that energetic molecules began leaking from these organisms into the environment (and still do). One of those highly energetic molecules is oxygen, O_2.

Not all organisms had developed the ability to absorb light, however, but the increasing presence of oxygen gave these latter individuals a new source of energy as well. Life all across the planet took a huge leap forward, as a clear distinction emerged between two lifestyles that are still in existence today. The first lifestyle is that of the **autotrophs**, which are the organisms that get their energy directly from a primary source (i.e., sun or inorganic energy source). Modern day examples include plants and algae. The second lifestyle is that of the **heterotrophs**. They get their energy by eating autotrophs and/or their byproducts. In effect, the autotrophs pull inorganic energy into the reactions of life, and the heterotrophs freeload off of them. Modern examples of the freeloading heterotrophs include herbivores, carnivores and politicians.

136

The important point to understand is that nearly all the energy that fuels life starts with the sun, gets captured by autotrophs and then gets used by heterotrophs. Even if one heterotroph makes a living by eating only other heterotrophs, (such as a carnivore), the energy that it gets still originated from the sun. No community of organisms can survive without autotrophs being part of the picture somewhere along the line.

The absorption of light energy and its storage within biological molecules is called **photosynthesis**. When autotrophs do photosynthesis, they take low-energy molecules (CO_2 and water)[32] and use energy to combine them in a series of reactions to make sugar. The overall reaction goes as follows:

$$6CO_2 + 12H_2O + light \longrightarrow 6O_2 + C_6H_{12}O_6 + 6H_2O$$

In other words, six carbon dioxides, twelve waters and the energy from light are mixed together in a tiny biological pot to make one sugar (specifically **glucose**, $C_6H_{12}O_6$). In addition to the glucose, six oxygen molecules and six of the original twelve water molecules pop out. This is how plants make oxygen to fill up the environment.

So why would a plant do this? What is the advantage of having a sugar sitting around? Well, the sun isn't there 24 hours a day. Cloudy days happen too. Sugar is highly energetic and represents the plant's attempt to save for a raining day. At night and during other low-light conditions, the plant cannot rely on photosynthesis to drive its reactions, so it uses the sugars it made earlier to continue living. All the plant needs to do is recombine the sugar with oxygen and water, run the reaction in reverse, and *TADA* the energy pops back out, only this time the energy is *not* in the form of light; it's chemical energy. This energy powers the plant. We use sugar in the same way. We breathe in oxygen and drink water, and combine these things with high energy molecules like sugar that we've eaten to acquire the energy we use to drive the reactions we need to keep us alive.

At this point, I should point out a very important fact. Plants do not release oxygen for our benefit. They don't do it because they want to be nice. They release oxygen because they can't hold it all in after it is produced. Plants use oxygen just like we do. It is useful to them, but they create so much of the stuff, that there isn't a need to be stingy. Not only that, but excess oxygen can interfere with the synthesis of glucose. Too much oxygen can be a bad thing. So, no, plants aren't being friendly to us.

Photosynthesis: The Light Reactions

Photosynthesis starts with the "light reactions," so named because it requires light to drive it. Light enters the plant's many chloroplasts (remember that from the last unit?) and hits molecules called **chlorophyll**. There are many kinds of chlorophyll actually, but the general form is shown in the next picture. The structure looks like a little satellite dish, with a single magnesium ion stuck in the middle. . . locked there, with dreams and ambitions, plans for greatness, all ruined by the wickedness of the chlorophyll. Chlorophyll looks green, so that means it must like green light, right? Sadly, no. Chlorophyll absorbs visible light mostly in the range of violets, blues and reds. The result is

---

[32] CO_2, H_2O, and N_2 are actually very common low-energy molecules found on planets having atmospheres. The reason is that carbon, oxygen, hydrogen, and nitrogen are common atoms, and when they form these molecules, they are very stable (e.g. happy in that form). The Earth is unique in that it has a LOT of O_2 in the atmosphere because of photosynthesis. O_2 is very energetic (e.g. unstable or unhappy). If plants died off, our atmosphere would quickly return to something similar to the other planets. In that regard, photosynthesis may be seen as one of the most important transforming agents of our planet. Therein lay one of the reasons for the great concern over deforestation.

that the green light bounces off unabsorbed. This is the reflected light that we end up seeing, making leaves look green.

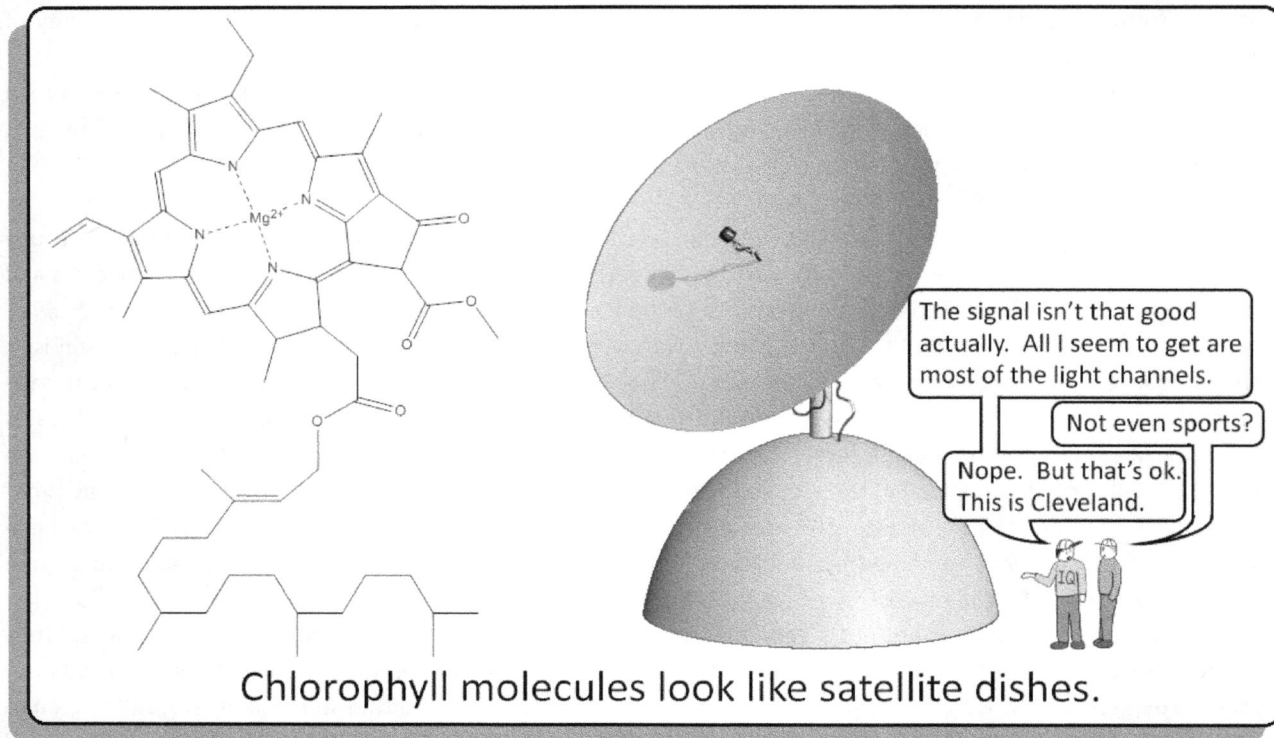

The signal isn't that good actually. All I seem to get are most of the light channels.

Not even sports?

Nope. But that's ok. This is Cleveland.

Chlorophyll molecules look like satellite dishes.

As it turns out, chlorophyll isn't the only molecule working in the craft of absorbing light. Chlorophylls have other molecules called **carotenoids** that help in the process. Together with these accessory pigments, the chlorophyll and carotenoids make up what's known as a **light harvesting complex, (LHC)**. A bunch of these magnesium-mistreating baddies are bunched together to make a **photosynthetic unit**. Within each unit, the numerous light harvesting complexes gather up energy from the light, and transfer it to a centralized area of the photosynthetic unit for the next step in the process. This centralized area is known as the **reaction center**.

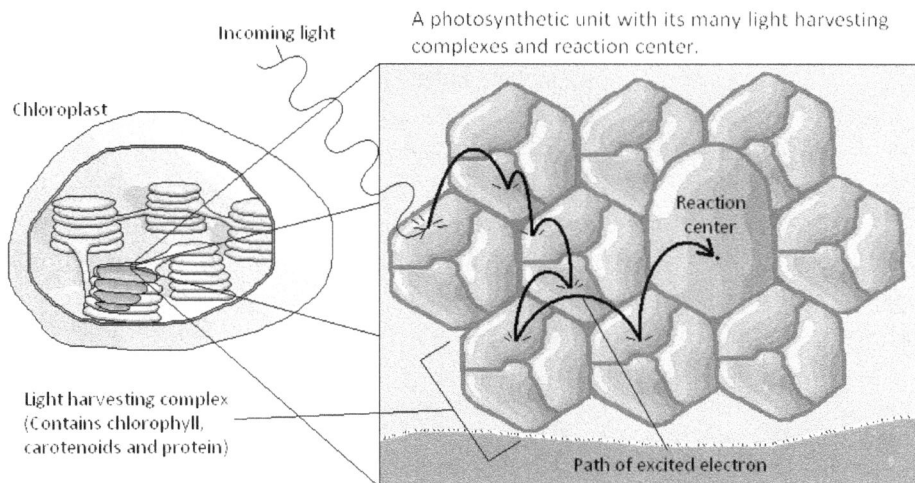

Chloroplast

Incoming light

A photosynthetic unit with its many light harvesting complexes and reaction center.

Reaction center

Light harvesting complex (Contains chlorophyll, carotenoids and protein)

Path of excited electron

This last image also gives a simplified take on how light is absorbed. In essence, the light of specific energy comes in and bumps an electron from its molecular orbital shell from within the magnesium-holding part of the chlorophyll. This electron becomes quite energetic and bounces along the pigments like a hot potato being passed around, until finally the electron arrives at the reaction center. From this point, the electron uses the reaction center as a launching point to hop to a number of other molecules, each jump releasing some of this energy in a controlled manner until the electron finally arrives onto a molecule called **NADP$^+$**. This entire event happens twice so that two electrons collect on the NADP$^+$. Electrons have negative charges, so the two electrons transform NADP$^+$ into NADP$^-$. This minus charge attracts H$^+$, which combines to make it **NADPH**. NADPH then floats away. It has more energy than just regular NADP$^+$, so it takes this energy away to other parts of the cell. The following image illustrates the concept. Here, the photosynthetic unit is designated "P700," a name that refers to the wavelength of light that excites it. The names of a few other proteins are included as well.

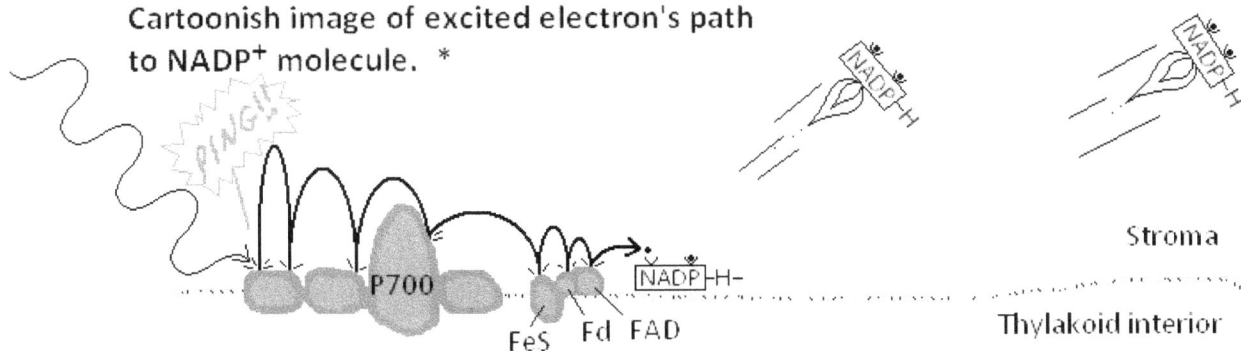

Cartoonish image of excited electron's path to NADP$^+$ molecule. *

PING!!

P700

FeS   Fd FAD

NADP-H-

NADP-H

NADP-H

Stroma

Thylakoid interior

* In reality, the electron does not actually bounce along the membrane structures, but rather travels *through* the structures. Depicting the electron as a bouncing ball makes the process easier to remember though.

139

Missing electron!

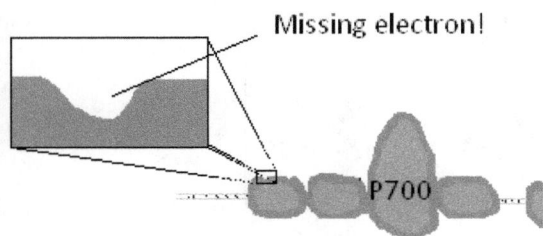

The story of the chlorophylls isn't quite over just yet, however. For each electron that gets excited by light (bouncing around and off the photosynthetic unit and onto $NADP^+$), there is a hole left behind where the electron was.

These holes need to be filled, or else the process will eventually come to a halt! To help fill these holes, modern plants use excited electrons from a *second* photosynthetic unit (called P680) to fill the first. This second photosynthetic unit is slightly different than the first one, because it requires slightly more energetic light to excite electrons. Otherwise, it works pretty much the same way, with electrons moving from site to site as in this picture:

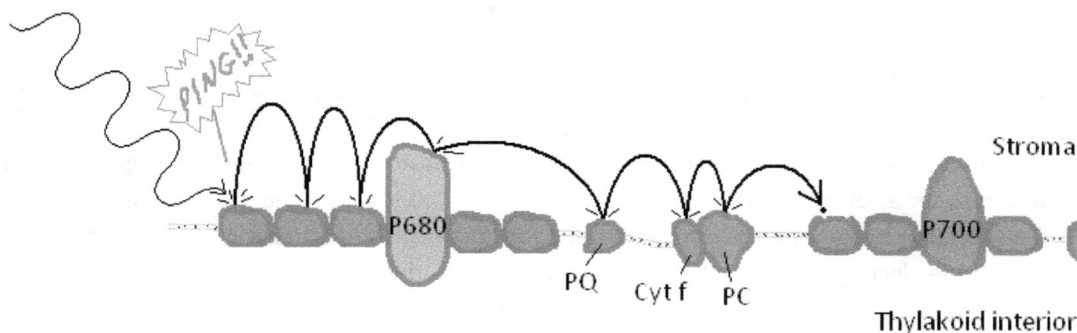

This, however, leaves the second photosynthetic unit deprived of electrons. To solve the problem, the P680 complex steals electrons from water, one for each time light bounces its own electrons down the line toward $NADP^+$.

Ultimately, this causes water to split up into one oxygen atom and two protons, as shown to the left.

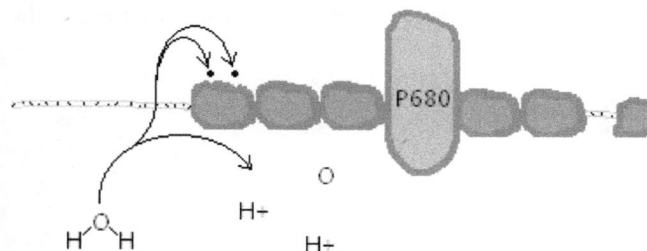

The lone oxygen that is created in the process brings us back to some of the basic chemistry we talked about earlier in the book. Lone oxygen atoms are not very happy because they do not have a full valence shell. Luckily, the photosynthesis process happens at an incredibly fast speed so that the lone oxygen does not have to wait long before another oxygen atom is created. The two oxygen atoms pair up to make molecular oxygen,[33] O_2. Plants are not airtight systems, however, so some of the O_2 that is created escapes into the air. This innocent little O_2 leak is the reason our atmosphere is so radically different than other planets.

The story of photosynthesis isn't quite over, though. After all, it wasn't necessarily the plants' goal to make O_2. Plants do photosynthesis to capture the sun's energy in chemical form, so we need to cover a few more details. Well, in addition to pushing electrons onto an $NADP^+$ molecule, (affectively charging the molecule with energy), the

---

[33] As we'll see in the next unit, molecular oxygen, O_2, isn't very happy either, but the two atoms pair up to temporarily solve their problem of not having a full valence shell. This little marriage of convenience doesn't last very long.

photosynthesis process uses the energy of the excited electron to pump protons across the membrane. Putting it all together, it would look like this:

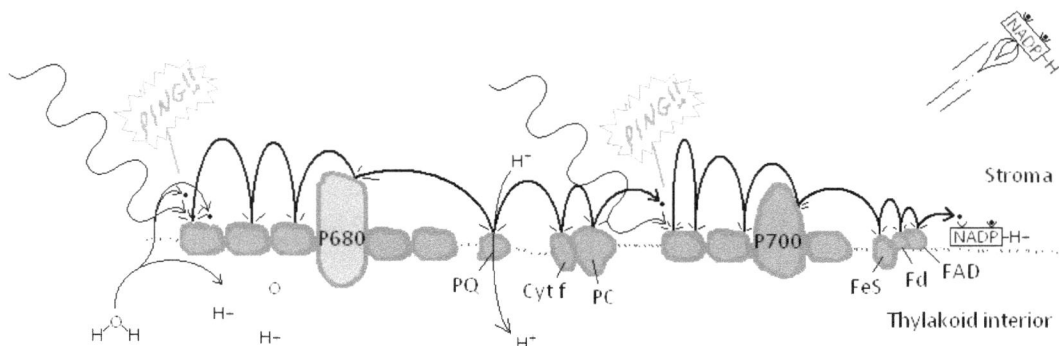

We are down to the homestretch. If you look at the last image, you'll notice that a bunch of H^+ ions are created/transported inside the thylakoid (the hollow pancakes inside the chloroplast) during the photosynthesis process. This leads to a very heavy overcrowding of H^+ in the thylakoid interior. If you recall from our discussion on concentration gradients (Unit 12), anytime you get an overcrowding of any chemical, that chemical will want to move in the direction that makes less crowding. That means the H^+ inside the thylakoid are unhappy and want out. This is where the plant uses a very creative trick: it doesn't let the H^+ out unless they pay a toll in the form of energy. That energy is used to convert ADP (a low-energy molecule) into ATP (a high-energy molecule) during a process called **phosphorylation**, which basically means "adding a phosphate." The highly energetic ATP floats away and helps power the rest of the plant.

Simplified chloroplast with crowded thylakoid interiors

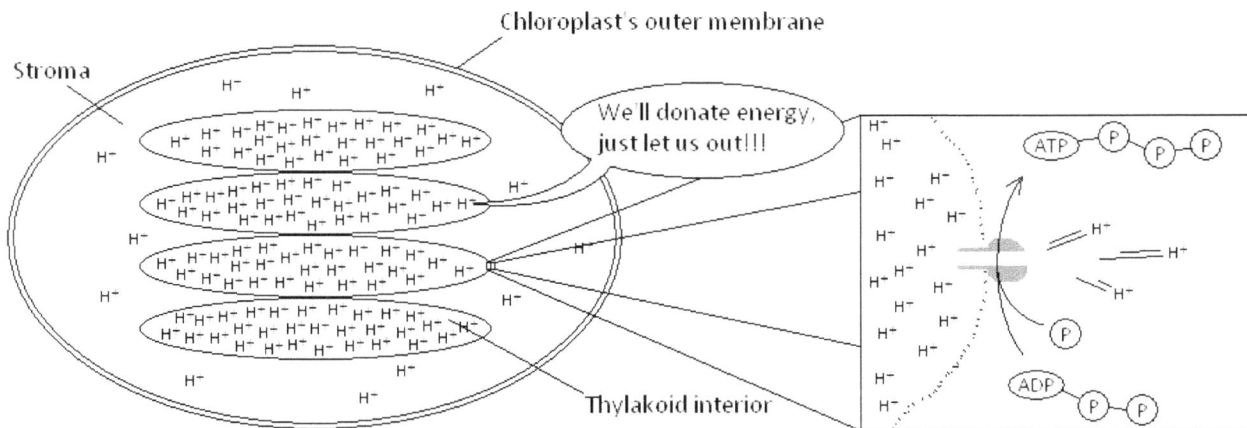

141

Photosynthesis: The Dark Reactions (The Calvin Cycle)

The next part of photosynthesis is referred to as the "Dark Reactions." This is not because they occur in the dark, but because they do not require light. They take place at any time. The general idea is that ATP and NADPH, (the two high-energy products from the previous set of reactions), are used to drive a cycle that puts together glucose. This is called the **Calvin Cycle**.

Step 1:

The first step in these reactions involves the combination of a CO_2 molecule with the five carbon sugar called **ribulose bisphosphate**, or **RuBi**. This makes a six carbon molecule that rapidly breaks into two **3-phosphoglycerate** molecules.

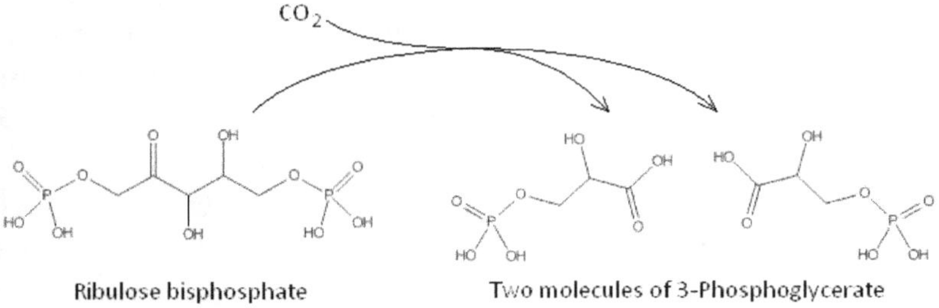

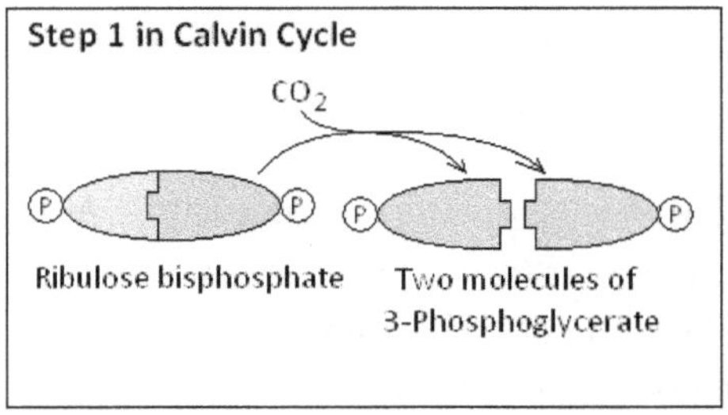

Ribulose bisphosphate   Two molecules of 3-Phosphoglycerate

Step 1 in Calvin Cycle

Ribulose bisphosphate   Two molecules of 3-Phosphoglycerate

Step 2:

The next step in the cycle involves the phosphorylation of 3-phosphoglycerate by ATP, forming **1,3-bisphosphoglycerate** and ADP (reaction shown below). ATP is a very energetic molecule, so any time you see ATP being used (rather than produced) in a reaction, think "here is where energy is required to keep the process going." So, in this case, the principle tells us that converting 3-phosphoglycerate to 1,3-bisphosphoglycerate requires energy. This also means that 1,3-bisphosphoglycerate is more energetic than 3-phosphoglycerate.

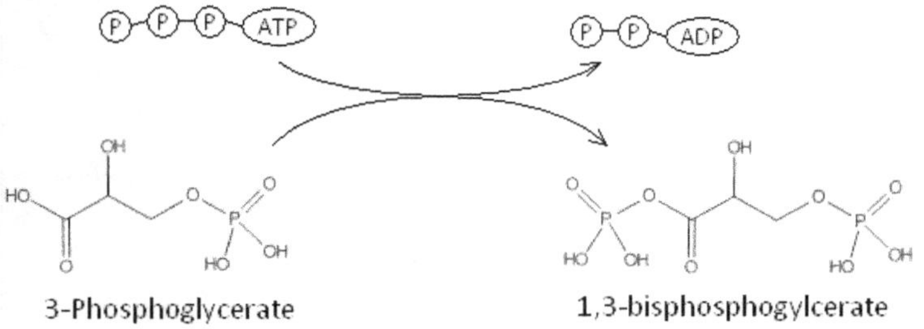

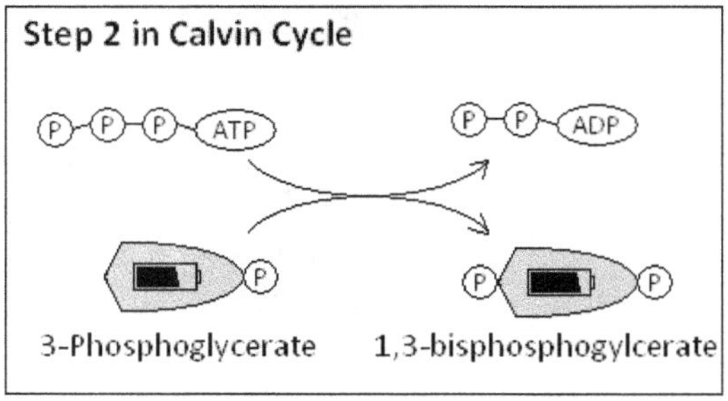

3-Phosphoglycerate   1,3-bisphosphogylcerate

Step 2 in Calvin Cycle

3-Phosphoglycerate   1,3-bisphosphogylcerate

Step 3:

After the creation of the energetic 1,3-bisphosphoglycerate molecule, a molecule of NADPH swoops in to convert it into **glyceraldehyde 3-phosphate**. Like ATP, NADPH is a highly energetic molecule made during the light reactions of photosynthesis. Also like ATP, any reaction that involves NADPH is a reaction that requires energy input. Below is an image of this reaction. Notice from it that NADPH has a hydrogen when it begins the reaction, but none as it floats away when the reaction is done. The "missing" hydrogen atom was attached to the glyceraldehyde 3-phosphate molecule. That leads us to another general pattern in these kinds of reactions: NADPH gives a hydrogen atom and energy to the molecules it reacts with.

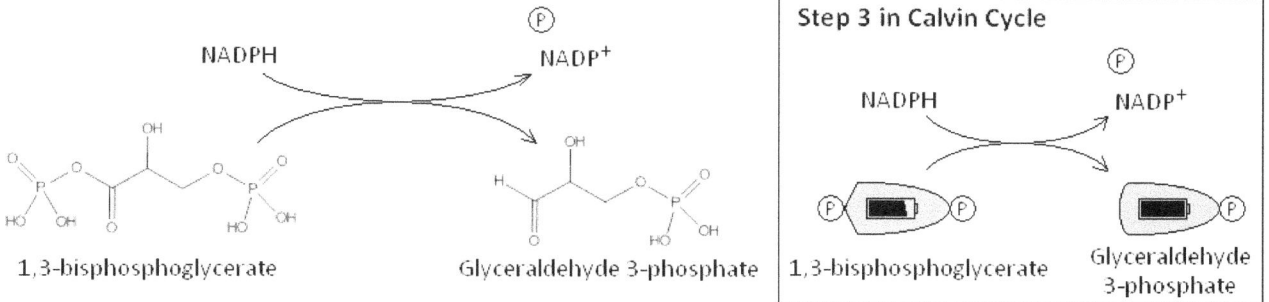

1,3-bisphosphoglycerate   Glyceraldehyde 3-phosphate

Step 3 in Calvin Cycle

1,3-bisphosphoglycerate   Glyceraldehyde 3-phosphate

By this point, the glyceraldehyde 3-phosphate molecule that was created has been the recipient of the energy of both ATP and NADPH. That makes glyceraldehyde 3-phosphate *very* energetic. The plant takes two of these little balls of energy and makes sugar with them.

Two Glyceraldehyde 3-phosphate molecules → Glucose (the plant's energy for a rainy day)

Our discussion of the Calvin Cycle is not quite over though. At the beginning of the cycle, we started with ribulose bisphosphate, our "RuBi" molecule. The plant has to recreate this molecule somehow, or it can't make any more of these reactions work. So, let's see how the plant recreates RuBi.

Step 4:

As it turns out, not all of the very energetic glyceraldehyde 3-phosphate molecules are used to make sugar. Each time a bunch get made, the chloroplast takes 5 of them and mixes them together in a complex set of reactions to recreate 3 RuBi molecules. Three more ATP molecules are required to do this. Here is what it looks like:

Step 4 in Calvin Cycle (many complex reactions)

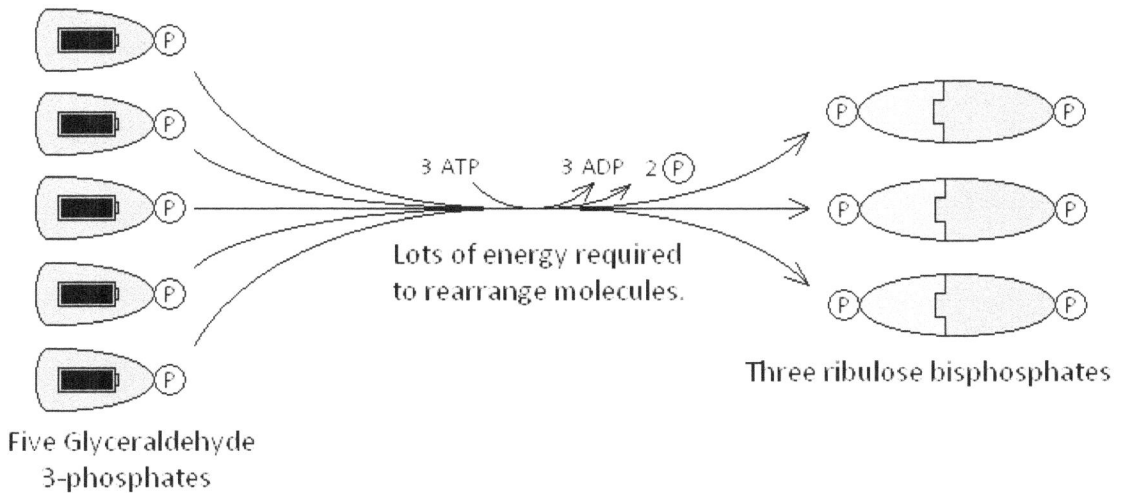

3 ATP   3 ADP   2 (P)

Lots of energy required
to rearrange molecules.

Three ribulose bisphosphates

Five Glyceraldehyde
3-phosphates

It is natural at this point for some confusion to creep into the situation, usually because the complex organic chemical structures complicate things. Also, you might be thinking something like: if it takes five glyceraldehyde 3-phosphates to make only three RuBi molecules, aren't we rapidly running out of molecules?

To make sure we clear up possible confusion, let's switch entirely to the little cartoonish symbols and go back over the reactions and keep track of the number of molecules. If you look at Step 1, each RuBi makes two smaller molecules. If you follow these molecules through Steps 2 and 3, each one becomes one glyceraldehyde 3-phosphate:

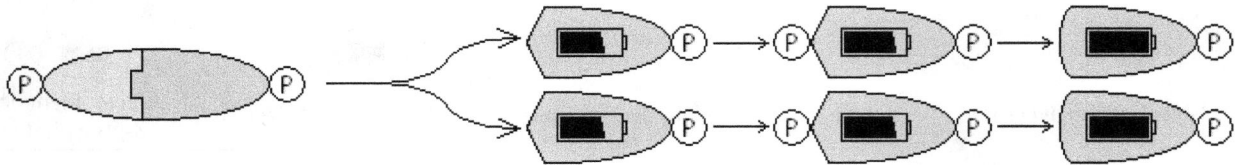

Or, in summary:

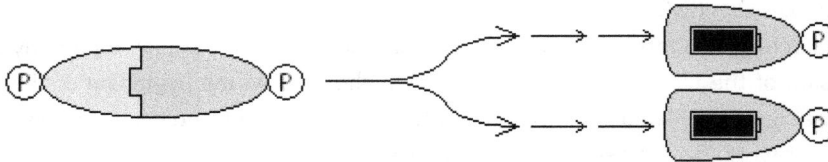

As this last image shows, each starting RuBi makes two glyceraldehyde 3-phosphates. Now imagine that we start the cycle with *three* RuBi molecules. Since each one ultimately makes two glyceraldehyde 3-phosphate molecules, we would end with six glyceraldehyde 3-phosphates at the end. We then can take five and remake three RuBi to do the cycle again.

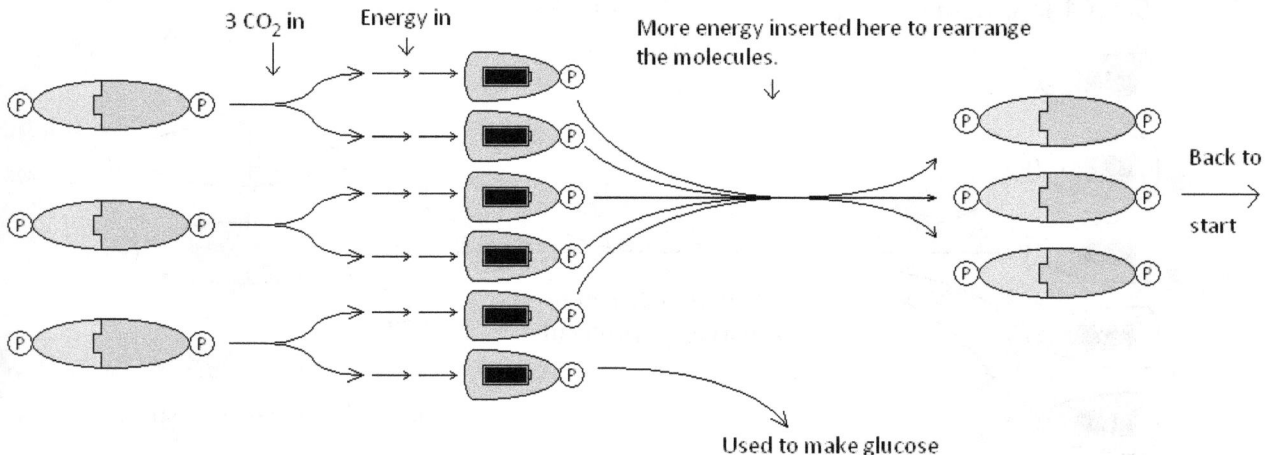

As this last image shows, the Calvin Cycle starts with three RuBi, absorbs three CO_2 as well as a bunch of energy, and creates one extra glyceraldehyde 3-phosphate that is used to make sugar. The rest gets turned back into three molecules of RuBi to start the cycle again. Here is a summary of the Calvin Cycle:

144

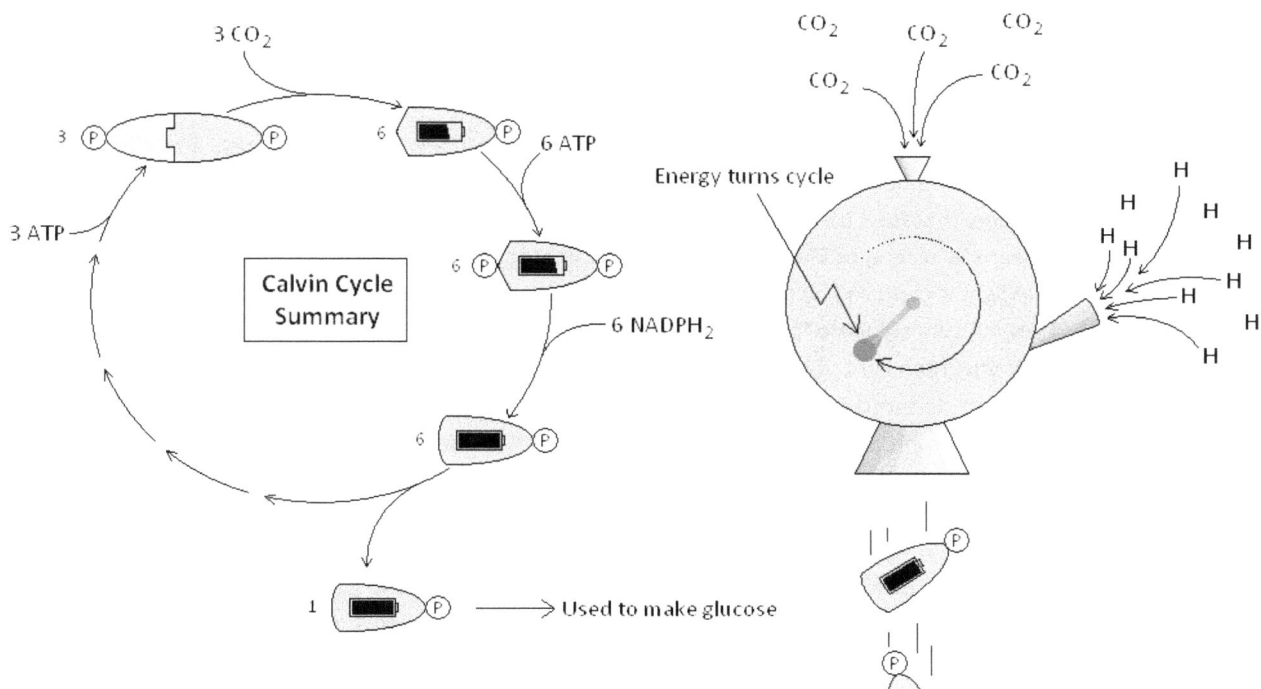

So to summarize our entire discussion of photosynthesis: the energy from sunlight is used to take CO_2 and water, and combines them together to make O_2 and glyceraldehyde 3-phosphate. This last molecule is used to make glucose. Here is the overall summary:

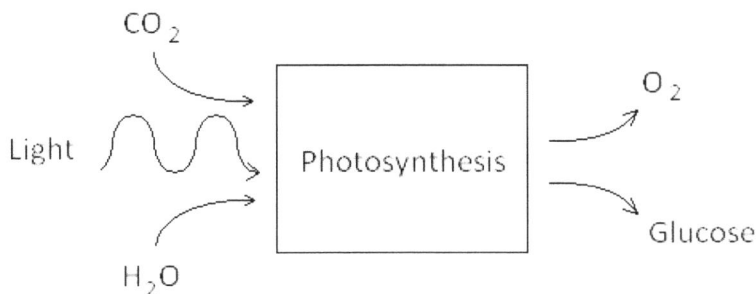

From this image, we see that CO_2 and water from our atmosphere is converted to O_2 and glucose. This simple innocent appearing little process is the reason that our planet has so much more oxygen and so much *less* CO_2 than other planets. It's how almost all life on Earth gets its energy.

In the next few units, we'll shift our attention to the chemistry that lets plants and animals pull the energy back out of glucose.

Summary

In this section, we talked about energy and photosynthesis.

- Energy is the ability of one thing to change another. There are many forms of energy, and almost all are relevant to the processes of life at one point or another.

- Nearly all the energy that fuels life on Earth starts with sunlight. "Light" is part of a spectrum of energy waves called electromagnetic waves, or EM waves. The visible light that we see is only a part of that spectrum.

- The sun's visible light is captured by plants in photosynthesis. Plants convert this energy into chemical energy as glucose. They consume glucose for energy whenever there isn't enough light to get energy.

- Plants do not absorb green light. Green light bounces off and gives plants their color.

- Photosynthesis is divided into two sets of reactions: the light reactions and the dark reactions. The light reactions require the sun's light. The dark reactions take the energy that was captured during the light reactions and converts it into chemical form. Another name for the dark reactions is the Calvin Cycle. These reactions don't necessarily take place in the dark, but are named that way because they don't require light.

- The light reactions begin when a chlorophyll molecule is hit by incoming light. Chlorophyll is found inside a protein complex called a light harvesting complex, or LHC. These are found packed together with other LHCs in the thylakoid membranes of chloroplasts. The entire collection of LHCs is called a photosynthetic unit, which has a special "reaction center." When the light hits chlorophyll, one of its electrons gets excited and bounces through the complex of LHCs until it hits the reaction center. Once there, the electron moves through other membrane proteins and finally arrives at an empty $NADP^+$ molecule. When the molecule captures two such electrons, it becomes NADPH and is used to fuel the Calvin Cycle.

- Each LHC that donates an excited electron to a $NADP^+$ molecule now has an electron deficit. This is filled by an analogous process taking place in a nearby similar set of LHCs. These LHCs now have an electron deficit. To solve the problem, this second set of LHCs steal electrons from water, and the oxygen atom that results pairs up with other oxygen atoms to make O_2 gas. Since plants are not gas tight, some of the O_2 is always leaking out into the environment.

- In addition to the energetic NADPH molecules that are made, the process of the electron bouncing through the membrane proteins causes a H^+ gradient to be produced inside the thylakoid interior. These crowded H^+ ions move through a channel in the membrane in order to escape back out, and in the process they provide energy needed to convert ADP into ATP.

- In the Calvin Cycle, CO_2 is pulled from the air and combined with ribulose bisphosphate. In the series of reactions that result, six molecules of glyceraldehyde 3-phosphate are produced. Five are recombined to recreate the starting ribulose bisphosphate molecules. One of the glyceraldehyde 3-phosphates is set aside and combined with another glyceraldehyde 3-phosphate (created during the next cycle) to make glucose, which is stored for later use when energy is needed. The NADPH and ATP generated in the light reactions are used to fuel the Calvin Cycle.

Unit 19

Oxidation and Reduction

In the last unit, we talked about energy and how it gets stored in glucose during photosynthesis. In the chapters following this unit, we will be talking about how energy is released from glucose. Until now, however, we haven't really commented much on why one molecule has more energy than another. We'll need to address this if we are going to have a deeper understanding of life and its energy patterns. The discussion centers on oxygen, which has a vital role in the movement of energy in living material. Oxygen is constantly jockeying for electrons in a process called **oxidation** and **reduction**, and this competition fuels practically all living reactions around you.

The oxygen molecule, O_2, or O=O, is a very unstable molecule. Oxygen likes to hog electrons, but if it's attached to *another* oxygen, neither win in the tug-of-war for electrons. This is a downright unhappy situation, leading to a very short molecular marriage. The O_2 breaks down whenever given the chance, exchanging partners for another more generous atom as soon as possible, and this releases the "tension" as energy. The fact that O_2 is so unstable is why you don't see much O_2 on other planets, like Venus and Mars. Here on Earth, we have plants that produce it during photosynthesis, so there is a large supply of unhappy O_2 in the air. This miserable oxygen is constantly butting into the business of other molecules, eyeballing neighboring chemicals for a chance to jump in to bonds and hog electrons. For example, when oxygen sees a carbon bonded to hydrogen, it sees an opportunity to be much happier. . .

Oxygen is *much* happier when it mixes with carbon and hydrogen, because it is easy to pull electrons from them. The more oxygen gets attached to C and H, the more energy gets released. In fact, that is precisely what

happens when things burn. Oxygen gets into bonds and hogs all the electrons, releasing a LOT of energy in the form of heat and light.

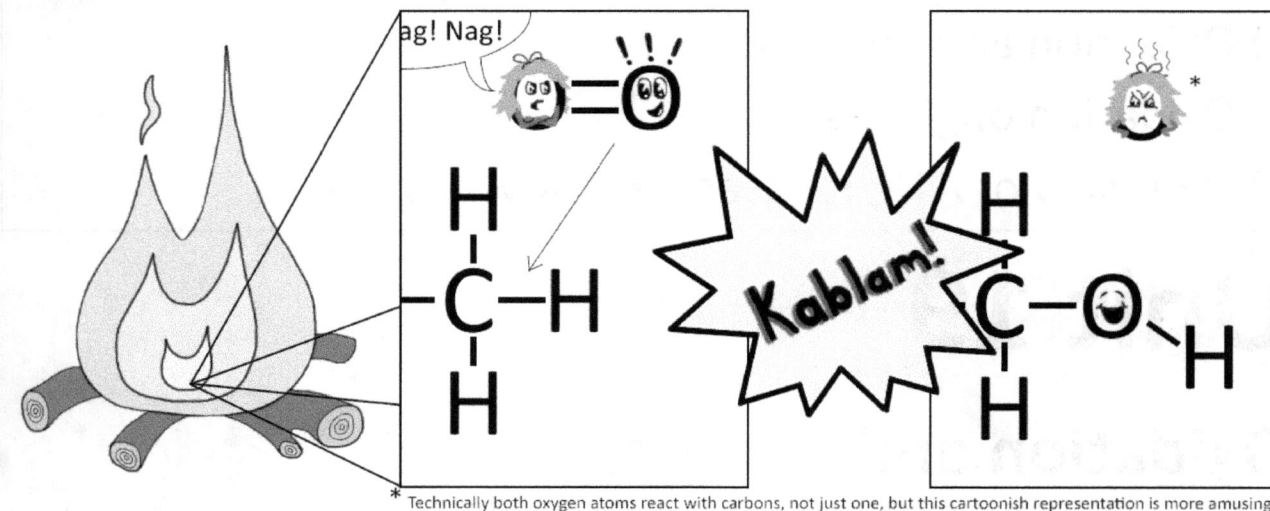

* Technically both oxygen atoms react with carbons, not just one, but this cartoonish representation is more amusing.

We say that the more oxygen that is attached to a carbon, the more **oxidized** the carbon is. The more oxidized it is, the more energy has been released. Conversely, this also means that the more oxidized a carbon is the less energy it has left to release. In other words, carbon molecules that have few oxygen have high potential energy. Carbon molecules that have a lot of oxygen have low potential energy.

For example, here is what it would look like when butane oxidizes:

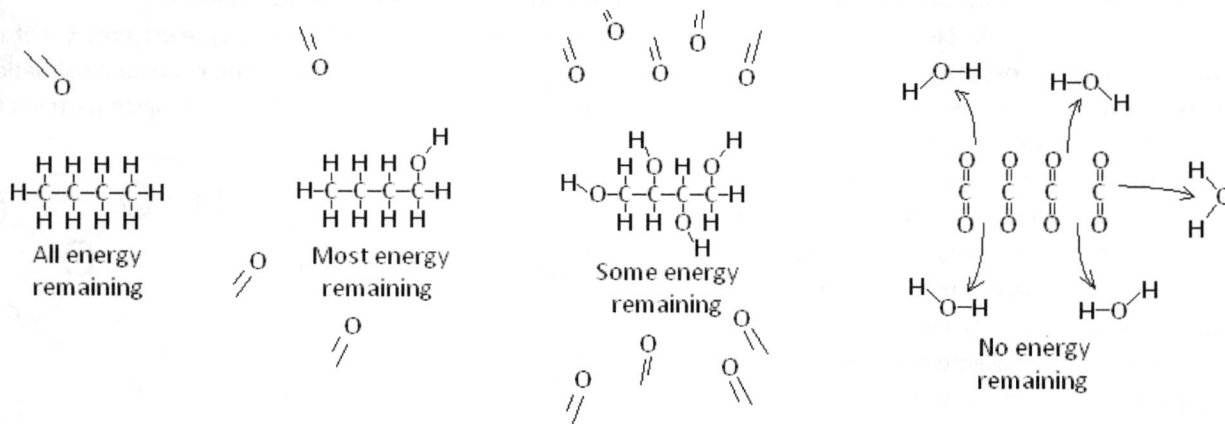

The oxygen keeps invading until the butane becomes so saturated with oxygen that is it broken into four CO_2 and five H_2O. At this point, butane has become fully oxidized and can release no more energy. The word "oxidized" is actually an unfortunate word, because oxygen isn't the only molecule that can cause this to happen. More globally, the word "oxidized" means that the atom has an electron pulled away from it, resulting in a positive or partial positive charge.

148

$$C\text{—}H \qquad C\text{—}O \qquad Na^+ \; Cl^-$$

Nonpolar   "Oxidized"   "Reduced"   "Oxidized"   "Reduced"

Polar   Ionic

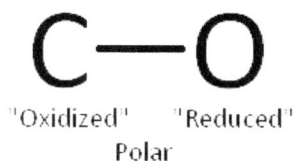

An atom becomes *oxidized when electrons are pulled away from it.* That is at the heart of the concept. Oxygen oxidizes carbon because it hogs electrons away from it. So, what do we say about the atoms that pull electrons toward them? We say that they become **reduced**. These atoms have electrons pulled toward them, and have a negative charge or partial negative charge.

Gain an electron ----> "Reduction"
Lose an electron ----> "Oxidation"
Mnemonic: LEO says GER

Ok, that said, let's look at glucose. It has six carbons. To remove as much chemical energy as possible, we have to oxidize it to the point where it is broken down into individual CO_2 and water molecules, much like what happened to butane. A biological cell could, in theory, just throw the glucose into a fire and let O_2 rip it to shreds, releasing all that energy, but the energy released would be uncontrolled (and dangerous), increasing the temperature and boiling the cell. That strategy won't work! The cell needs to oxidize the glucose one step at a time, carefully allowing the energy to be released in other forms. That way, much less energy gets lost as uncontrolled heat. How should the cell store these bits of carefully released energy? The answer is simple. It uses a number of other molecules that act as "middlemen" in the process. These middlemen take the electrons one at a time *from* glucose and give them *to* oxygen. This allows energy to be released more slowly and carefully. During this highly controlled process, the middlemen take these little bits of energy and store them in a much more useful and practical form: ATP.

Hey! Where did that electron go?

Electron

Energy given to ATP

149

ATP is the short-term, day-to-day energy currency of the cell. In fact, if glucose were a gold brick, ATP would be a $10 bill and much more practical for pushing around reactions or moving materials. No one pays for things with gold bricks after all. We might store our wealth in gold, but we pay for things with smaller value denominations.

What makes ATP capable of being an energy molecule? Let's look at it for a moment and see.

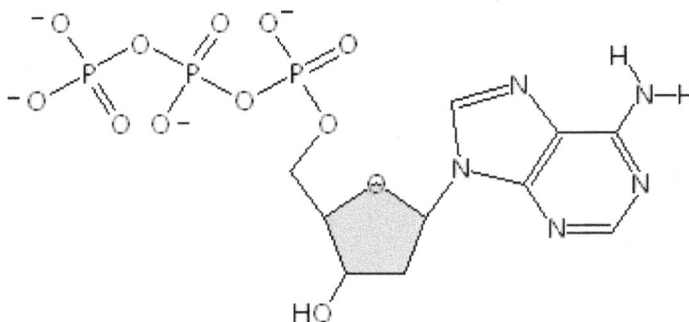

Adenosine triphosphate, "ATP"

Notice that ATP has three phosphate groups in it, strung up in a line. These phosphate groups are not happy in this situation, because each one is negatively charged. Negative charges repel, so each phosphate begrudgingly bonds to another, but you can bet they are counting down the time for when they can break up and leave. As such, ATP is lively and energetic. The cell makes ATP by taking the energy that is released when glucose is oxidized to force a phosphate group to add to the end of ADP in the following reaction:

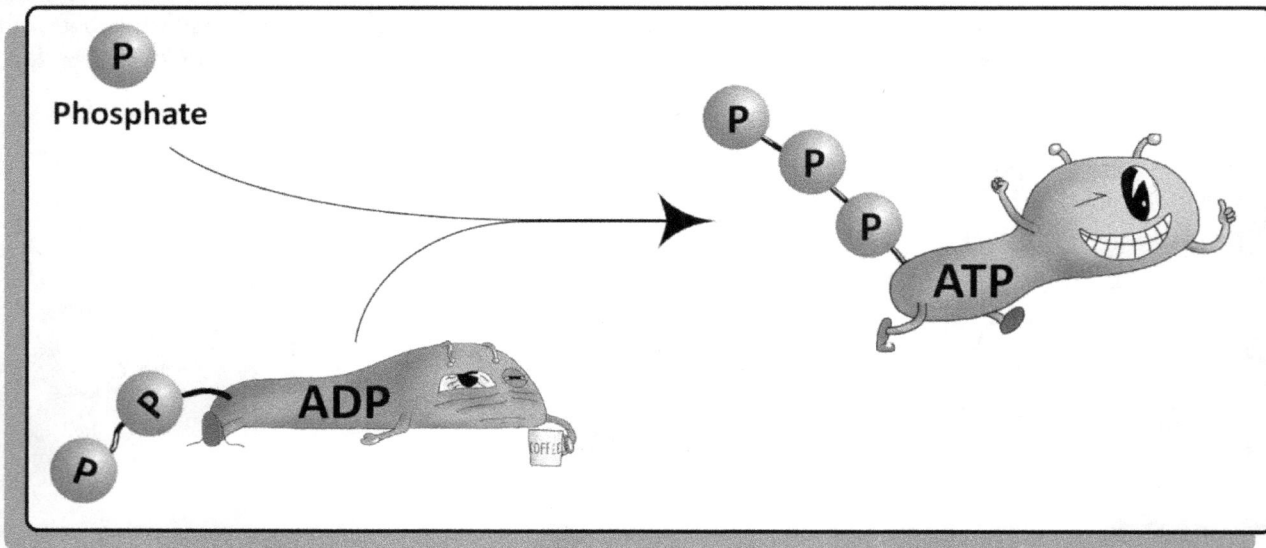

There are other forms of short-term energy currency in the cell, however. For example, energy can be stored as GTP (molecules similar to ATP), CTP (again similar), NADH and $FADH_2$ molecules.[34] These other forms of short-term energy currency are used for somewhat specific purposes, whereas ATP is basically all-purpose. This helps a cell divide up the kinds of energy that it stores for short-term use. We'll discuss these molecules as they appear again later.

At any rate, the punch-line for this discussion goes as follows: Glucose represents the storage of lots of energy (much more than is useful to be released at once). So, the cell oxidizes it slowly, and the energy that is released gets stored in short-term currency, mostly as ATP. ATP has an amount of energy that is much more practical for the cell to use on a regular basis. ATP is the all-purpose energy currency. GTP, CTP, and other forms of short-term energy molecules are used for more specific purposes. Cells only oxidize as much glucose as they need to keep a regular supply of short-term energy currency around. The rest of the glucose gets set aside for later use.

Now that we have the basic process down, we'll spend the next two units seeing the details. . .

Summary

In this section, we talked about oxidation, reduction, and an overview of ATP.

- Oxidation occurs when electrons are pulled from an atom. These atoms become positively or partial positively charged.
- Reduction occurs when electrons are added to an atom. These atoms become negatively or partial negatively charged.
- Oxygen is commonly the atom that oxidizes organic molecules, though any molecule that is more electronegative than carbon may also oxidize organic molecules.
- When oxygen oxidizes carbon and hydrogen, energy is released. Oxygen likes electrons and releases energy when it gets them.
- When oxygen fully oxidizes carbon and hydrogen, CO_2, H_2O and all the energy is released.
- Fire occurs when O_2 oxidizes organic material in an uncontrolled fashion, releasing all the energy as heat and light.
- Living systems control the oxidation process by putting a middleman between the organic molecule and oxygen. The middleman takes the electron from the organic molecule and gives it to oxygen. This process releases energy in a controlled manner and stores it as ATP.
- ATP is the basic all-purpose energy currency of the cell. It is a more practical form of energy than glucose. Glucose is a storage form of energy, whereas ATP is the active form of energy that the cell uses for day-to-day chemical reactions.
- The cell uses a few other energy molecules as well, such as GTP, CTP, NADH, and $FADH_2$. These molecules have limited use, however, and are restricted to only a few processes in the cell.

---

[34] NAD^+ = nicotinamide adenine dinucleotide, related to NADP from photosynthesis. FAD = flavine adenine dinucleotide. NADH and $FADH_2$ are their reduced forms.

Unit 20

Respiration, Part 1: Glycolysis

Having seen photosynthesis and the processes of oxidation and reduction, we are now in a position to talk about how the biological world (including plants) breaks down glucose, a complex set of reactions known as **respiration**. In this process, we do the reverse of the overall reaction that went into making the glucose, namely:

$C_6H_{12}O_6 + 6O_2$ ----> $6CO_2 + 6H_2O$ + energy

This happens in two sets of reactions. The first is a more primitive set of reactions known as **glycolysis**. In glycolysis, a small amount of the energy and a little water is released from glucose, transforming the glucose into **pyruvic acid** (also called **pyruvate**). The second set of reactions is a cycle that breaks the pyruvate down into CO_2, more water and a LOT more energy. This cycle of reactions has many names: Krebs cycle, the citric acid cycle, and the tricarboxylic acid cycle. We'll refer to it as the citric acid cycle.

Here is an overview of respiration:

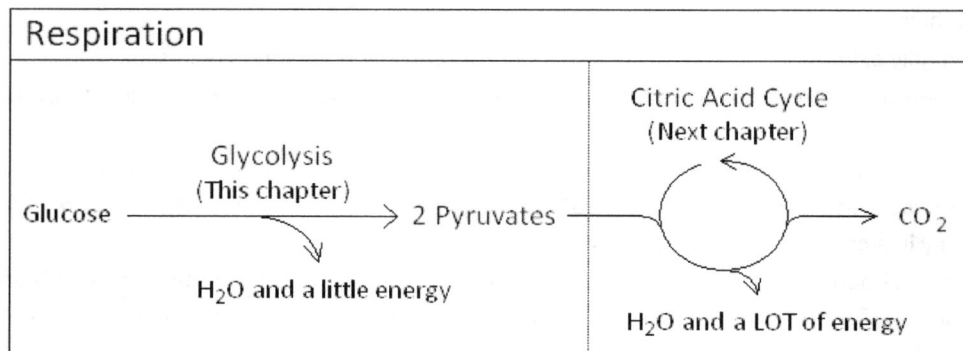

If you look at the overall reaction, glucose is broken down into CO_2, water, and energy, just like the butane was in the last chapter. The net effect is the same as if O_2 just went in and ripped glucose apart, but as we noted in the last chapter, the cell has to be careful to release the energy in an organized fashion. If the cell lets O_2 have free reign to mingle with the carbon and hydrogen, too much energy will be released too quickly and in a disorganized

152

fashion. In that sense, glycolysis and the citric acid cycle are the mechanisms the cell uses to keep this energy harvesting under control. We will focus on glycolysis in this unit, and then look at the citric acid cycle in the next.

Glycolysis

Glycolysis means roughly "splitting glucose." During the course of glucose oxidation, one glucose gets broken in half, and the two parts are then converted into two pyruvate molecules. During the course of these reactions, a small amount of energy is harvested by the cell. In glycolysis, O_2 is not used to complete the oxidation. Rather, a molecule called NAD acts as the middleman, taking electrons from glucose and transporting it to where oxygen is. That story picks up in the citric acid cycle, which we'll discuss later. In addition to the NAD, a net total of two ATP are produced in glycolysis, but to understand how this happens, we'll need to examine the details.

There are ten reactions in the glycolytic pathway. The first three reactions basically prepare the molecule for being split in half, which happens in the fourth reaction. The remaining reactions pull a small amount of energy from the pieces, and spit out two pyruvate molecules at the very end.

Let's examine these ten steps. The following images are an overview of the reactions, each step labeled numerically, 1 through 10. After these images, a summary is displayed as well.

Step 1:

In the first reaction, the enzyme **hexokinase** takes one ATP and uses it to attach a phosphate onto glucose, making **glucose-6-phosphate**. The ATP is converted into ADP and floats away. Remember, the reaction ATP --> ADP releases energy, and in this step the energy is invested in the glucose molecule when the phosphate is added.

You may be wondering what the point of this step is, considering it *spends* ATP rather than making it. This is a small investment of energy (and not the only one) that the cell will eventually make back before glycolysis is over.

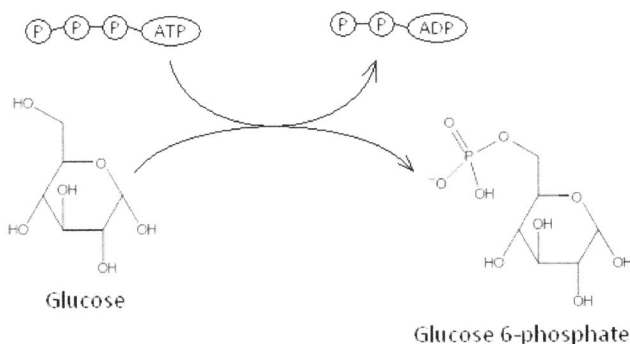

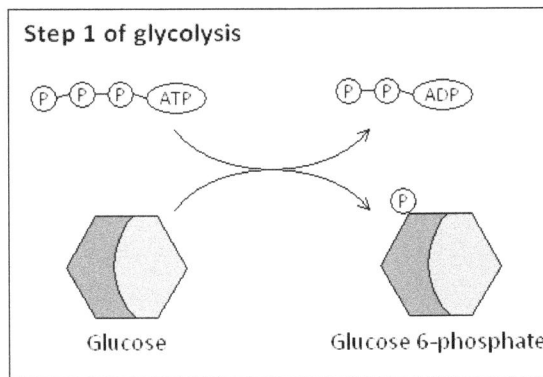

Glucose

Glucose 6-phosphate

Step 1 of glycolysis

Glucose

Glucose 6-phosphate

Step 2:

In the second step, the glucose-6-phosphate is rearranged internally into **fructose-6-phosphate** by an enzyme called **phosphoglucoisomerase**.

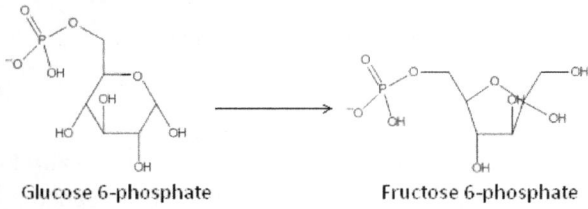

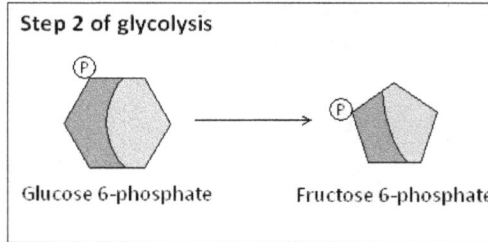

Glucose 6-phosphate → Fructose 6-phosphate

Step 2 of glycolysis

Glucose 6-phosphate → Fructose 6-phosphate

Step 3:

Next, we take this rearranged molecule and stick another phosphate on it. The enzyme that does this is **phosphofructokinase**. The molecule that results is **1,6-bisphosphate**

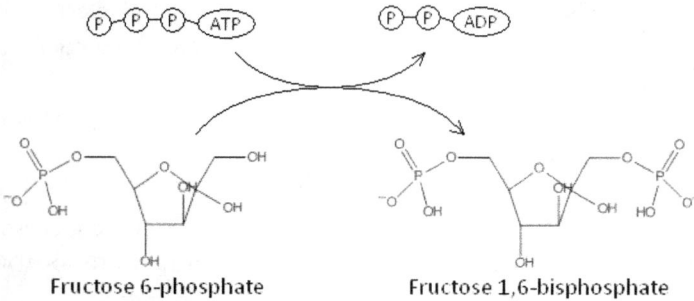

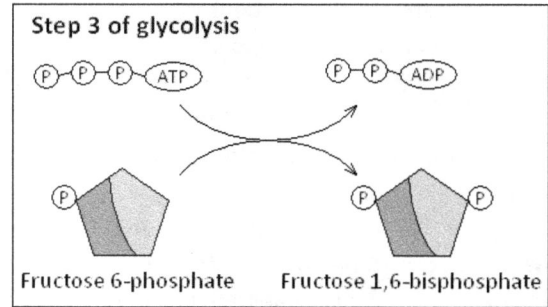

Fructose 6-phosphate → Fructose 1,6-bisphosphate

Step 3 of glycolysis

Fructose 6-phosphate → Fructose 1,6-bisphosphate

Again the cell invests another ATP at this point. So far, the cell is two ATP in debt, but it won't stay that way.

Step 4:

In reaction 4, we finally break the fructose 6-phosphate molecule in half using an enzyme called **aldolase**. This creates two molecules, each with a phosphate on the end. The two molecules are not identical, however. The first is called **dihydroxyacetone phosphate**, and the second is called **glyceraldehyde 3-phosphate**. The glyceraldehyde 3-phosphate is the same molecule that is produced during photosynthesis (Unit 19).

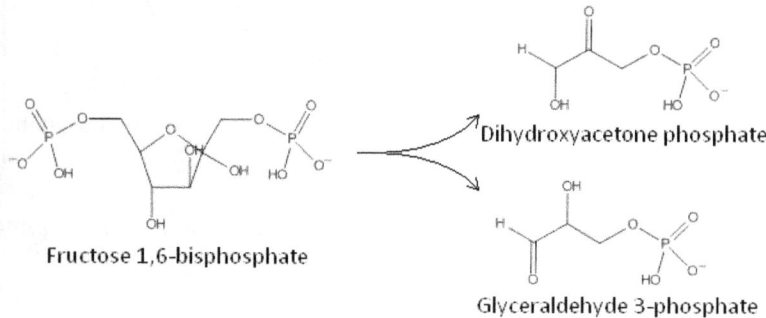

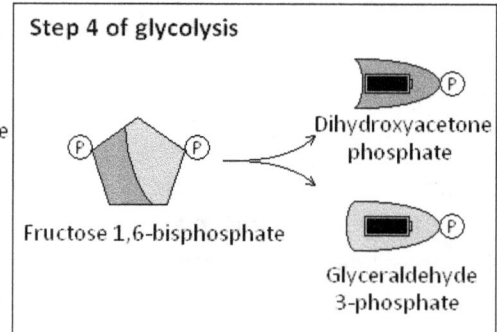

Fructose 1,6-bisphosphate → Dihydroxyacetone phosphate / Glyceraldehyde 3-phosphate

Step 4 of glycolysis

Fructose 1,6-bisphosphate → Dihydroxyacetone phosphate / Glyceraldehyde 3-phosphate

Step 5:

Next an enzyme called **phosphotriose isomerase** converts dihydroxyacetone phosphate into yet another glyceraldehyde 3-phosphate.

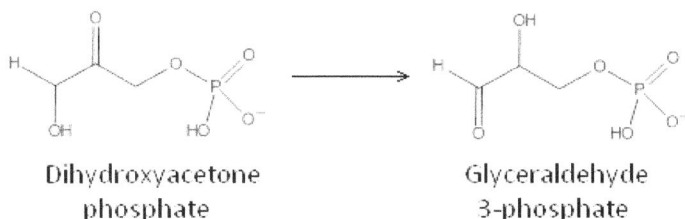

At this point, it is important to understand that we now have *two* copies of glyceraldehyde 3-phosphate: one was created when the fructose 6-phosphate was split in half, and the other was created by converting the dihydroxyacetone phosphate molecule. That means that all the following reactions happen *twice* per the glucose molecule that originally enters into glycolysis.

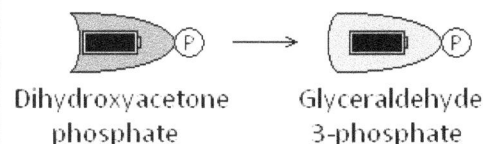

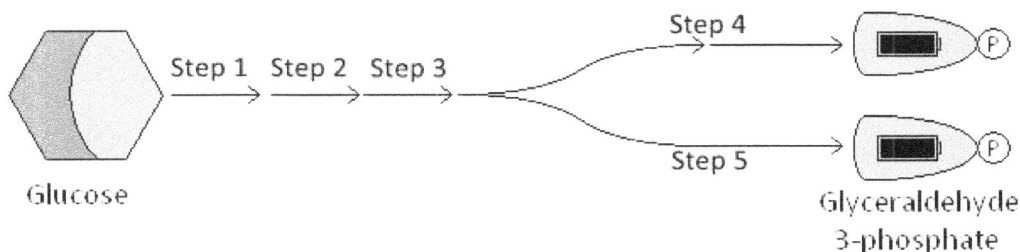

If you'll recall from the discussion of photosynthesis, plants put two glyceraldehyde 3-phosphates together to make glucose in an effort to *store* energy. Considering that, it shouldn't be surprising that we can get two of these molecules back out again by breaking down the glucose to *release* energy for use.

Step 6:

Here is where the action begins, so hold on to your hats! In the sixth step each glyceraldehyde 3-phosphate is oxidized. From our discussion on oxidation, this means an electron is pulled from the molecule. This electron is given to a molecule called NAD^+, a relative of the $NADP^+$ we encountered when we studied photosynthesis. By receiving the electron in this step, the NAD^+ becomes reduced to NADH and is more energetic. This is the first energetic molecule that emerges from the glycolysis reactions. It looks like our initial investment of two ATPs is starting to pay off. Anyway, the reaction also adds another phosphate to the glyceraldehyde 3-phosphate. The product that results is **1,3-bisphosphoglycerate**.

The enzyme that performs this step is called **glyceraldehyde 3-phosphate dehydrogenase**. The word dehydrogenase refers to an enzyme that participates in a reaction that pulls hydrogen from a molecule. If you look at the image that follows, a phosphate is added, and an H^+ is freed. The release of that H^+ is what the name "dehydrogenase" indicates.

155

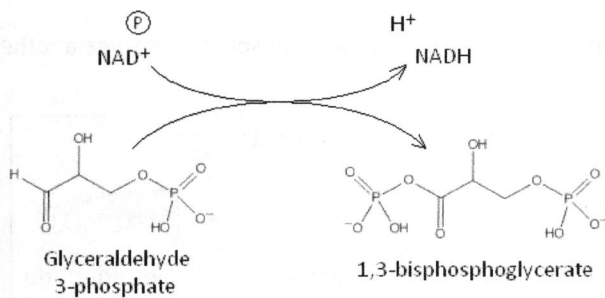

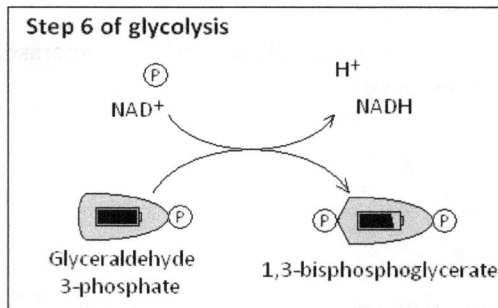

Glyceraldehyde 3-phosphate → 1,3-bisphosphoglycerate

Step 6 of glycolysis

Step 7:

We are now ready to generate the first ATP! The 1,3-bisphosphoglycerate is an unhappy molecule with two phosphates stuck on it. It has enough energy that it gives one of these phosphates to ADP, creating ATP. The molecule that forms is **3-phosphoglycerate**. Remember that Steps 6 and 7 happen twice for every glucose we use, because two glyceraldehyde 3-phosphates were made from the original glucose. That means we've made back our initial investment of ATP. So far, we've netted two NADH energy molecules. Yay!

The enzyme that performs this step is called **phosphoglycerate kinase**.

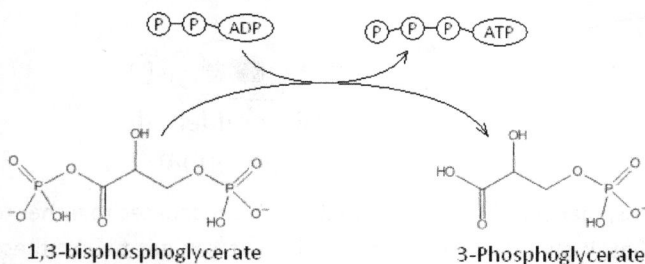

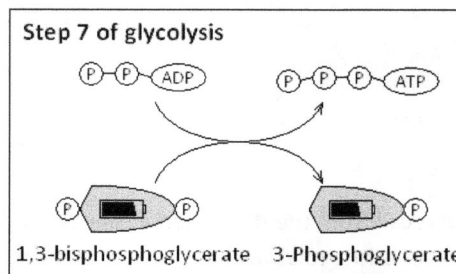

1,3-bisphosphoglycerate → 3-Phosphoglycerate

Step 7 of glycolysis

Step 8:

Now, we do an internal rearrangement, using the enzyme **phosphoglyceromutase**. This changes the 3-phosphoglycerate into **2-phosphoglycerate**.

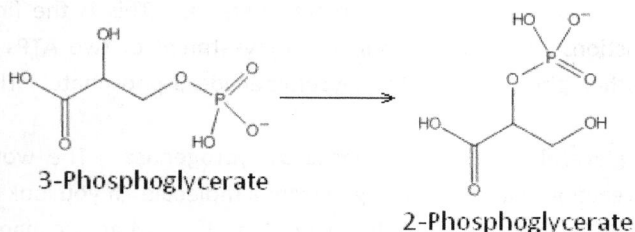

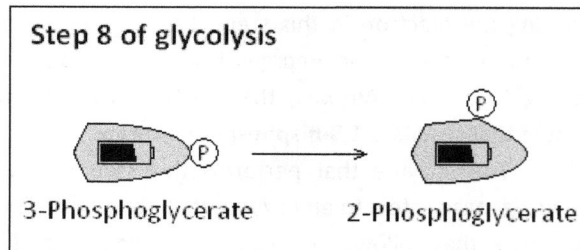

3-Phosphoglycerate → 2-Phosphoglycerate

Step 8 of glycolysis

156

Step 9:

Then, we do yet another rearrangement. The enzyme **enolase** performs this task. Notice that a water molecule is produced here. This is our first evidence of the overall oxidation end products: CO_2 and water.

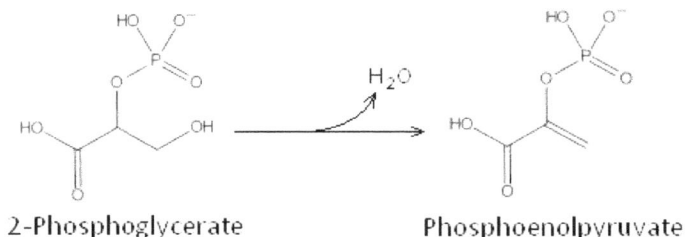

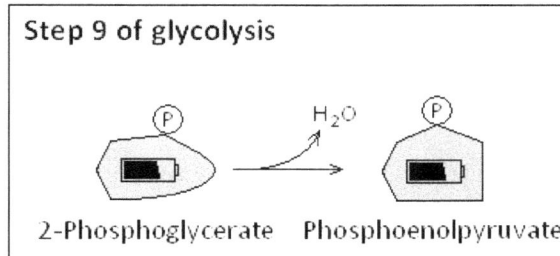

Step 10:

Now, after all that mess, we are finally ready to reap the rewards of glycolysis. The enzyme **pyruvate kinase** pulls the last of the phosphates from the **phosphoenolpyruvate** to turn ADP into ATP. Remember, this happens twice per glucose, so this step yields another two ATP, one for each phosphoenolpyruvate. Our initial investment of two ATP was already paid back, so these new two ATP are pure profit! Woohoo! The final molecule that we get is pyruvate.

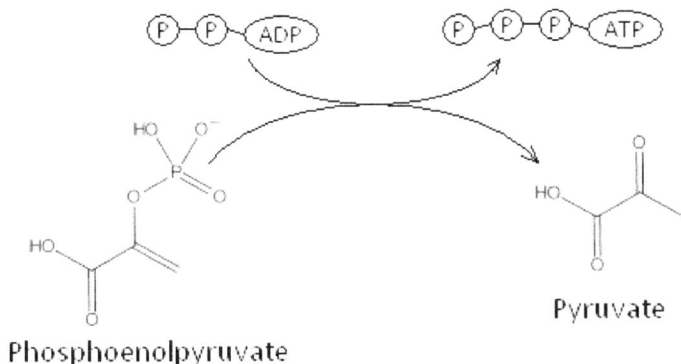

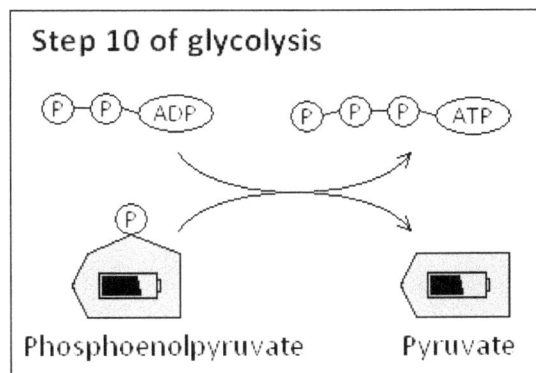

Glycolysis summary:

Let's put all these reactions together into one picture to see how they work. If you examine the image below, you'll see glucose (the first hexagon shape) going through various changes. These changes ultimately require the input of two ATP in order to get the reactions started. So, that puts the living system in "energy debt" with negative two ATP in its account, so to speak. After the glucose gets broken into the smaller pieces, each individual piece eventually releases two ATP and one NADH, earning the living system four ATP *back*, as well as two molecules of NADH. To summarize, the living system spends two ATP and gets four back, so it earns two overall. The smaller

157

picture in the bottom right corner of the following image shows the net effect: one glucose, two NAD, and two ADP enter into glycolysis, yielding two pyruvate, two ATP, and two NADH.

Summary of Glycolysis

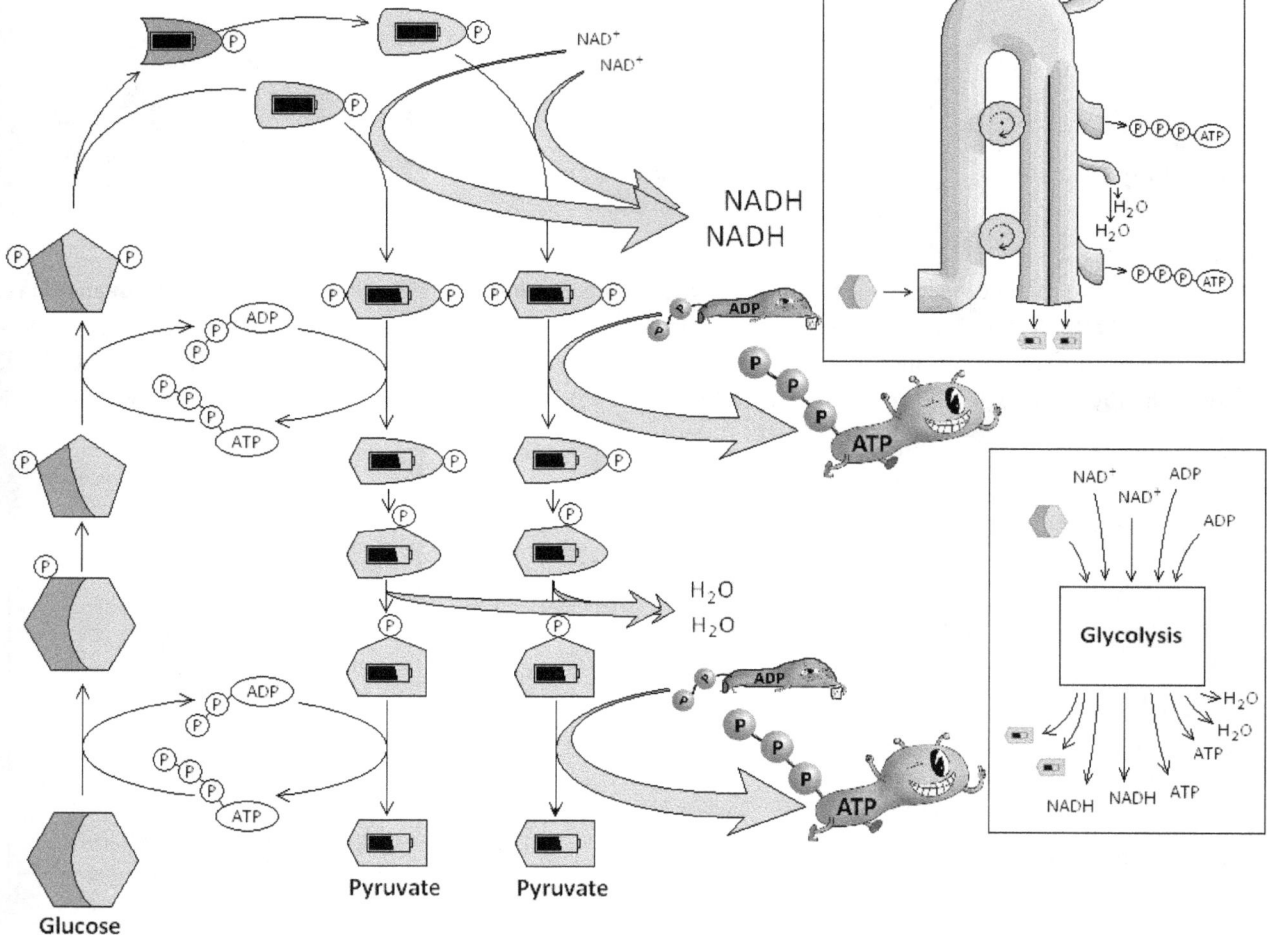

The story of how living systems get energy out of glucose is not over yet though. The two pyruvate molecules that result from glycolysis still have a lot of energy left in them! From here, the two pyruvates and the two NADH molecules move on to the next set of reactions, the citric acid cycle, where the rest of the energy will be released.

NAD/NADH Balance and Lactic Acid

There is, however, one last thing we need to touch upon before we conclude the unit. During the course of glycolysis, the living system converts NAD^+ into NADH. The NADH molecule is very energetic, and when it is used in other reactions, it gives up its energy and is converted back into NAD^+. It then returns to glycolysis to get charged up again. Unfortunately, although NADH is energetic, it is not a very convenient form of energy for the living system to use, not like ATP is anyhow. If ATP was a $10 dollar bill, NADH would be a special token that can only be used at three convenience stores across the country.

The point is: there are only a few reactions that use NADH as an energy source, and if these reactions do not need energy, the NADH would just sit there collecting dust and get in the way. If they just sit there, it means they do not get converted back into NAD^+. Therein lays the real problem: the cell could run out of NAD^+! Glycolysis MUST have NAD^+ or it can't proceed, so the cell must always have a way to convert NADH back into NAD^+ or else it can't get any more energy out of glucose. If glycolysis stops, the living system will die.

As it turns out, there are two standard ways in which NADH is converted back into NAD^+. The first (and more preferable way) occurs during the citric acid cycle, which requires the presence of O_2. The second way is for NADH to react with the newly made pyruvate, turning it into **lactic acid**. The reaction looks like this:

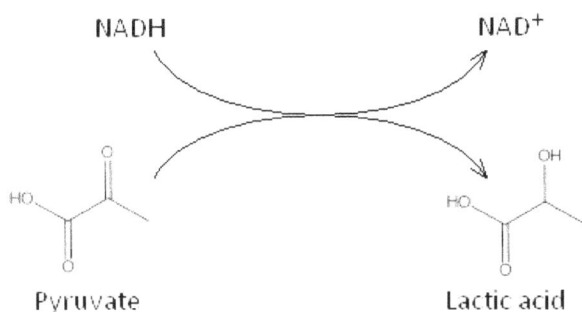

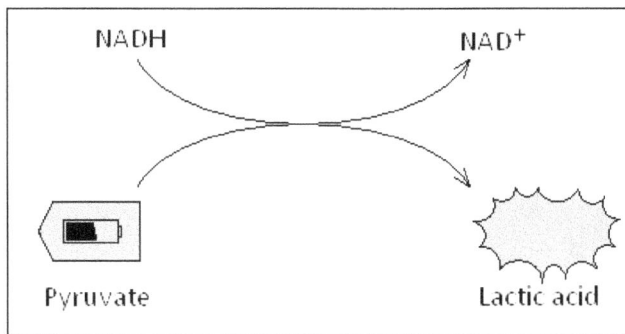

When O_2 is present, both pyruvate and NADH will proceed together into the citric acid cycle, where a *great* deal of energy is released. When this is the case, respiration is called **aerobic respiration**. When O_2 is *not* present, pyruvate and NADH react together. When this happens, respiration is called **anaerobic respiration**, and it doesn't produce nearly as much energy.

Net Result of Glycolysis with O_2 Present

"Aerobic Respiration"

P—P—ADP   P—P—P—ATP

P—P—ADP   P—P—P—ATP

NAD+
NAD+

Glycolysis
(O_2 present)

NADH
NADH

H_2O
H_2O

Pyruvate

Pyruvate

Glucose

To the citric acid cycle for roughly 30 more ATP!

Net Result of Glycolysis with O_2 Absent

"Anaerobic Respiration"

P—P—ADP   P—P—P—ATP

P—P—ADP   P—P—P—ATP

Glycolysis
(O_2 absent)

H_2O
H_2O

Lactic acid

Glucose

Lactic acid

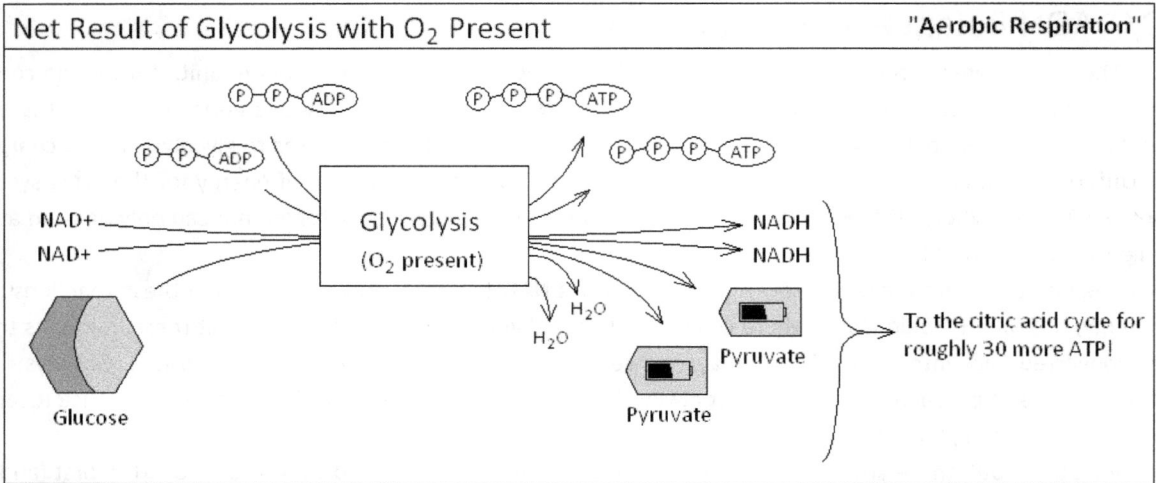

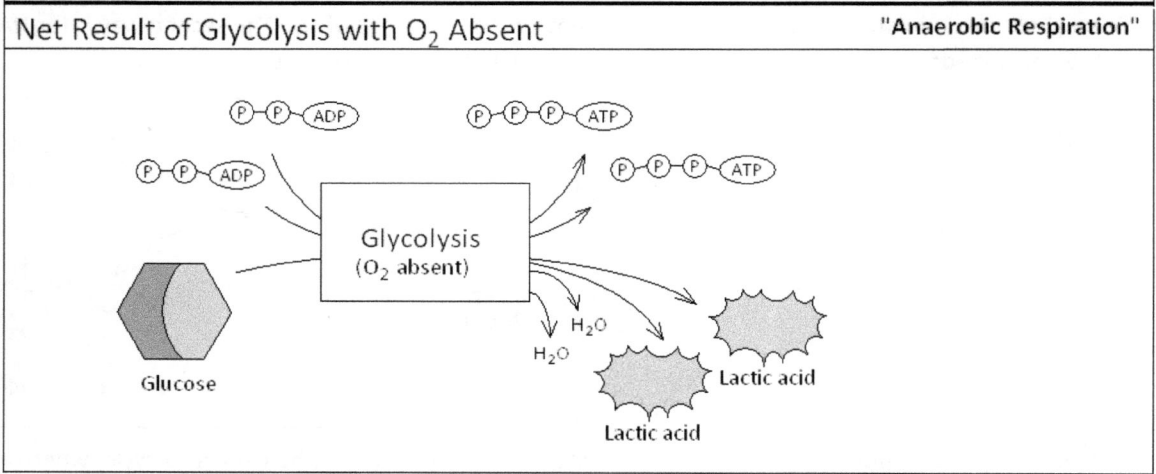

Due to the nature of anaerobic respiration, lactic acid forms any time the body is deprived of O_2, such as during suffocation, intense exercise, or in the moments following clinical death. When the body dies, for example, the heart stops beating and blood stops flowing. When it flows, blood delivers oxygen to the body. So, when the blood stops pumping, all the cells around the body begin suffocating. Since they can't get oxygen, they cannot do oxidative respiration and switch to reacting NADH and pyruvate together. This makes a lot of lactic acid. Since anaerobic respiration produces far less energy, some of the energy dependent organs, such as the brain, begin to rapidly shut down.

When you exercise vigorously, you begin using energy much faster than the blood can deliver oxygen to the muscles. In order to continue working, your muscles – now deprived of the necessary oxygen – switch to reacting NADH with pyruvate. This creates lactic acid and is why you get that familiar "burn" when exercising.[35]

---

[35] The soreness you get days later is not due to lactic acid though. It is due to microscopic tears in the muscle that are healing.

Summary

In this unit, we talked about glycolysis and lactic acid production.

- Glycolysis means "breaking sugar" and it is the first part in a set of reactions called respiration. The second part of respiration is the citric acid cycle, discussed in the next unit.
- Glycolysis breaks glucose in half and partially oxidizes it in a controlled set of ten reactions. Two ATP, two NADH, two water molecules and two pyruvates are created.
- If oxygen is present, the pyruvate and NADH move on to the next set of reactions, the citric acid cycle, where much more energy is harvested. This version of respiration is called aerobic respiration.
- If oxygen is not present, the pyruvate is reacted with NADH to make lactic acid and NAD^+. This frees up NAD^+ to participate in another round of glycolysis. This version of respiration is called anaerobic respiration and it produces far less energy than aerobic respiration.

Unit 21

Respiration, Part 2: Citric Acid Cycle (a.k.a. Krebs Cycle or Tricarboxylic Acid Cycle)

In the last unit, we talked about how glucose is transformed into two pyruvates during the ten reactions of glycolysis. We also touched on the issue of keeping NAD^+ and NADH in balance when there is no oxygen to be the acceptor of the electron that NADH carries around. Without oxygen, the pyruvate that is made is turned into lactic acid.

In this section we talk about what happens in the **citric acid cycle**, where the rest of the energy in pyruvate is converted into water and CO_2. Together, the citric acid cycle and glycolysis achieve the reverse of photosynthesis:

Photosynthesis
$$6CO_2 + 6H_2O + \text{energy (as light)} \text{-----}> 6O_2 + C_6H_{12}O_6$$

Respiration (glycolysis + citric acid cycle)
$$6O_2 + C_6H_{12}O_6 \text{---->} 6CO_2 + 6H_2O + \text{energy (chemical)}$$

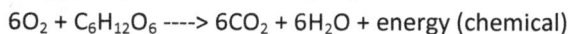

The summary of these two reactions is: 1) autotrophs do photosynthesis, storing the energy of light in glucose, and 2) both autotrophs and heterotrophs then take the glucose and break it down to get the energy back out.

Citric Acid Cycle

The Introductory Reaction

At the end of glycolysis, glucose has been cracked in half and turned into two pyruvate molecules. From here, these two pyruvates are taken into the mitochondria (see unit 17) where the citric acid cycle awaits them.

Introductory Step:

The first reaction to take place is the addition of a chemical side group called **Coenzyme A** (CoA) to pyruvate. CoA has a fairly complex structure and is related to nucleic acids, as shown to the right. CoA was introduced in Unit 11. Its addition to pyruvate is catalyzed by the enzyme **pyruvate dehydrogenase**, as shown below:

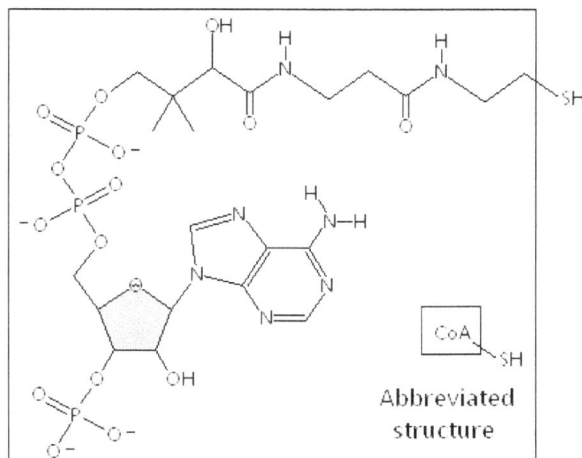

Coenzyme A

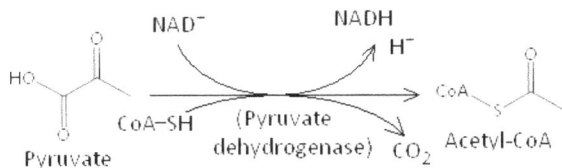

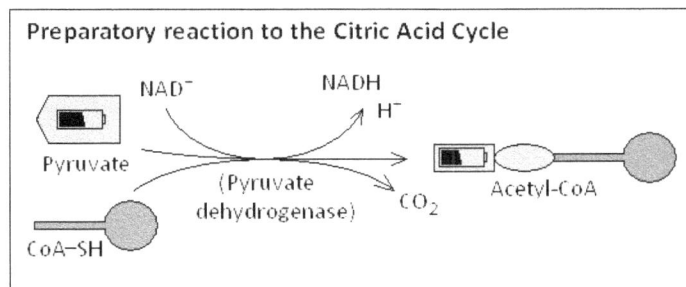

As a side note: pyruvate dehydrogenase is a heavy-hitter in courses like biochemistry and on standardized exams, especially in the medical profession. If that is your chosen goal,[36] then you may as well get used to this enzyme.

Anyway, as you'll notice from this reaction, one of the CO_2 in pyruvate gets set free. Since two pyruvates per glucose are created, two CO_2 total are set free at this stage. Remember that glucose started with six carbon atoms. This means that we've carried out complete oxidation (gotten the most energy possible) out of two of the six carbons in glucose. The citric acid cycle frees the remaining four.

Acetyl-CoA is the product of this initial step. For practical purposes, acetyl-CoA is the official start of the citric acid cycle. It is worth pointing out that acetyl-CoA is also produced during the breakdown of most lipids and many proteins as well. Basically, the living system recycles the energy it took to make the lipids and proteins by using them to make acetyl-CoA and/or pyruvate that go into the citric acid cycle to get broken down into CO_2, H_2O and energy. In

---

[36] Oh you poor soul...

that sense, the citric acid cycle is not limited to the breakdown of glucose, but is a universal set of reactions that uses many of the cell's chemicals.

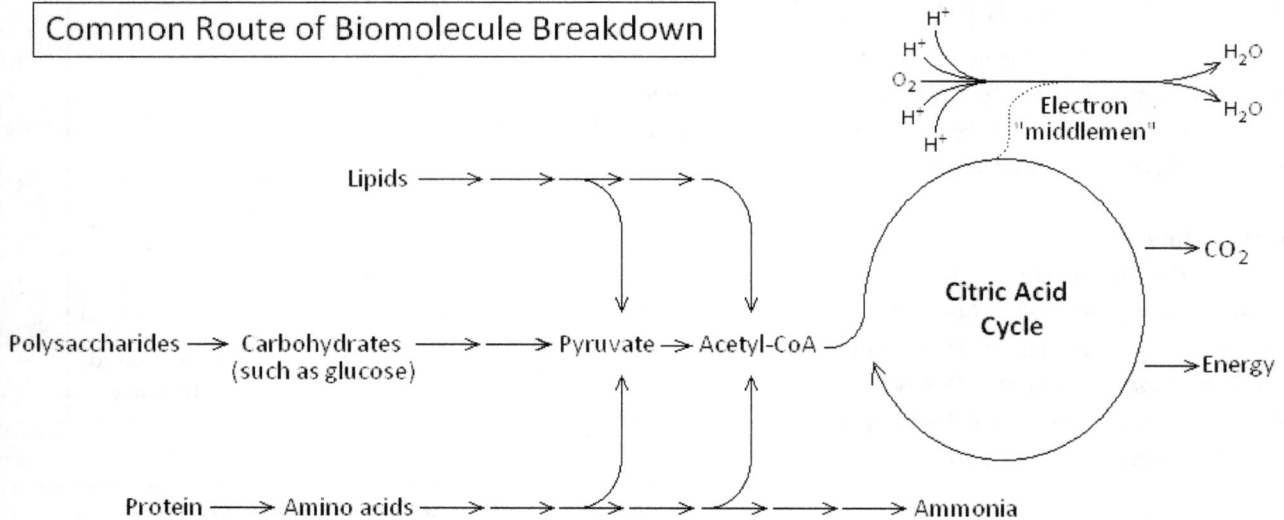

Common Route of Biomolecule Breakdown

As this image shows, the citric acid cycle makes energy, CO_2 and electron "middlemen." These middlemen represent the majority of the actual energy to emerge from the citric acid cycle. These molecules carry electrons to oxygen, and this process unlocks a great deal of energy, as we'll see. Otherwise, the actual amount of energy that is produced from the citric acid cycle (labeled as just "Energy" in the picture) is very minimal.

You'll also notice from the image that when protein is broken down to enter into the citric acid cycle, ammonia is produced as a side product. The chemical formula for ammonia is NH_3, and this chemical gets made because of the nitrogen in amino acids. The citric acid cycle does not handle nitrogen compounds, so this chemical gets put off to the side to be removed by other means. In animals, such as humans, it is removed in the urine.

The Citric Acid Cycle Reactions

As with the Calvin Cycle and glycolysis, the individual reactions are listed here in sequence. Some introductory biology courses will not require you to memorize the reactions, but you may find this listing helpful if you have to study the reactions more in depth at a later time. Otherwise, a summary image is given at the bottom as well as a brief discussion of the overall effects of the citric acid cycle.

Step 1:

Once acetyl-CoA is formed, (regardless of which biomolecule it started from), it is now ready to enter into the citric acid cycle. In the first reaction, the acetyl part of the acetyl-CoA is added onto a pre-existing molecule called **oxaloacetate**. This addition forms **citric acid**, which gives the cycle its name. The enzyme that performs this step is **citrate synthase**.

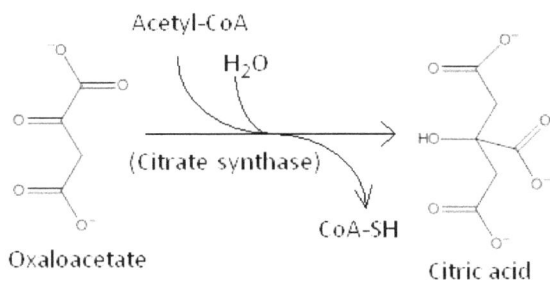

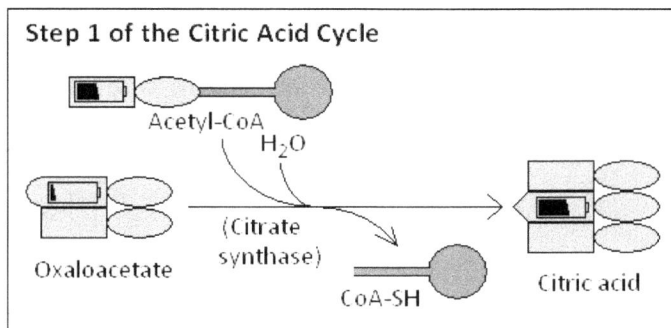

Oxaloacetate

Citric acid

Step 1 of the Citric Acid Cycle

Step 2:

The next step is catalyzed by the enzyme **aconitase** and is a simple rearrangement of citric acid to **isocitric acid**. Apart from this minor rearrangement, nothing else of much excitement happens here.

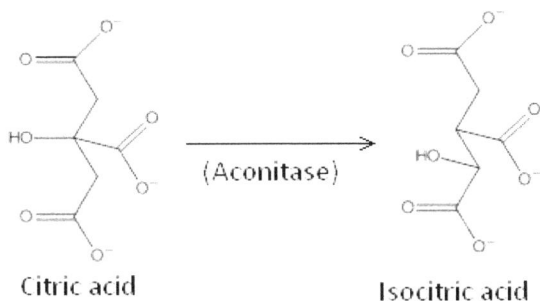

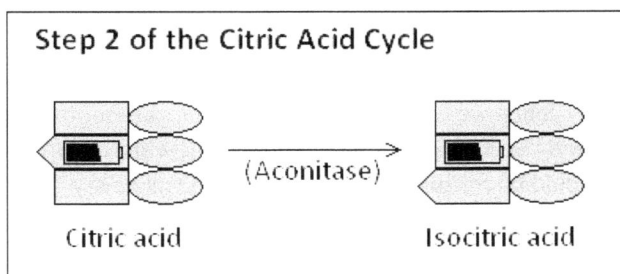

Citric acid

Isocitric acid

Step 2 of the Citric Acid Cycle

Citric acid

Isocitric acid

Step 3:

The next step in the cycle is performed by the enzyme **isocitrate dehydrogenase**. In this reaction, one CO_2 is released from isocitrate to make **α-ketoglutarate**. One NADH is created in the process. Remember that this happens once per pyruvate. Since there are two pyruvate made from each glucose, this step ultimately releases *two* CO_2 and *two* NADH molecules per glucose. This means that two of the final four carbon atoms are now accounted for. We only have two carbons left to release!

The NADH that is produced is one of the electron middlemen. It is highly energetic, and once made, it floats off to accomplish its task of making energy by giving electrons to oxygen.

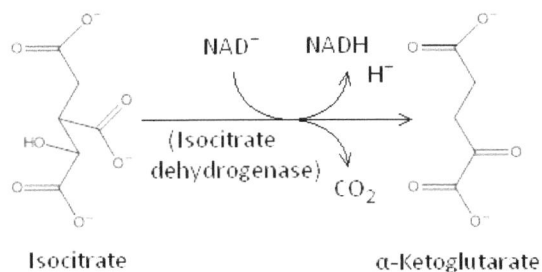

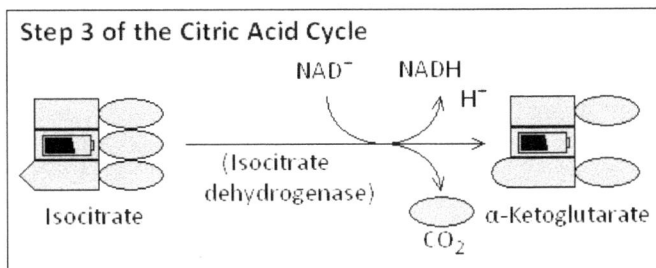

Isocitrate

α-Ketoglutarate

Step 3 of the Citric Acid Cycle

Isocitrate

α-Ketoglutarate

165

Step 4:

Step 4, catalyzed by **α-ketoglutarate dehydrogenase**, takes α-ketoglutarate, removes CO_2 and attaches CoA to make **succinyl-CoA**. Once again, this happens twice per glucose, so this accounts for the last two carbon atoms.

In addition to freeing the last carbon dioxide, this step also produces NADH, which floats off to make energy like the NADH in the previous reaction.

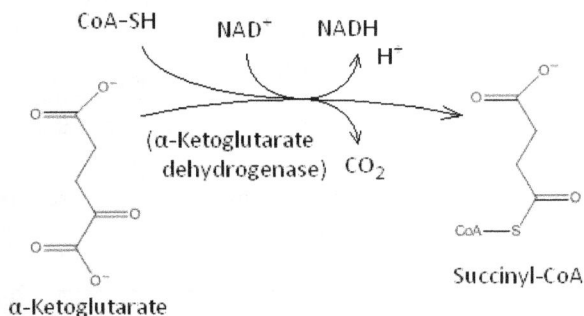

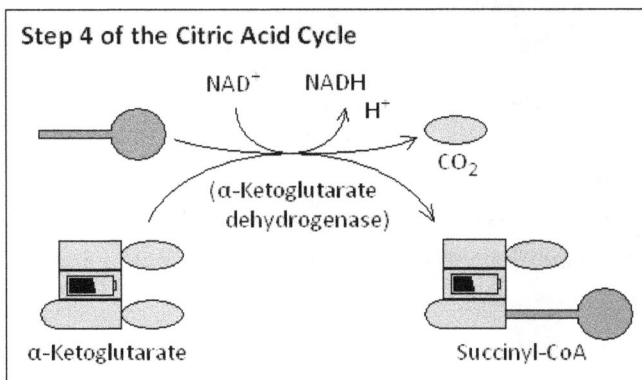

Since we've accounted for the last carbon, then it's entirely possible that the story of the citric acid cycle is over, and I've just filled the rest of this unit with random letters that just happen to make sentences. Well, for as nice as it would be for the story to be over, we still have a very big stumbling block to take care of: we need to regenerate oxaloacetate so we can run the cycle again! Oddly enough – as it turns out – the process of regenerating the oxaloacetate manages to squeeze a *large* part of the remaining energy left over from the pyruvate. It's weird – the carbons are gone, but the energy is still left lurking, like the ghost of glucose past. So let's keep going and see what kind of molecular brouhaha we can discover.

Step 5:

In step 5, **succinyl-CoA synthase**, takes the succinyl-CoA that was created during the last reaction and pulls the CoA from the molecule to make **succinate**. This process releases a little energy, which the citric acid cycle uses to make GTP. If you'll recall, GTP is like ATP; it is an energy molecule, but with less universal use. Shortly after GTP is made, another reaction rapidly converts it into ATP. Since this happens twice per glucose, this step produces two total ATP.

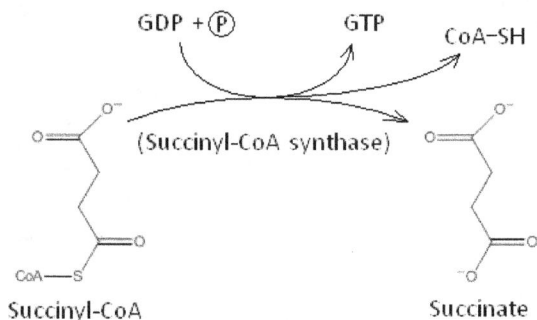

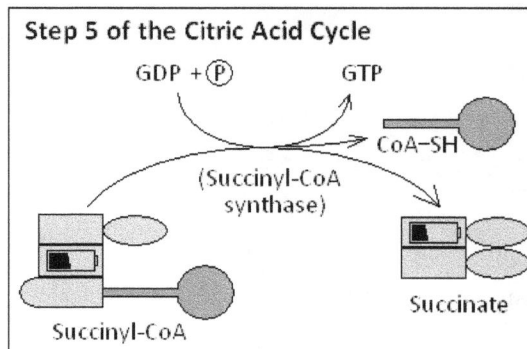

Step 6:

The next step to occur is the conversion of succinate to **fumarate**, which is catalyzed by the enzyme **succinate dehydrogenase**. This reaction creates one $FADH_2$ which is an electron middleman like NADH.

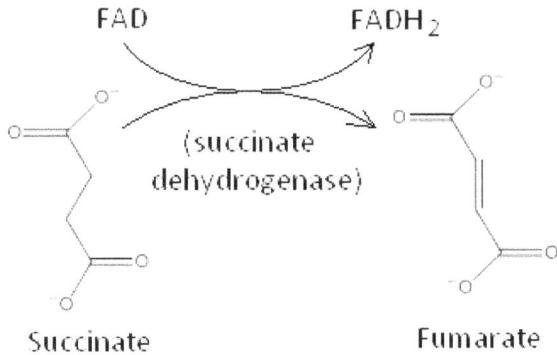

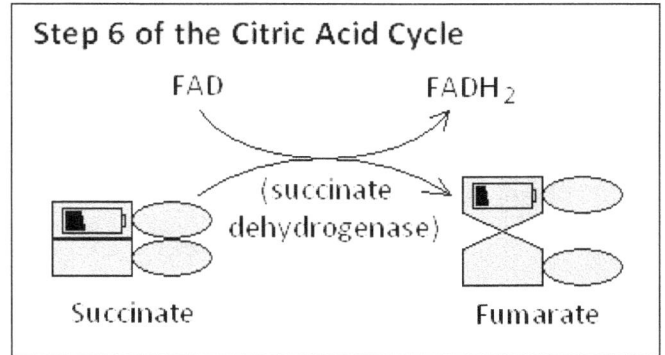

Step 7:

In step 7, the enzyme **fumarase** catalyzes the addition of water to fumarate to make malate. Apart from this, the reaction is otherwise unexciting.

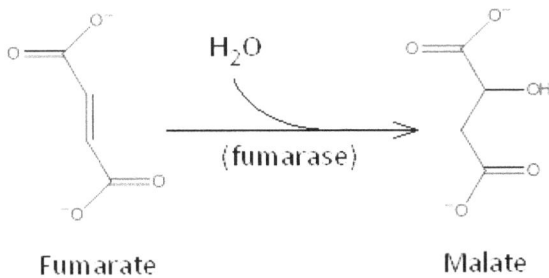

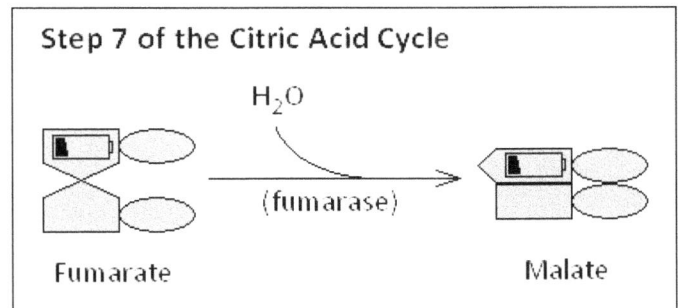

Step 8:

Step 8 is the last reaction in the cycle. This step makes the original oxaloacetate molecule (needed in step 1) so that the cycle can happen again. This reaction is catalyzed by **malate dehydrogenase**, which makes another molecule of NADH in the process as well.

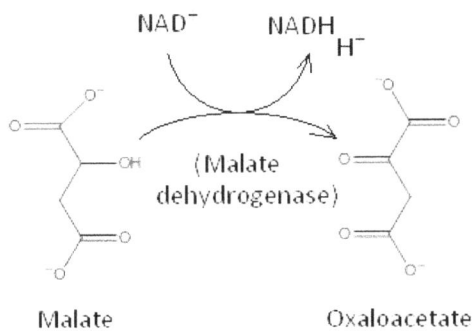

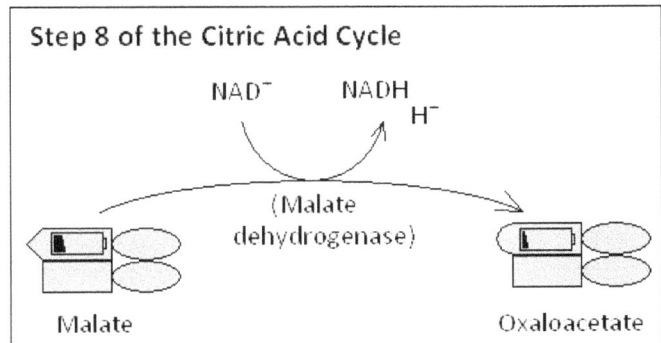

167

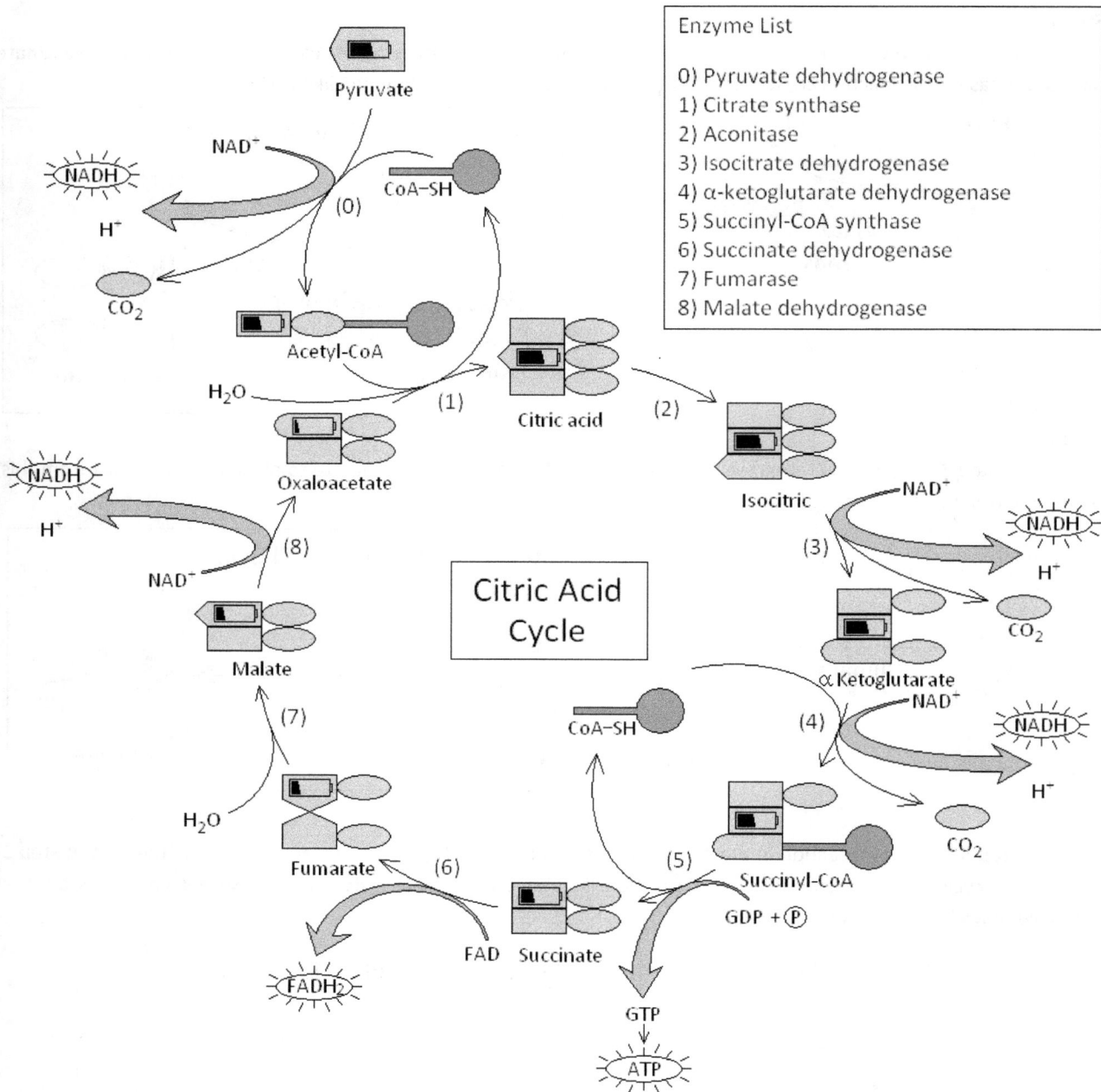

Enzyme List

0) Pyruvate dehydrogenase
1) Citrate synthase
2) Aconitase
3) Isocitrate dehydrogenase
4) α-ketoglutarate dehydrogenase
5) Succinyl-CoA synthase
6) Succinate dehydrogenase
7) Fumarase
8) Malate dehydrogenase

Summary:

     Let's take an overview of the reactions of the citric acid cycle. Pyruvate is initially converted to acetyl-CoA during the introductory reaction. This releases one CO_2. Within the next three steps of the cycle, another two CO_2 are freed. Two pyruvates are made per glucose, so ultimately six CO_2 are released during the introductory and first three steps of the citric acid cycle. That accounts for all six carbon molecules in the starting glucose.

In addition to the freed carbon dioxide, each round of the citric acid cycle (along with the introductory step) makes one ATP (converted from GTP), four NADH and one $FADH_2$. This gets doubled, since two pyruvates are made per glucose.[37] All the NADH and $FADH_2$ that is made then participates in the last bit of reactions that make energy from oxygen's greedy interest in electrons. That story is soon to follow.

Below is a more cartoonish version showing the overall reaction. As it shows, one pyruvate gets crunched up in a machine to make three CO_2, four NADH, one $FADH_2$ and one ATP.

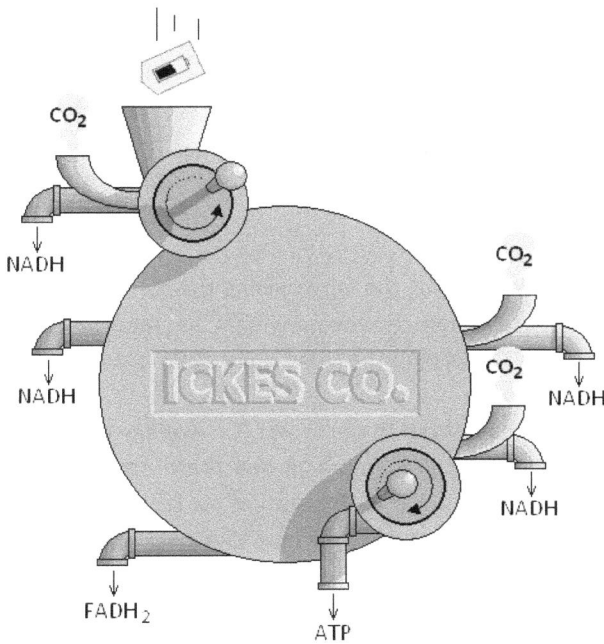

Electron Transport Chain

We are now at the last part of the energy formation process of respiration. This is a set of reactions called the **electron transport chain** that are *very* similar in principle to the process taking place in chloroplasts. To see how this is so, have a look at the image at the top of the next page. The electron transport chain involves a series of proteins that are imbedded in the inner membrane of mitochondria. Between this inner membrane and outer membrane is a space, and a LOT of H^+ are jammed within it. Like the process in chloroplasts, the mitochondria create this overcrowding by "bouncing" electrons along membrane protein islands.[38] To fully appreciate how this happens, recall from our discussion on oxidation and reduction that oxygen loves electrons. In science, this is the same thing as saying that *electrons love oxygen*. NADH and $FADH_2$ drop off their electrons onto protein islands, and as the electrons urgently move from protein to protein in attempt to get to oxygen, the mitochondrion uses this intense energy to push H^+ ions across the membrane and into the crowded space. This cruel overcrowding causes the H^+ to want out, very badly. To get out, they pay an "energy toll," which generates ATP from ADP and phosphate.

---

[37] It may seem like I'm pounding the concept of "two pyruvates per glucose" into the ground, but teachers and professors love to trick students with this on exams. It is critically important to understand that once the glucose is broken in half during glycolysis, *everything* that follows, including the citric acid cycle, happens *twice*.

[38] Once again I must offer a disclaimer: the electrons do not technically "bounce" along the proteins in the membrane, but rather travel through them in a highly energetic fashion. The image of bouncing electrons provides us with a much more memorable idea of what is happening though.

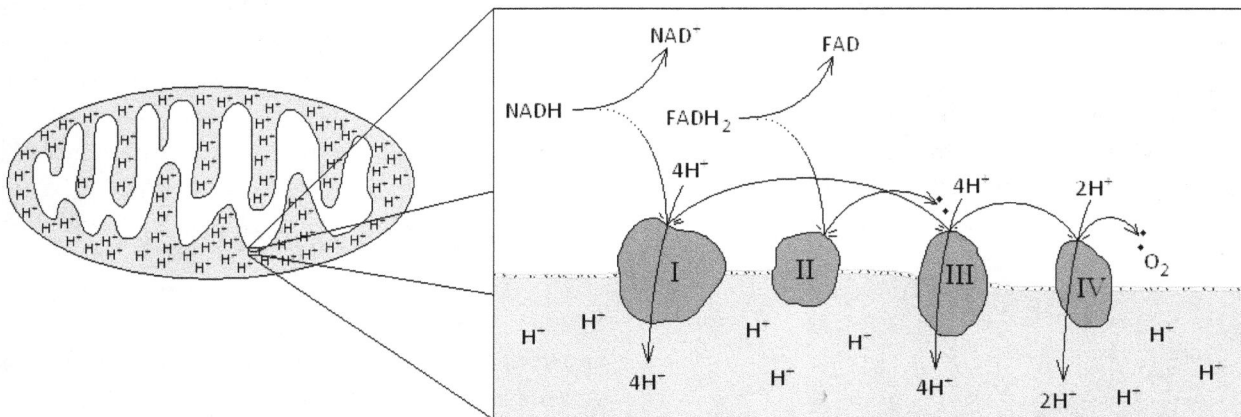

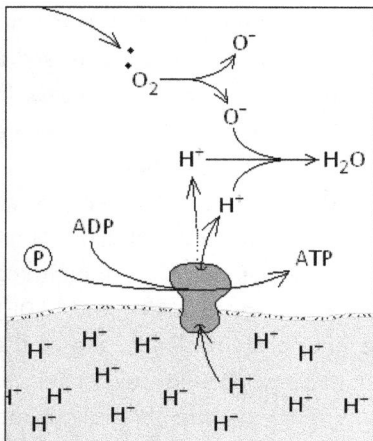

As the image on the left shows, the O_2 receiving the electrons breaks apart into two O^- ions, and these rapidly recombine with H^+ to create water. At the same time, the energy of H^+ escaping the inner space gets used to create ATP.

Just how much ATP is made from this process? On average each NADH molecule pushes ten H^+ across the cell membrane, and these protons create three ATP as they escape back out. Each $FADH_2$ pushes six H^+ across the cell membrane, and these create two ATP as they escape back out. Looking back over glycolysis and the citric acid cycle, we can now begin to take a tally of all the ATP that is created from a single glucose molecule. From glycolysis, we get a net of two ATP plus two NADH. Each NADH is worth three ATP. So, with the use of the electron transport chain (which requires O_2) we get (2 ATP + 2×3 ATP) = 8 ATP total. During the course of the introductory step and subsequent reactions of the citric acid cycle, (which happens twice per glucose), we get another two ATP, another eight NADH, and two $FADH_2$. This amounts to (2 ATP + 8×3 ATP + 2×2 ATP) = 30 ATP total. The grand total ATP harvested from glucose during glycolysis and the citric acid cycle is 8 ATP + 30 ATP = 38 ATP.

Glycolysis ----> 2 ATP + 2 NADH ----> 2 ATP + 6 ATP =		8 ATP
Citric acid cycle ----> 2 ATP + 8 NADH + 2 $FADH_2$ = 2 ATP + 24 ATP + 4 ATP =		<u>30 ATP</u>
		38 ATP

If oxygen is not present, the entire process comes to a screeching halt at the end of glycolysis, yielding only two ATP and two lactic acid molecules. The radical difference in energy is the reason we must breathe oxygen.

The Grand Summary

The formation of water and storage of energy in ATP molecules brings to an end the complex processes of photosynthesis and respiration. The first process stores the energy as glucose, and the second pulls it back out and puts the energy into ATP for the cell to use as general currency. Carbon dioxide and water were used to make the glucose, and as the glucose is broken down, CO_2 and water are released once again. The net result is as follows:

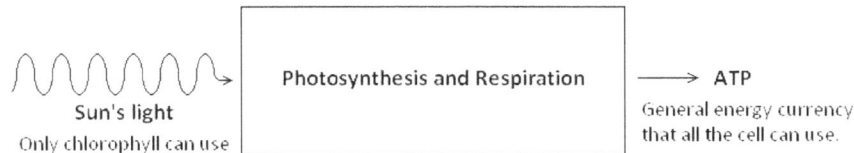

Sun's light
Only chlorophyll can use

Photosynthesis and Respiration ⟶ ATP

General energy currency
that all the cell can use.

Summary

In this section, we discussed the citric acid cycle and the electron transport chain.

- The two pyruvates that result from glycolysis are each converted to acetyl-CoA, which is the entry point for the citric acid cycle. One CO_2 from each pyruvate is released at this point.
- Pyruvate and acetyl-CoA are chemicals that can be made from the breakdown of amino acids and fatty acids, as well as from glucose. This makes the citric acid cycle and the electron transport chain the common fate of many of the cell's molecules.
- The citric acid cycle creates CO_2, a few ATP and a LOT of electron middlemen. These electron middlemen, (NADH and $FADH_2$), carry electrons to oxygen. Electrons want to be with oxygen just as bad as oxygen wants to be with electrons. So, NADH and $FADH_2$ force electrons to do work in making a LOT of ATP in order to get to oxygen.
- Steps two and three of the citric acid cycle remove the remaining two CO_2 from each starting pyruvate. This accounts for all the carbon originally locked in glucose.
- During the course of the citric acid cycle, one GTP is produced, but is rapidly converted to ATP.
- A total of four NADH and one $FADH_2$ are created in the citric acid cycle and its introductory step.
- The electron transport chain is the last step in the energy harvesting process. It takes place in the inner membrane of the mitochondrion. The electrons from NADH and $FADH_2$ are brought to the proteins of the electron transport chain, and these electrons move from protein to protein in a desperate attempt to get to the oxygen waiting at the end. This energy pumps H^+ across the membrane and builds a large H^+ gradient.
- The H^+ that get pumped across the inner membrane gather in the space between the inner membrane and outer membrane. They are very unhappy in this state and push through a protein canal that extracts their aggravation energy and uses it to make ATP.
- Each NADH makes roughly three ATP. Each $FADH_2$ makes roughly two ATP.
- The grand total energy that can be extracted from glucose is ~38 ATP, assuming oxygen is present to allow the electron transport chain to work. If oxygen is not present, only two ATP and two lactic acids are made.
- By the time this respiration process is finally complete, CO_2 and water are released. This is the reverse of what happens in photosynthesis, in which CO_2 and water are consumed.

Unit 22

Replication

In the early units of this book, we talked about the stuff from which life is made, and then we got our feet wet with a little biochemistry when we discussed how life gets its energy. Now, we turn our attention to life's ability to reproduce. DNA is at the center stage of this particular drama, so let's focus our attention back on this information molecule to get us started on the discussion.

As you'll recall, DNA is a long chain of nucleotide monomers. There are four different kinds of nucleotides: A, G, T and C, shown here:

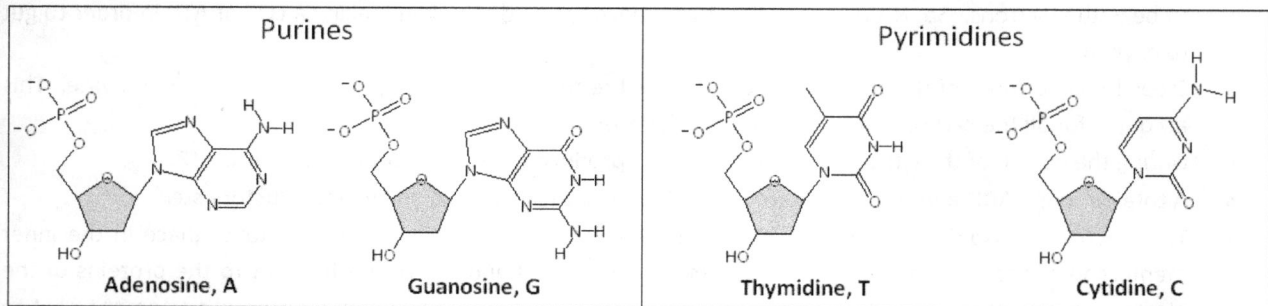

Purines: Adenosine, A — Guanosine, G. Pyrimidines: Thymidine, T — Cytidine, C.

The carbons in the gray sugar ring are numbered from 1' to 5' (read as "one prime" to "five prime"). The phosphate of the nucleotide is attached to the 5' position:

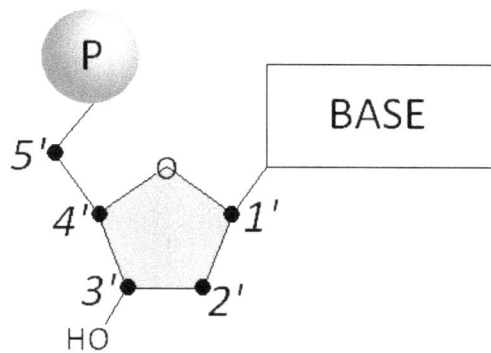

When nucleotides are strung together to make DNA, the numbering system helps describe the directionality of the polymer that results.

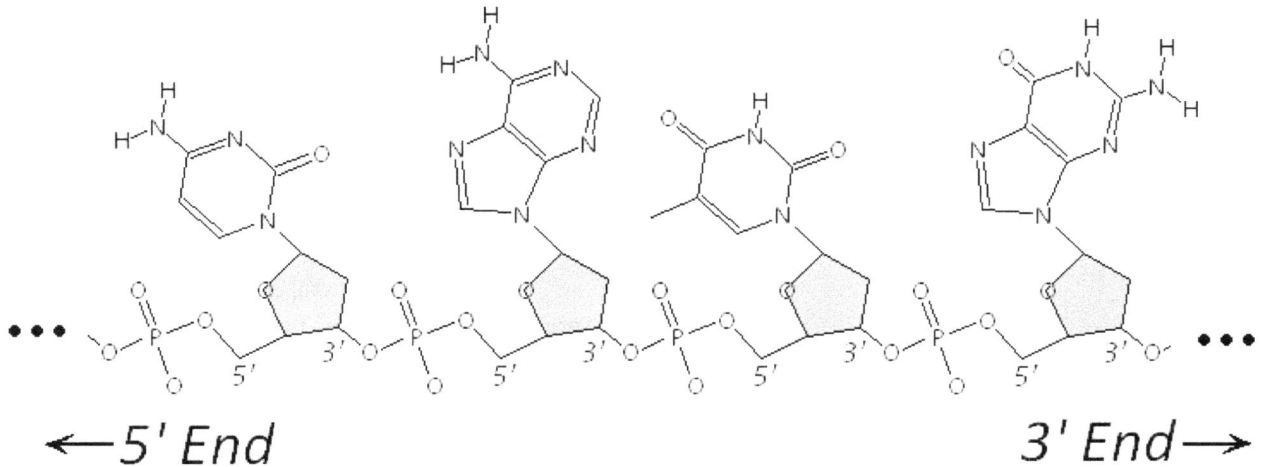

←5' End   3' End→

Each chain of nucleotides has a complementary chain that attaches to it, side-by-side, like a long ladder, each rung being a nucleotide-pair. The opposing strands of DNA run in opposite directions. One strand goes from 5' to 3' while the other is 3' to 5'.

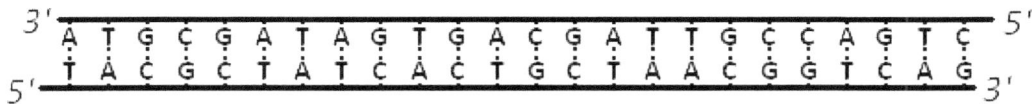

3' A T G C G A T A G T G A C G A T T G C C A G T C 5'
5' T A C G C T A T C A C T G C T A A C G G T C A G 3'

In addition to opposing directions of the strands, the nucleotide pairs have a very special nature. If an A is on one side, then a T is on the other. If a G is on one side, then a C will be on the other. This special interaction occurs through hydrogen bonds. A and T share two hydrogen bonds, while G and C share three. That said, this DNA "ladder" is also twisted, resulting in a complex three-dimensional shape called a double helix.

173

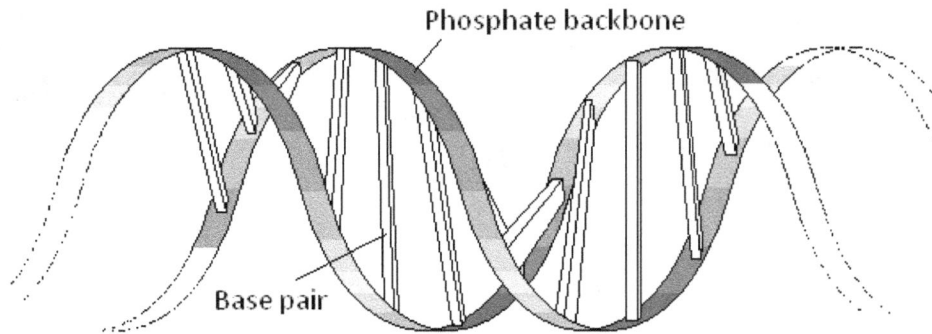

Phosphate backbone

Base pair

The DNA helices of living organisms typically contain hundreds of thousands of these nucleotide pairs. In bacteria, these long DNA ladders are circular with the ends attached to one-another, while in eukaryotes, such as humans, the DNA has two free ends (called **telomeres**). The difference between prokaryotic and eukaryotic DNA goes a little further, though, because eukaryotic DNA is also wrapped tightly around protein clumps called **histones**. We touched upon this briefly in our chapter on cell anatomy. Five histone proteins are connected together to make a larger protein complex, and the DNA is wound around this complex. Taken together, the entire thing is called a nucleosome (see unit 17 for illustration). Either way, we call one full chain of DNA, (whether looped or straight), along with whatever proteins that help package it, a **chromosome**. Bacteria typically have only one chromosome. Eukaryotes, on the other hand, differ in the number of chromosomes they have, although their chromosomes generally come in pairs. Humans, for example, have 46 chromosomes, or 23 pairs. The complete set of chromosomes that an organism has is called its **genome**. So, most bacteria (for simplicity) have one chromosome in their genome, while humans have 46 chromosomes in their genome.

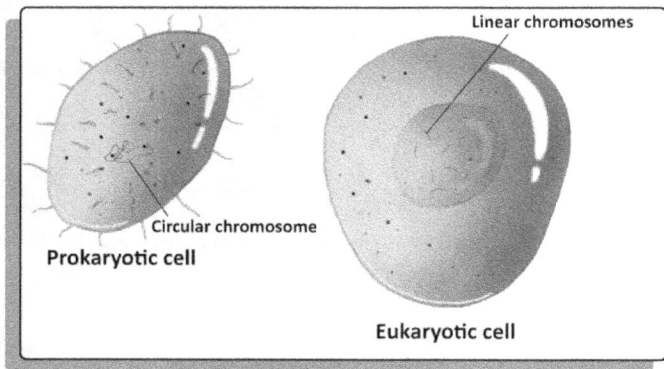

Linear chromosomes

Circular chromosome

Prokaryotic cell

Eukaryotic cell

That said, let's talk about the complementary nature of the double helix strand some more. Suppose we are out walking around and we happen to stumble across only one strand of a DNA helix. The complementary strand is missing for some reason, so that the strand we find looks like half of a ladder. Because of the interesting complementary nature of DNA base pairs, we already know what the other half of the ladder should look like! Wherever there is a G, the other side will have a C. Wherever there is an A, the other side will have a T, and so forth. In other words, one side of the DNA helix contains the information for the other, and vice versa. This fact is important in the replication of DNA. As a cell prepares to divide, it slowly pulls apart the DNA helix and makes a new copy of both halves using the complementary information. When this is done, the cell ends up with two copies of the original chromosome. Life apparently had copy machines long before we did!

DNA Replication

There are subtle differences between how prokaryotes and eukaryotes replicate their DNA, but the complex geometry of the helix imposes the same problem either way. Let's say we separate the helix altogether so that the two halves are floating free in the cell. The two halves are complementary and will want to reconnect (**anneal**), but do you think they anneal just like they were before? Will the two halves come back together perfectly? Theoretically, it is possible, but the odds are actually quite remote. The problem is that any given A doesn't know *which* T it was originally bound to, and similarly with the G's, C's and T's. Instead of annealing perfectly, two other events are much more likely to occur: 1) The two halves of the DNA would clump together in a huge tangle, or 2) a strand could double over so that its own A's bind to its own T's, its C's bind to its G's, and so on. Either way, you get a complete mess. The DNA strands in living systems are just too long to be pulled apart entirely.

This leads to a very important rule: No DNA strand can be left without a complementary strand. If the cell needs to work with the DNA base pairs, it separates only a small part of the helix and then puts it back together after it's done.

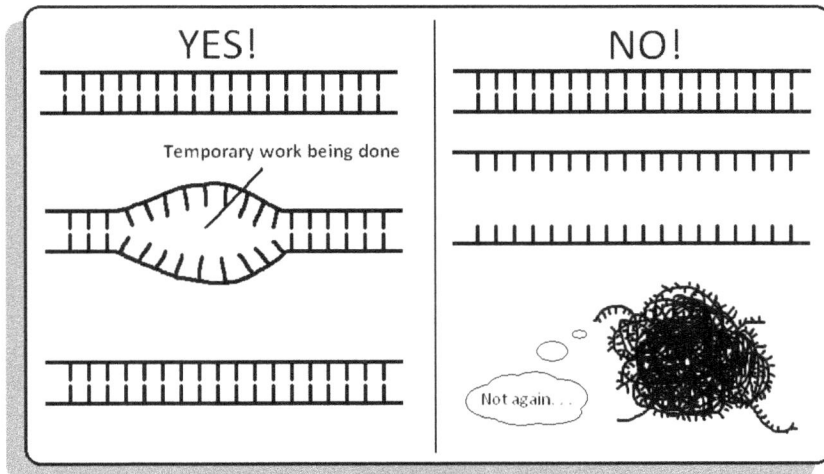

This rule remains true for the replication process as well. Replication begins at an origin of replication, where proteins bind and slowly pull a small section of the strands apart. If the cell has to start somewhere on the DNA chain to begin replication, how does it know where in the chain to begin? Well, DNA has specific start sites that typically have a LOT of A's and T's. A-T pairs bond with only two hydrogen bonds, whereas G-C pairs bond with 3. This means that A-T bonding is weaker than G-C bonding. It also means that a long stretch of A-T's will be easier to pull apart than a long stretch of G-C's. So, a collection of proteins called the **pre-replication complex** binds to the origin of replication and pulls this long stretch of A-T's apart so that replication can begin.

175

After the pre-replication complex completes this work, the copying machinery comes in and begins inserting the new nucleotides to both of the original halves of helix, moving in opposite directions as it does so. In this manner, neither of the original strands is left with long stretches of unpaired nucleotide bases. To see how this all works, let's start with a very simplistic model of the replication machinery, and slowly add complexity to the images as the details emerge.

Our highly simplistic first view of DNA replication.

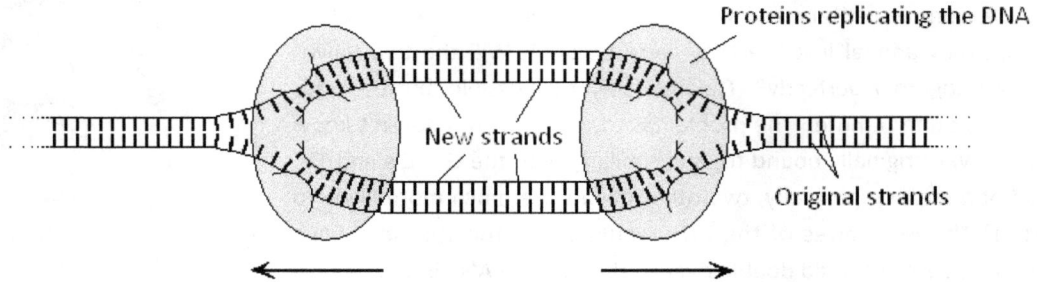

The actual separation of DNA strands is accomplished by a protein called **DNA helicase**, but here we are presented with a problem. Separating DNA strands is not as easy as it sounds. The two strands are wound around each other in a helix. That means if you try to pull the two strands apart, they will still be spiraled around each other and get tangled. In order to fully separate them, you must *untwist* them as well. The following image shows this process.

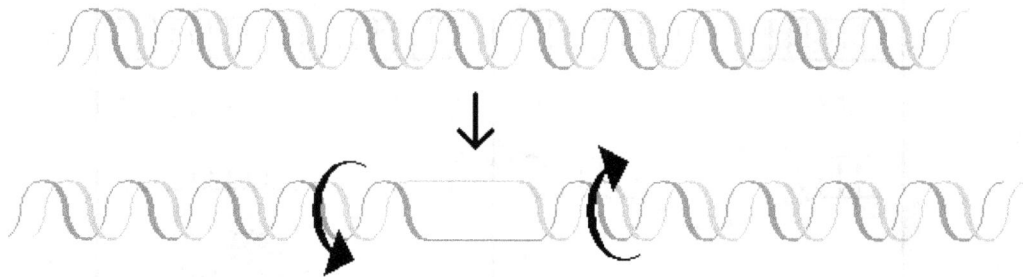

The job of untwisting the DNA helix falls to a protein called **DNA gyrase.**[39] Together with helicase, the two proteins are used to separate DNA strands anytime the cell needs to do work with the base pairs, such as during replication or in the initial stages of protein production (discussed in later chapters).

So once helicase breaks open the DNA at the starting spot and the gyrase untwists the DNA, we are ready to create the new DNA strands from the unraveled single strands. The protein responsible for creating the new strands is

---

[39] Actually, gyrase doesn't technically "untwist" DNA. It cuts the backbone of one strand, spins that strand around the other, and then seals up the backbone again. The effect, however, is the same as if the protein untwisted the DNA. The reason it doesn't simply untwist the helix is that it would have to untwist it all the way to the end, which could be millions of nucleotides away. That would be very slow and tedious work.

called **DNA polymerase**.[40]   One DNA polymerase moves along the helix in one direction, while another DNA polymerase protein moves in the other direction. In front of each DNA polymerase are helicase and gyrase.

A better but still simplistic view of Replication

Replication starts. . .

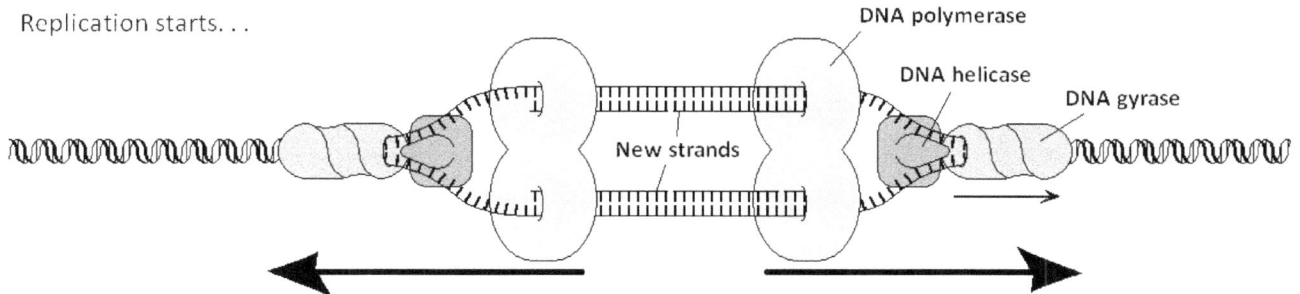

DNA polymerase

DNA helicase

DNA gyrase

New strands

. . .and goes in both directions

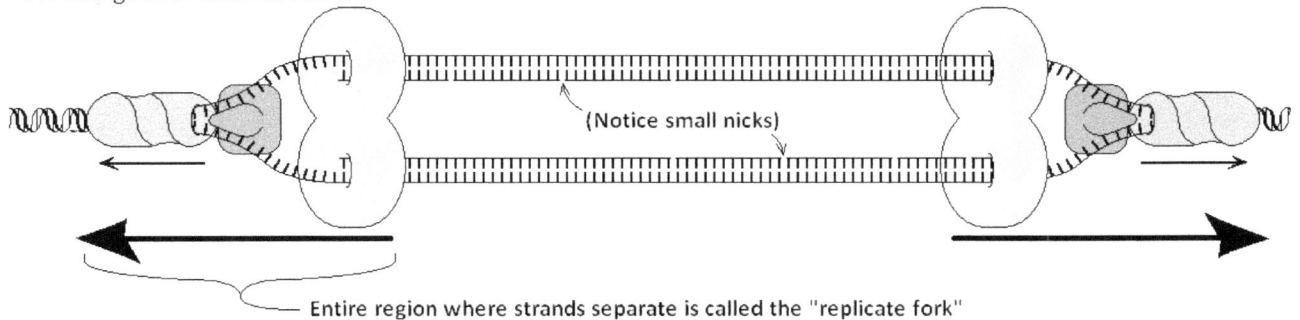

(Notice small nicks)

Entire region where strands separate is called the "replicate fork"

The last image shows little nicks in some of the backbones of the new strands. To see why these happen, we'll need to dig a little deeper into the process. As it turns out, the generation of the new strands is a little more complex than our initial overview would suggest.

DNA polymerases work by reading the old strand from the 3' to 5' direction.   As we saw above, complimentary strands face in opposite directions. So, if the polymerase reads the *old* strand in the 3' to 5' direction, it adds the *new* strand in the 5' to 3' direction. In other words, it adds new nucleotides onto the 3' end of the new strand.

---

[40] Each DNA polymerase is actually made of two circular polymerase "holoenzyme" proteins, each shaped like a donut. The DNA is threaded through the centers, like two strings running through a figure eight.

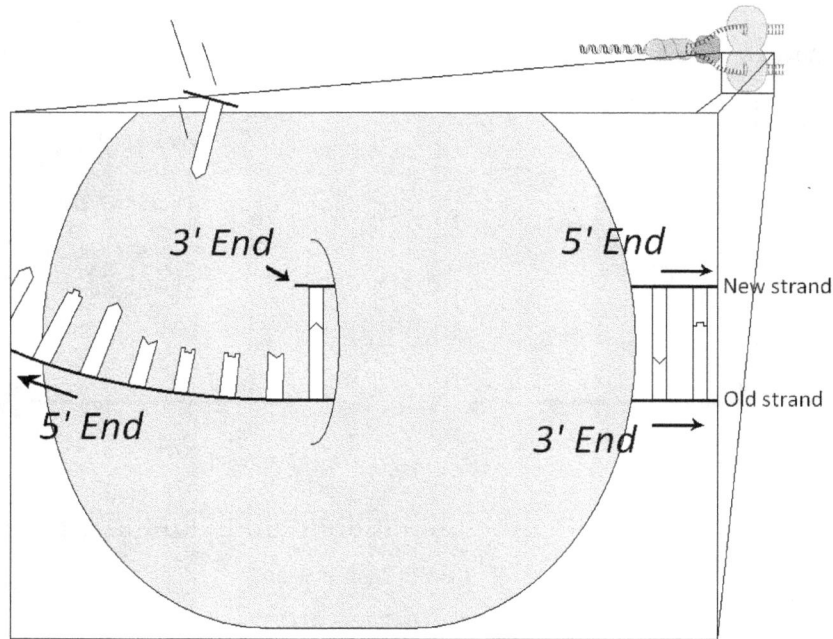

3' End

5' End

New strand

5' End

3' End

Old strand

A problem arises because only one of the old strands is facing the 3' to 5' direction. That means if a polymerase is scooting along the helix, one strand will be facing backwards. Since polymerase works by reading old strands in the 3' to 5' direction, it must loop this backwards strand around and read it in the other direction:

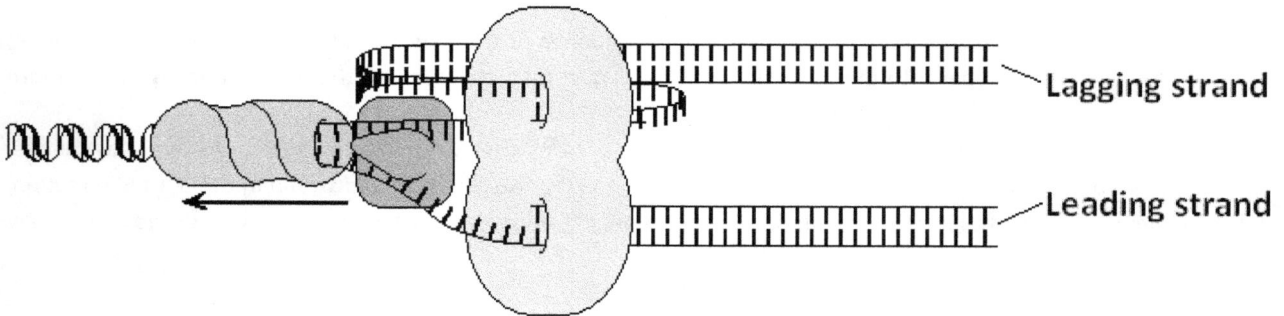

Lagging strand

Leading strand

By looping it around, the polymerase is affectively able to read the strand in the same direction as the other strand: 3' to 5'. This strange looping process seems quite bizarre and is rather difficult to grasp the first time it is encountered. Conceptually, it is hard to visualize just how the mechanism would work as the polymerase moves down the helix. Here is a cartoon to help show the process. In the cartoon, think of the polymerase as a bead and the DNA as a string. The bead slides down the string until it hits the bend in the string, at which point you need to pull more string through the hole in the bead to have enough slack to keep sliding it.

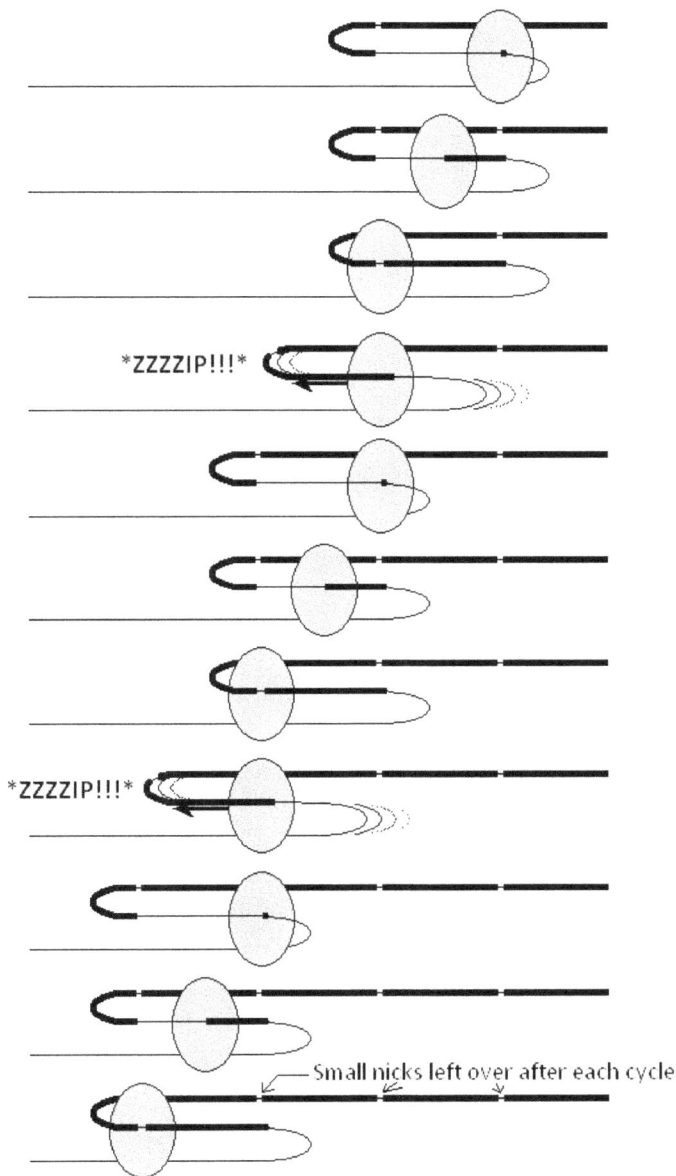

ZZZZIP!!!

ZZZZIP!!!

Small nicks left over after each cycle

In the image, the DNA polymerase slowly moves left. Just as the DNA polymerase bumps into the looped strand, the strand is threaded forward. The DNA polymerase continues moving left until it nearly bumps into the strand again, and so forth. This gradual ratcheting movement is a reasonable approximation of how it occurs in the cell.

Sadly, there is one more problem that needs to be addressed when it comes to the synthesis of a new DNA from the stubborn 5'-3' looped strand. If you look at the close-up picture of the DNA polymerase several pictures above, you'll note that DNA polymerase adds new nucleotides to the 3' end of the *previously* placed nucleotides. If there are no previously placed nucleotides, DNA polymerase will stall. *This is a problem*. If you look again at the cartoon to the left, each time the thread ZZZZIPs forward, it appears as though the DNA polymerase has to start a new thick line from scratch. There is one more piece of information, however, before we have the *completed* picture: before the polymerase starts its work on this looped strand, a protein called **DNA primase** lays down a temporary stretch of RNA nucleotides, called an **RNA primer**. That primer gives the DNA polymerase a 3' end upon which to add new nucleotides. With this correction, the individual parts of the above cartoon would look something like the next image.

179

From this process, the new strand that results is a series of short DNA stretches called **Okazaki fragments**,[41] which are separated from each other by small nicks (places where the phosphate backbones are not sealed) and short sequences of temporary RNA nucleotides. To complete the synthesis of this rag-tag strand, **DNA ligase** seals up the nicks in the backbone and then another type of DNA polymerase[42] replaces the RNA primers with DNA nucleotides.

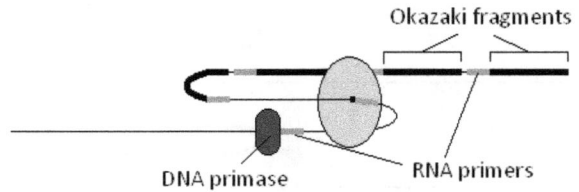

Okazaki fragments

DNA primase

RNA primers

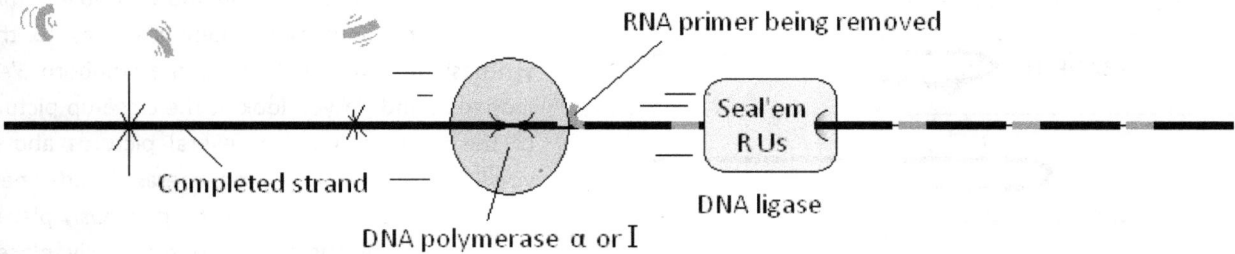

RNA primer being removed

Seal'em R Us

Completed strand

DNA ligase

DNA polymerase α or I

Overview

Let's put it all together. Two DNA polymerase molecules attach to the DNA near the origin of replication, where a set of proteins called the pre-replication complex have gathered to mark the spot for replication to start. Once attached, the two DNA polymerases move along the helix in opposite directions. The helix itself is pulled apart and untwisted by helicase and gyrase. As each polymerase moves along, it begins making new strands using the old strands as templates, matching the nucleotides with their complementary bases. The 3'-5' old strand is used in a straight forward fashion, and new nucleotides are added sequentially without difficulty. This creates the leading strand. The 5'-3' strand is facing in the opposite direction, however, so it has to be looped around the polymerase and threaded forward periodically. Since the polymerase requires an existing 3' end of a previously placed nucleotide, a protein called DNA primase inserts a temporary sequence of RNA nucleotides, called RNA primers. New nucleotides are then attached to the 3' end of these primers. The new strand is thus made in short lengths called Okazaki fragments. Each fragment is separated by little nicks, where the backbones have not yet been bonded together. Finally, after this is finished, DNA ligase seals up the nicks in the backbone, and another type of DNA polymerase replaces the RNA primers with DNA nucleotides. This completes the lagging strand synthesis.

---

[41] Although they may appear small in the cartoons of this chapter, Okazaki fragments are actually several hundred to a thousand base pairs long, depending on the species. Eukaryotic Okazaki fragments are usually much shorter than prokaryotic Okazaki fragments.

[42] The protein is named DNA polymerase α in eukaryotes and DNA polymerase I in prokaryotes. Generally speaking, while the patterns of many biological processes are often the same between eukaryotes and prokaryotes, the names and features of the proteins that execute the individual cellular mechanisms can vary quite a bit. Also, in some cases, proteins can have multiple names. For example, DNA gyrase (the protein that untwists the helix) is also known as topoisomerase. It can make learning the material difficult, so keep the possibility of name variation in mind.

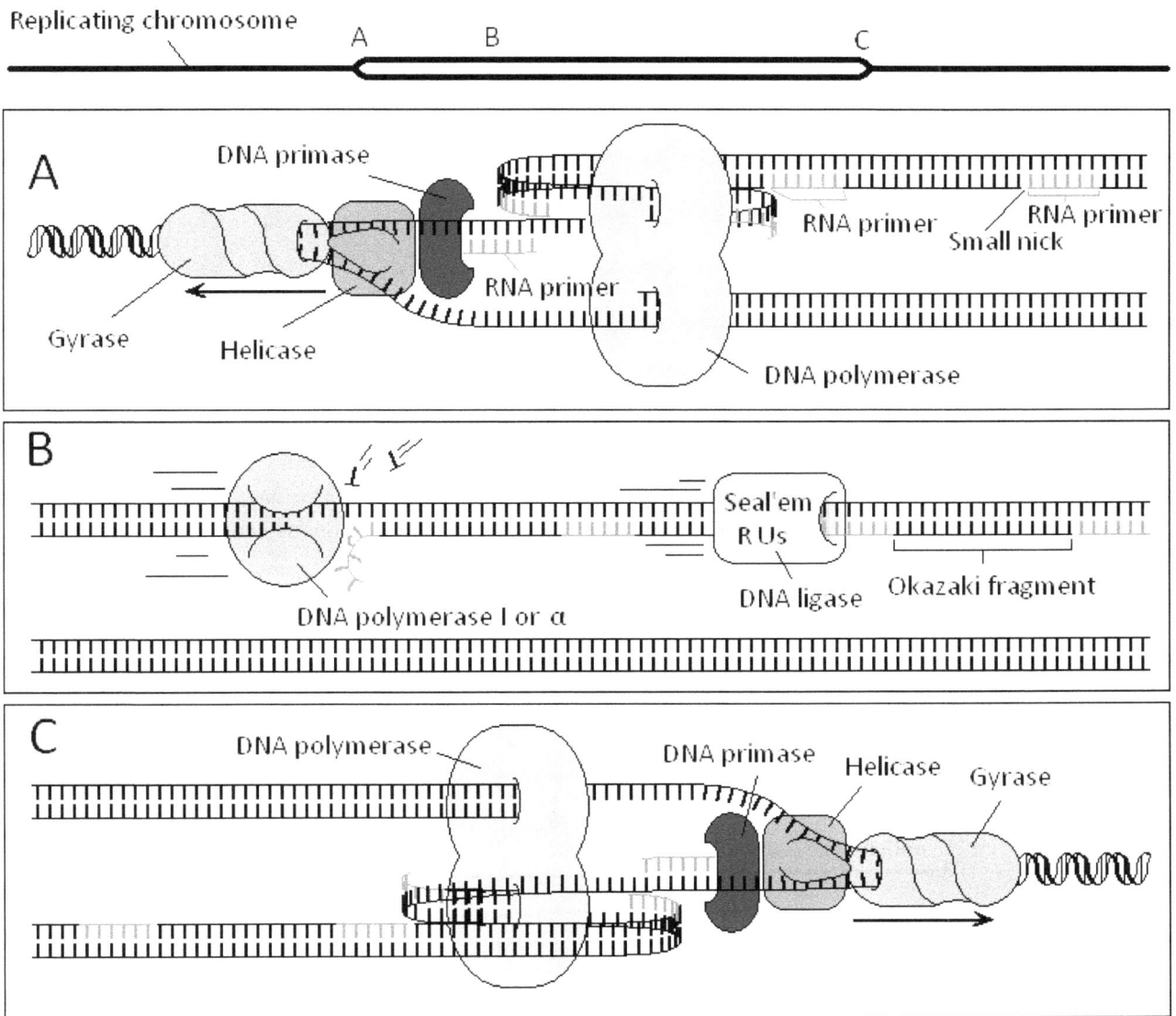

A — DNA primase, Gyrase, Helicase, RNA primer, DNA polymerase, RNA primer, Small nick

B — DNA polymerase I or α, Seal'em R Us, DNA ligase, Okazaki fragment

C — DNA polymerase, DNA primase, Helicase, Gyrase

Replication: Prokaryotes vs Eukaryotes

The primary difference in DNA replication between eukaryotes and prokaryotes is in the number of origins of replication used. Because prokaryotic DNA is circular, bacteria typically have only one origin of replication. Replication starts there and races in both directions around the circular track until they meet each other. At that point, each of the old strands has a completely new complementary strand.

Replication of prokaryotic chromosome

Replication starts . . . and spreads both ways . . . until two new chromosomes are made.

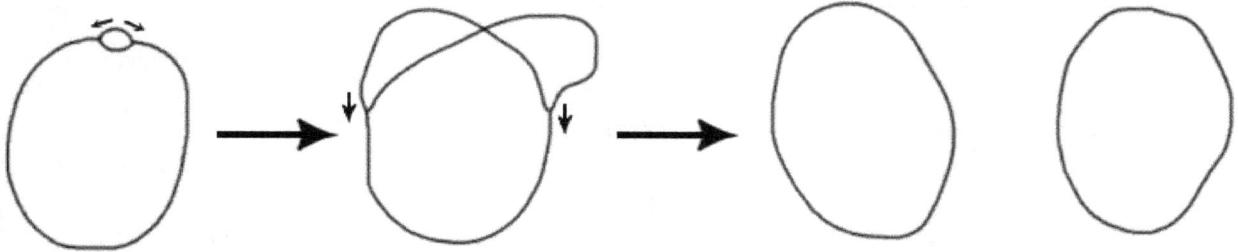

Eukaryotic chromosomes generally have many more than one origin of replication. Why? First, eukaryotic DNA is wound around histones. Supercoils are also present, so that unpacking the DNA is a considerably slower process. Our bodies are choked-full of cells that divide at a rapid rate, so they need to be able to replicate their DNA quickly! That means the DNA needs to be copied from multiple places at the same time, or the process would take FOREVER to get done! Therefore, eukaryotes have multiple origins of replication. One last thing to point out is that the origins or replication are always somewhere within the chromosome and *never* on the ends.

Replication in eukaryotes

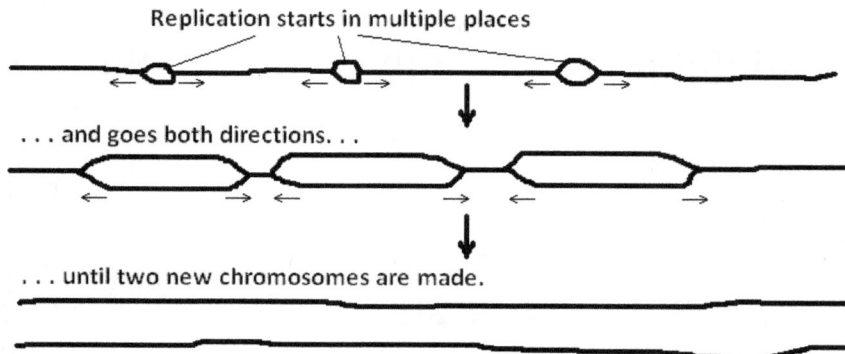

Replication starts in multiple places

. . . and goes both directions. . .

. . . until two new chromosomes are made.

Summary

In this section, we talked about DNA replication.

- DNA is comprised of four different base pairs: adenosine (A), guanosine (G), thymidine (T) and cytidine (C). A's pair with T's through two hydrogen bonds, and C's pair with G's through three hydrogen bonds.
- The strands of DNA are paired up in a helix, which looks like a twisted ladder.
- Anytime the cell needs to access the base pairs, it separates the two strands very minimally to reduce the chance of errors when the strands come back together. The cell never leaves a strand by itself because it will not fit together again correctly with the other strand.

- DNA replication begins at an origin of replication, which is a long stretch of A-T pairs. A set of proteins called the pre-replication complex binds to the origin of replication, marking it as a place for replication to begin.
- Next, two DNA polymerase proteins attach and begin making new strands, using the old strands as templates. A's are matched with T's, and G's are match with C's.
- Two polymerase proteins move in opposite directions to make new strands, like a rope cut down the center.
- The DNA strands are difficult to separate since they form a twisted helix. In order to open the helix, it must be untwisted as well as pulled apart. The job of untwisting the DNA falls to DNA gyrase. The job of pulling apart the exposed base pairs comes to DNA helicase. Together, they move along the helix in front of DNA polymerase, pulling it open so the polymerase can access the base pairs of the old strands.
- DNA polymerase reads old strands in the 3' to 5' direction, and makes the new strands in the 5' to 3' direction. Only one of the old strands is facing in the 3' to 5' direction. This strand is paired with a new strand (the leading strand) quite easily. The other old strand is facing backwards. It must be looped around the polymerase and read in the opposite direction. As the polymerase moves along, this backwards strand is periodically ratcheted forward. The result is a new strand being made in fragments, separated by little nicks where the backbone of the new strand has not yet been sealed. This fragmented strand is called the lagging strand, and each new fragment is called an Okazaki fragment.
- DNA polymerase requires the 3' end of a previously placed nucleotide in order to lay down a new nucleotide. Since the looped strand is facing backwards, there are no previously placed nucleotides for polymerase to work with, so a protein called DNA primase puts a temporary strip of RNA down to provide the 3' end. This temporary strip is called an RNA primer.
- As DNA polymerase moves away, DNA ligase runs along the lagging strand, sealing up the backbone. Then, another form of polymerase (α in eukaryotes or I in prokaryotes) follows after the ligase, replacing the RNA primers with DNA nucleotides.
- The process of replication is quite similar in both eukaryotes and prokaryotes, except for the number of origins of replication. Prokaryotes typically have one circular chromosome, with one origin of replication, while eukaryotes have many linear chromosomes, each with many origins of replications. Eukaryotes have more origins of replication because their DNA replication takes longer to replicate; it wouldn't get done in time with only one origin of replication.

Unit 23

Protein Production, Part 1:

RNA Transcription

The Central Dogma of Biology

In the last section, we saw how DNA is replicated. Now, let's turn our focus to the primary importance of DNA: protein production. DNA stores protein blueprints. The proteins that are made from these blueprints manipulate the environment to increase the odds that the genome will make it to the next round of replication. In other words, proteins are the genome's tools for survival.

At this point, a nagging question might be playing on your thoughts: how did DNA make protein in the first place if it needed proteins to do it? How did it even replicate to begin with? Isn't this a chicken and egg scenario? Not quite, but that question is a natural one for many students to have. A full appreciation of the elegant answer requires a lot more understanding of biology. By the time we finish this book, we will be in a position to tackle the explanation.

The important thing about DNA is the *order* in which the nucleotides are attached to each other. This order is used as a means of storing a code. It's a lot like the letters in our language. Using only 26 letters and a few punctuation marks, we can express a limitless number of ideas.[43] All we have to do is arrange the letters in some meaningful fashion. DNA works in a similar fashion. Each triplet of nucleotides translates into one of the twenty amino acids, and the order of the triplets appearing in the DNA chain tells the cell the order in which to put together

---

[43] ... most of which are rubbish, if history tells us anything.

the amino acids. This order of amino acids then determines the structure and behavior of the protein (primary, secondary, tertiary and quaternary structure as discussed in our unit on Proteins).

Not all of the genome is used to make protein though. In fact, it is generally believed that most of the DNA, like our libraries and governments, is full of junk. You might, for example, find a short stretch of DNA that encodes for a protein and then go for thousands and thousands of base pairs that don't encode for anything whatsoever. All at once, you will again encounter another stretch of meaningful DNA, and so forth. It's like a book full of random letters and meaningless drivel, with an occasional sentence here or there.[44] The meaningful little stretches of useful DNA are called **genes**. The rest, quite simply, is called junk.

Like our written language, biological systems use a kind of punctuation as well. Genes are sandwiched between small sequences of nucleotides that tell the cell's biological machinery when to start reading and when to stop. The "start" is a recognizable set of base pairs called the **promoter**, which various proteins bind to in order to start the process of reading the gene. The "stop" punctuation, on the other hand, varies from eukaryotes to prokaryotes. We'll discuss this more later in this unit.

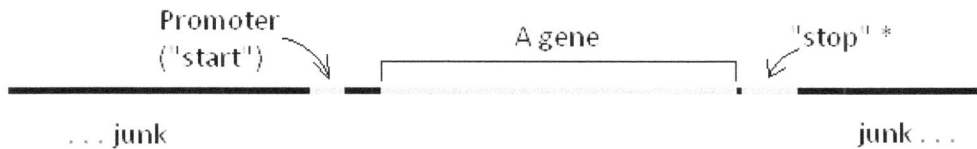

That's all fine and good, but how does the gene get put into action? Well, the first thing that happens is **RNA transcription**, which is a fancy name that basically means a small "copy" of the gene is made using RNA. During this process, the DNA helix is pulled open a little and the strand *opposite* the one having the meaningful base pairs is used as a template to put together an RNA strand, called a **messenger RNA**, or **mRNA**. Since the base pairs bond in a complimentary fashion, the mRNA that results will be an exact RNA copy of the strand having the gene.[45] The image below shows this idea. From this image, you can see that the strand that contains the encoded protein is called the **sense strand**. The strand that is complimentary to the sense strand is the **nonsense strand**.

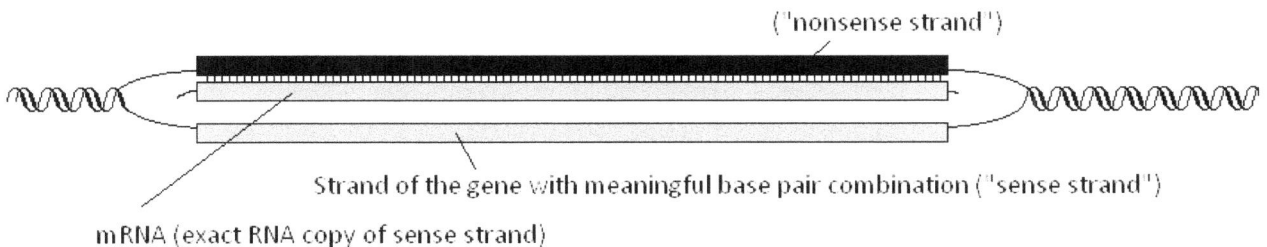

After this happens, ribosomes take these little RNA strands and use them to decipher the order of amino acids needed for the protein in question. Taken together with RNA transcription, this entire process, (DNA to RNA to

---

[44] . . .just like this particular book, for example.

[45] "Exact" is used somewhat lightly here. RNA is technically different than DNA, but the base pair sequence is the same as the sense strand, except that all T's are replaced with U's, as you'll see in the rest of chapter. I use the word "exact" to help facilitate the concept that the RNA transcript is the same functional equivalent of the sense chain.

protein), is dubbed the Central Dogma of Biology. For the rest of this chapter, we will cover the first part of the process, the transcription of RNA from DNA. Then, in the next chapter, we'll finish up the story by covering how protein is created from RNA.

RNA Transcription

If you'll recall, RNA is just like DNA. It is made from nucleotide monomers similar to those that make DNA, where the major differences are 1) the sugar unit on the nucleotides in RNA has an extra hydroxyl group (-OH) attached to it, and 2) thymidine is replaced by uridine. Uridine is very similar to thymidine, but with the methyl group missing, as shown to the right.

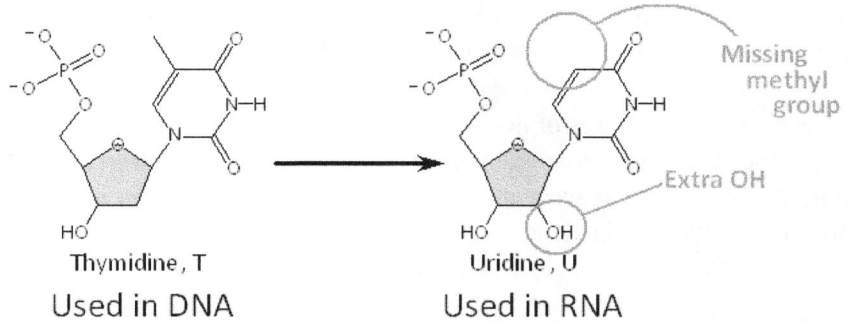

Thymidine, T
Used in DNA

Uridine, U
Used in RNA

Missing methyl group

Extra OH

These RNA nucleotides are strung together into strands (i.e., polymers), where the function of the strand depends, in large part, on the sequence of the nucleotides. RNA polymers are divided into many categories. The three largest RNA categories are messenger RNA (mRNA), which we encountered in the previous section, **transfer RNA (tRNA)** and **ribosomal RNA (rRNA)**. We will encounter transfer RNA in the next unit and will therefore put off discussion of it until that point. Ribosomal RNA are polymers that help make up the two gloves of ribosomes (see Unit 17). Here is a picture of how the three categories typically look.

Three kinds of RNA

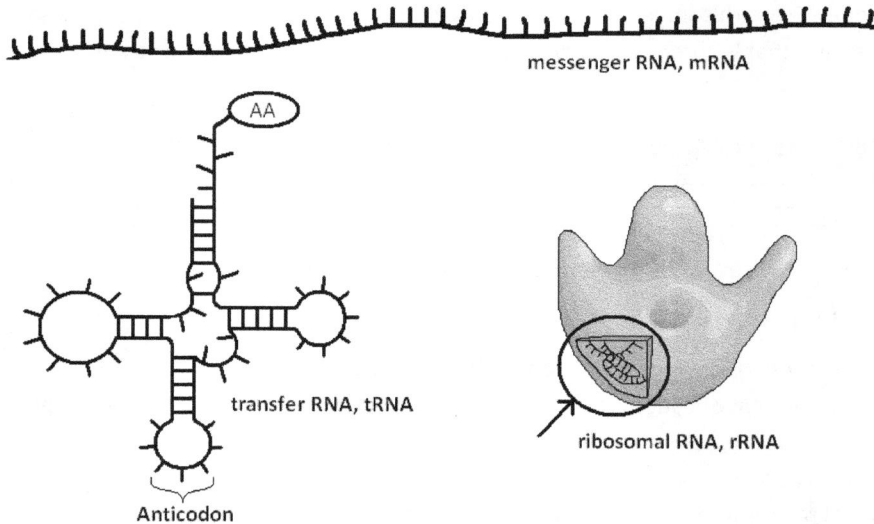

messenger RNA, mRNA

transfer RNA, tRNA

Anticodon

AA

ribosomal RNA, rRNA

Messenger RNA are the blueprints for proteins and are made in RNA transcription. At the very start of this process, an enzyme complex called **RNA polymerase** binds to a promoter, or "start" signal, sitting just before a gene.[46] Think of the promoter as the landing platform for the RNA polymerase complex. It has a very recognizable pattern, usually something like: TTGACA . . . . . .TATA, where the ". . . . ." represents a short stretch of base pairs that could be practically any combination. Once the binding of RNA polymerase to the promoter is complete, the structure begins moving along the DNA. As it does so, it reads the nonsense strand in the 3' to 5' direction, matching up the strand with RNA nucleotides that float by and enter an intake space on the top of the complex. If the RNA polymerase reads a G on the nonsense strand, it attaches a C to the growing messenger RNA. If it reads a T, it attaches an A. If it reads an A, it attaches a U (uridine instead of T, thymidine), and so forth. The growing messenger RNA emerges from a space in the top of the polymerase complex.

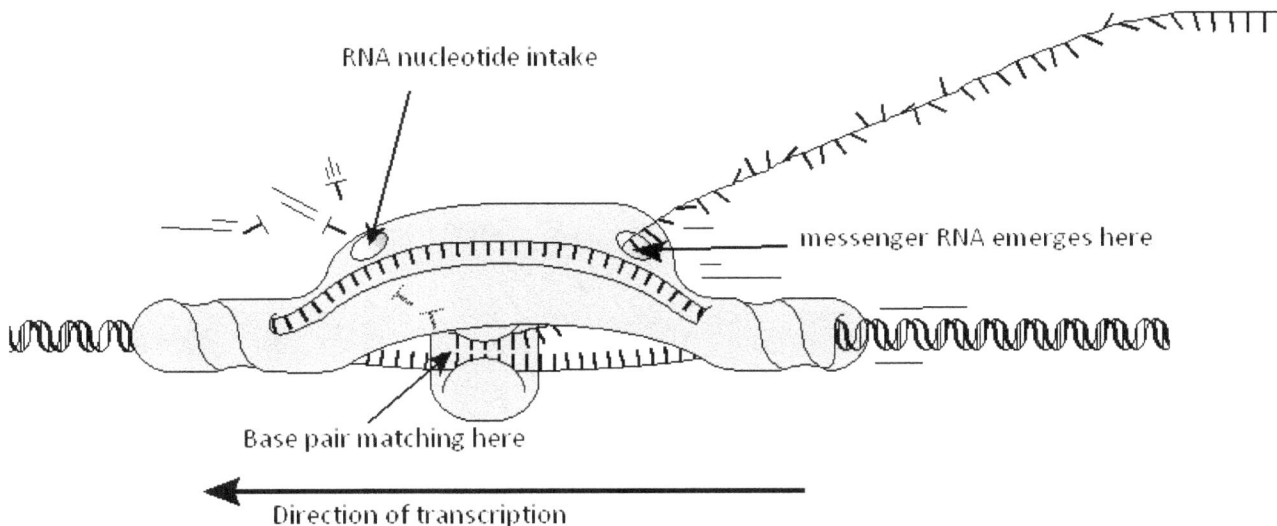

Just as in DNA replication, the problem of unwinding the DNA must be handled, or the DNA cannot be affectively separated without tangling. To handle this, the RNA polymerase complex incorporates DNA gyrase-like and helicase-like subunits that accomplish this job. As it moves along, these subunits separate and unwind the DNA, threading the sense strand to the side and out of the way. As the polymerase passes it winds the DNA back together again behind it. As always, the DNA strands are never left unpaired for very long. All told, it looks like a little bubble moving along a string.

At this point, you may be wondering why the mRNA that emerges from the RNA polymerase doesn't clump together into a jumbled ball. After all, that is what DNA and RNA strands do if left without a complimentary partner strand, and we *did* make a fuss about it in the last unit. The reality is that the mRNA *does* clump up here and there,

---

[46] There is a great deal of variability in this process, depending on the species, though the patterns are the same. Generally speaking, in eukaryotes RNA polymerase binding happens piecemeal. The proteins that are involved in the complex land separately, gradually forming the RNA polymerase complex. In prokaryotes, the RNA polymerase complex (made of many subunits) is more-or-less already formed when it binds to the gene promoter. Here, we avoid the minutia in attempt not to lose the forest for the trees.

but it also tends to have a lot of proteins (not shown in the picture for simplicity) interacting with it. These proteins, no matter their other roles, also prevent aggregation (clumping). Also, a small amount of RNA chain aggregation is not nearly as problematic as aggregation of an entire genome would be, so mRNA clumping is a minor issue.

At any rate, this process of mRNA creation continues until the "stop" signal is encountered. In prokaryotes, a stop could be signaled by one of two methods, depending on the gene and the species. One "stop" method, for example, is when a stretch of self-recognizing G-C rich base pairs is created in the growing mRNA. This self-recognizing area creates a hair-pin (an abrupt turn of the chain onto itself) in the mRNA, and this occurs while a stretch of A's are currently being read. As we saw in DNA, A's bind less tightly to their complementary bases than do G's and C's. As a result, the stretch of G-C in the mRNA will snap together and yank the forming end of the messenger RNA out of the RNA polymerase. This event is believed to cause the RNA polymerase complex to pop off the DNA strand.

Example of transcription termination

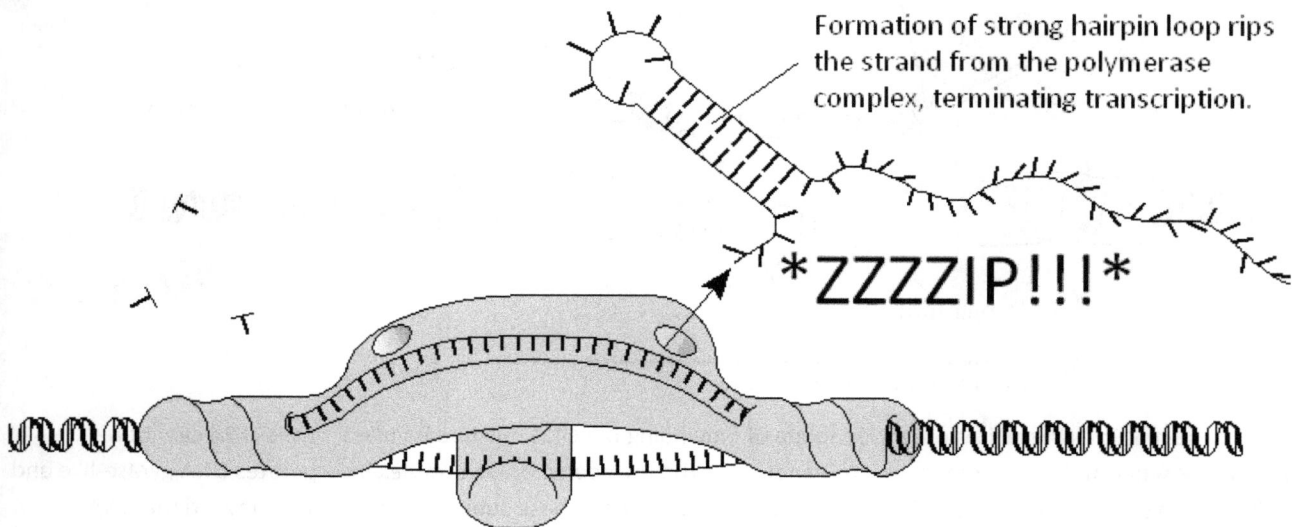

Formation of strong hairpin loop rips the strand from the polymerase complex, terminating transcription.

ZZZZIP!!!

Another way transcription ends in prokaryotes is when a protein called "Rho" attaches to the free end of the new mRNA and crawls toward the RNA polymerase complex. When it reaches the complex (where the mRNA is emerging from the polymerase), it causes the process to stop. How RNA transcription ends in eukaryotes, such as in you and me, is not fully understood.

After transcription termination, the fate of the mRNA differs in prokaryotes and eukaryotes. In prokaryotes, the mRNA proceeds more-or-less uninterrupted to the protein-making part of the process. In eukaryotes, the messenger RNA, or **pre-mRNA** as it is called at this stage, experiences a number of modifications before it is ultimately ready to serve as the protein blueprint. The first two changes that occur are 5' "capping" and **polyadenylation** of the 3' tail. The 5' end of the messenger RNA is capped with the addition of **7-methylguanosine**, which is basically an altered guanosine snapped backwards onto the front of the mRNA. Meanwhile, a stretch of 100-200 adenines is attached to the 3' end. This is called the **poly-A tail**.

Post-Transcriptional Modifications to mRNA in Eukaryotes

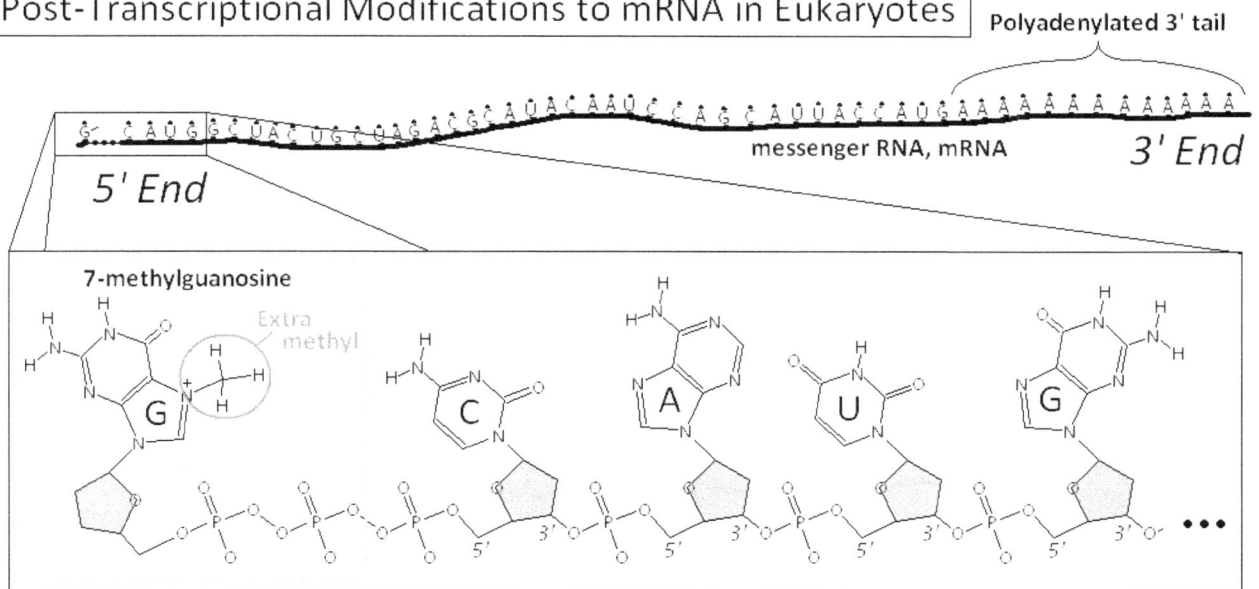

Polyadenylated 3' tail

5' End

messenger RNA, mRNA

3' End

7-methylguanosine

Extra methyl

G C A U G

After these modifications, short stretches of base pairs within the pre-mRNAs are often removed. It's like taking a string, snipping out small parts in the middle and gluing the rest back together. The pieces that are removed are called **introns**: they stay IN the nucleus. The pieces that are not removed are called **exons**: they ultimately get to EXIT the nucleus. This process is called **intron splicing**, and it happens when small RNA-protein complexes called **snRNP** (read as "snurps") recognize specific sequences in the mRNA. These snurps also combine with the other proteins and, collectively, the entire mess is called a **spliceosome**. After the formation of the spliceosome, the introns are removed (as **lariat loops**) and the ends of the exons are spliced together. After all the introns are removed, the pre-mRNA becomes a mature mRNA and is finally allowed out of the nucleus to play.

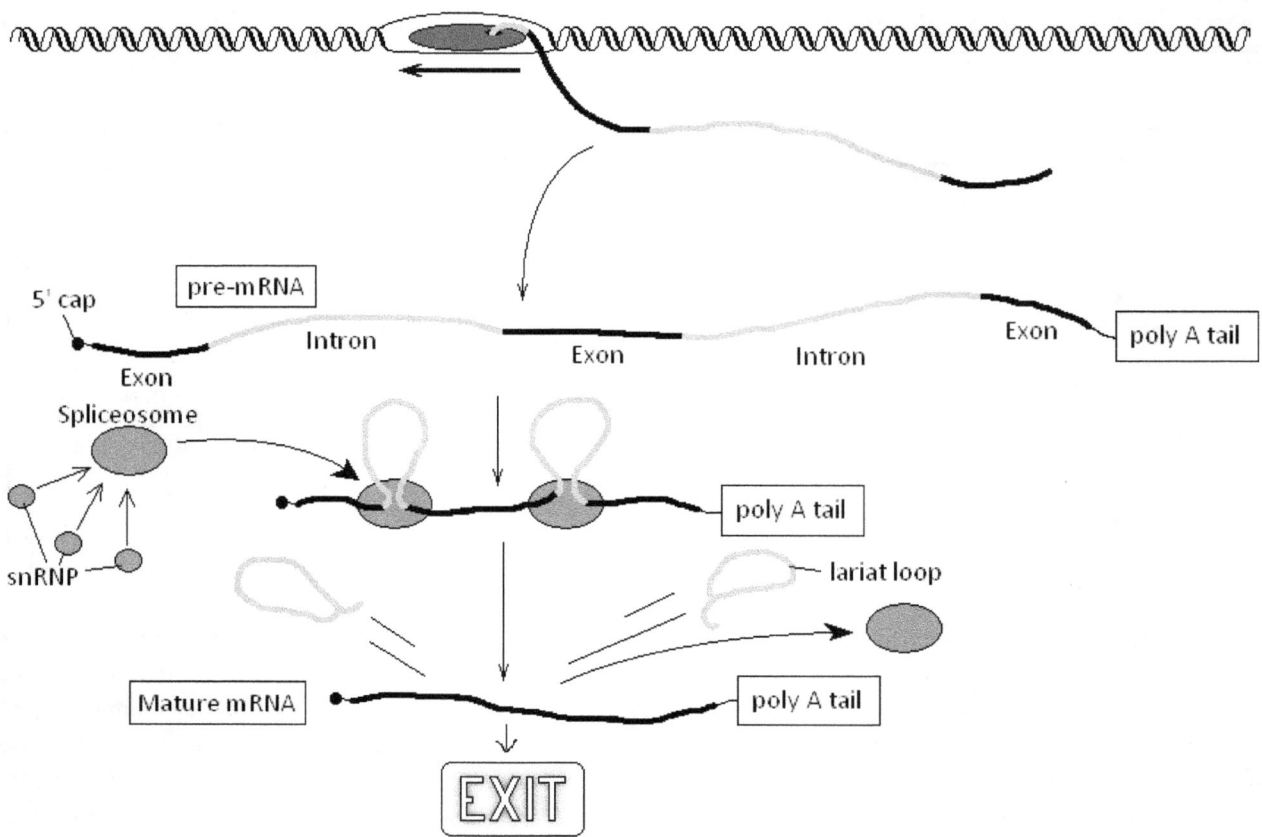

5' cap   pre-mRNA

Intron   Exon   Intron   Exon   poly A tail

Exon

Spliceosome

snRNP

poly A tail

lariat loop

Mature mRNA   poly A tail

EXIT

With that, we bring to a close the first half of the story of protein production. In the next unit, we'll finish the story and see how the mRNA interacts with ribosomes to make protein.

Summary

In this section, we discussed RNA transcription.
- The general theme of information flow from DNA to RNA to protein is called the Central Dogma of Biology. It is divided into two parts, occurring in order, RNA transcription and RNA translation.
- Most of the genome in any given species is full of random nucleotide sequences, called junk. Small stretches here and there contain the information for proteins. These small stretches are called genes.
- Genomes that contain the information for efficient proteins will be more successful in surviving to replication than genomes encoding for crummy proteins. Over time, the more efficient genomes will outnumber the less efficient ones.
- Genes have "start" signals in front of them called promoters. They also have "stop" signals as well, but these appear in a number of different variations.

- The process of protein production starts with RNA transcription.
- RNA transcription begins when a complex of proteins called RNA polymerase binds to a promoter. Once fully formed, the RNA polymerase moves along the gene, making mRNA from the nonsense strand.
- In prokaryotes, transcription stops when either a G-C hairpin loop forms and yanks the rest of the mRNA away from RNA polymerase or when a Rho protein crawls along the mRNA and reaches the RNA polymerase. Transcription termination in eukaryotes is not well understood.
- In eukaryotes, the mRNA are first known as pre-mRNA and experience a number of modifications before becoming a mature protein blueprint. In prokaryotes the mRNA proceed to the next step without any modifications.
- The first modification to pre-mRNA is 5' capping with a backwards-facing, altered guanosine called 7'-methylguanosine. The second modification is the addition of 100-200 adenines to the 3' tail of the pre-mRNA. This is called the poly-A tail.
- Lastly, a number of introns are removed from various points along the mRNA. These are removed from the mRNA by spliceosomes. The remaining mRNA pieces, called exons, are bonded together, resulting in the mature mRNA.

In this section we cover:

1) The Genetic Code

2) transfer RNA

3) RNA translation

4) Protein routing to endoplasmic reticulum

5) Speed of protein production

Unit 24

Protein Production Part 2:

RNA translation

In the last chapter, we discussed the first half of the Central Dogma of Biology, where messenger RNA is transcribed from DNA. The second half of the dogma is called **RNA translation** and is the focus of this chapter. It has this name because the cell's machinery takes the mRNA and translates the order of its nucleotides into the order of amino acids for the new protein. Before we tackle that process, however, let's talk for a bit about the genetic code as well as about transfer RNA (tRNA). An understanding of both is vital before moving on.

The Genetic Code

We spoke a bit in previous sections about the information that DNA holds. Let's zoom in a little more into how this works. It's all based on *combinations*. Consider a stretch of 3 nucleotides (a triplet). The first nucleotide can have one of 4 bases, so we start with 4 possibilities: U, A, C and G. Similarly, the second nucleotide can also have 4 possible bases, so together there are sixteen (i.e. 4 × 4) combinations available: UU, UA, UC, UG, AU, AA, AC, AG, CU, CA, CC, CG, GU, GA, GC, GG. After the third nucleotide, our total combinations increase to 64. In other words, since each of the three nucleotides can be one of four possible bases, there are a total of 64 possible triplets we can make.

So what, right? Well, the cell has only twenty kinds of amino acids. The trick living systems use goes as follows: they assign one amino acid for each triplet (hereafter called a **codon**), like say, AAA is the code for lysine, GGA for glycine, UAC for tyrosine, and so forth. With 64 possible codons, there are more than enough combinations to cover the twenty amino acids. In fact, there are so many codons left over that multiple codons can be assigned to a single amino acid (i.e., GGU, GGC, GGA and GGG all go to glycine). Here is the full list of the codon assignments (called the **genetic code**).[47]

		U		C		A		G
U	UUU	Phenylalanine, phe	UCU	Serine, ser	UAU	Tyrosine, try	UGU	Cysteine, cys
	UUC	Phenylalanine, phe	UCC	Serine, ser	UAC	Tyrosine, try	UGC	Cysteine, cys
	UUA	Leucine, leu	UCA	Serine, ser	UAA	Stop	UGA	Stop
	UUG	Leucine, leu	UCG	Serine, ser	UAG	Stop	UGG	Tryptophan, try
C	CUU	Leucine, leu	CCU	Proline, pro	CAU	Histidine, his	CGU	Arginine, arg
	CUC	Leucine, leu	CCC	Proline, pro	CAC	Histidine, his	CGC	Arginine, arg
	CUA	Leucine, leu	CCA	Proline, pro	CAA	Glutamine, gln	CGA	Arginine, arg
	CUG	Leucine, leu	CCG	Proline, pro	CAG	Glutamine, gln	CGG	Arginine, arg
A	AUU	Isoleucine, Ile	ACU	Threonine, thr	AAU	Asparagine, asn	AGU	Serine, Ser
	AUC	Isoleucine, Ile	ACC	Threonine, thr	AAC	Asparagine, asn	AGC	Serine, Ser
	AUA	Isoleucine, Ile	ACA	Threonine, thr	AAA	Lysine, lys	AGA	Arginine, arg
	AUG	Start, methionine, met	ACG	Threonine, thr	AAG	Lysine, lys	AGG	Arginine, arg
G	GUU	Valine, val	GCU	Alanine, ala	GAU	Aspartic acid, asp	GGU	Glycine
	GUC	Valine, val	GCC	Alanine, ala	GAC	Aspartic acid, asp	GGC	Glycine
	GUA	Valine, val	GCA	Alanine, ala	GAA	Glutamic acid, glu	GGA	Glycine
	GUG	Valine, val	GGG	Alanine, ala	GAG	Glutamic acid, glu	GGG	Glycine

Let's see a brief example of how this works, using a picture:

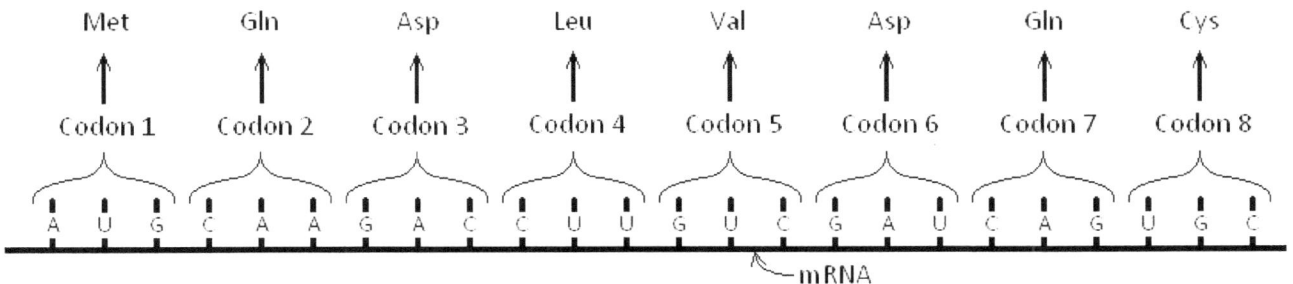

In this image, you can see that AUG causes a methionine to be placed in the first position. Then CAA causes glutamine to be placed next, and so forth. In this manner, the codons determine the order of the amino acids in the protein once the codes of the codons are translated.

---

[47] There are technically a few variations of this table depending on the organism, but most living systems use this one. Also, this table is listed in RNA form. Remember that the U's are T's in DNA. So, for example, encountering the DNA nucleotide triplet of TTG means leucine, ACT means threonine, and so forth.

Notice from the genetic code that the codon AUG represents both a start signal and methionine. Also, there are three codons that represent termination signals. These tell the cell when the end of the amino acid order has been reached. Do not confuse these "start" and "stop" codons with the start and stop signals we saw during RNA transcription. They are different processes. Before we can explore that further, however, we'll need to go over the details of transfer RNA.

Transfer RNA

We spoke briefly about tRNA during the last chapter. These molecules bridge the gap between the codons of the mRNA and amino acids, and their structure is critical in the process. Transfer RNAs are single-stranded chains of RNA nucleotides that have coiled up to form a tangled blob roughly shaped like a bent "t," shown to the right.

At the top is the amino acid. At the bottom is the **anticodon**, which consists of three RNA bases pointed down like a diverging pitchfork. During the course of RNA translation, the tRNA matches up its anticodon with the codon on the mRNA, and in doing so marks that spot for the amino acid. Once in place, another tRNA bearing the next amino acid lands on the next codon, as shown below.

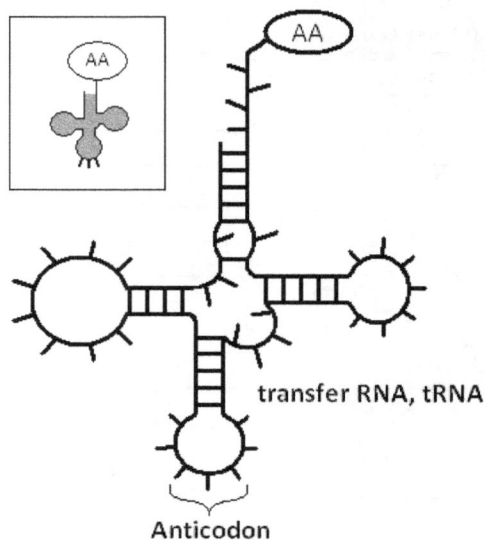

transfer RNA, tRNA

Anticodon

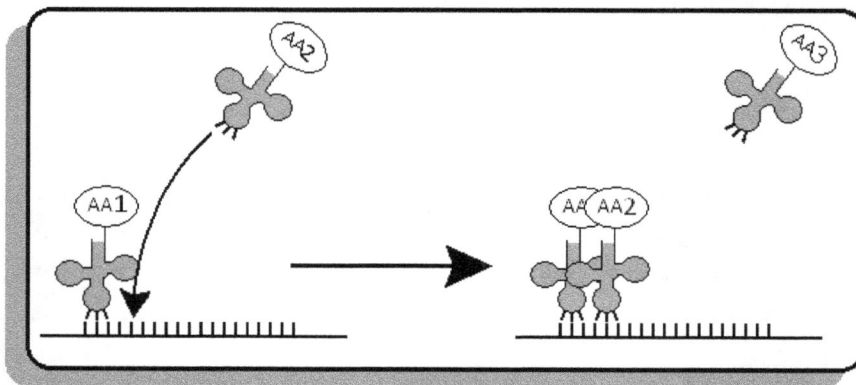

So, how many different tRNA's exist? Well, the answer depends on the species. Let's take a guess though. A quick look at the genetic code will tell you that there are 64 total codons (UUU, UUC, UUA, UUG . . . GGU, GGC, GGA, GGG). So our first guess is that 64 tRNA are necessary. Not quite. We have to subtract 3 because three of the codons are translated as termination codons; they don't code for amino acids. We're down to 61, which is a good guess. As it turns out, however, there are *less* than 61 types of tRNA in most living systems. Why?

The reason there are less than 61 tRNA is due to a process described by the **Wobble Hypothesis**. If you take another look at the picture of the tRNA, you'll notice that the anticodon is curved. This means that the first two bases can match up with the first two bases of the codon, but the last one has poor contact and contributes less distinction to the matching process.

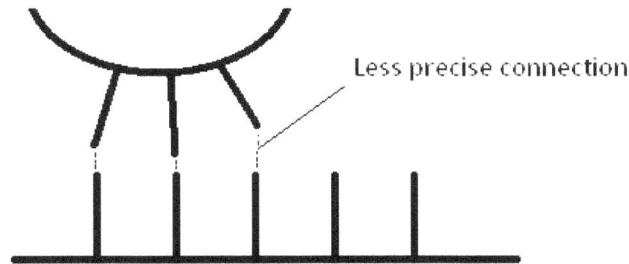

As a result, one anticodon might recognize up to four different codons, depending on the identity of the third base. This is why several codons code for the same amino acid in the genetic code. Because of this principle, the actual number of required tRNA drops to approximately 31, depending on the species. Some require more, and some require less, but 31 is an average number.

We are almost done with our preliminary discussion and will soon be looking at RNA translation. There is, however, one more question we need to ask: How do the amino acids get matched up with the right tRNA? After all, if proline, say, does not get correctly matched with the tRNA that has the anticodon GGG, it will not get correctly placed when the codon CCC appears on the mRNA. The job of making sure that the right amino acid is attached to the right tRNA comes down to a group of proteins called **aminoacyl tRNA synthetases** (think of this as amino acid-onto-transfer-RNA synthesis enzyme). With the help of these enzymes, the cell has a whole collection of tRNA molecules, each with their matching amino acids, just waiting for RNA translation to begin.

Initiation of Translation

We are now ready to talk about the initiation of RNA translation. This process is like catching a fly and caterpillar in your hands using bug bait (see image on the following page). If you recall, the two halves of a ribosome look like little hands. The smaller half, (or small subunit), starts the translation process by binding small proteins called **initiation factors** (i.e., the bug bait). Once these factors bind, the small unit is then ready to attract the mRNA (the caterpillar) as well as the first transfer RNA, (the fly). This tRNA is carrying methionine. The mRNA attaches at a very specific place on the small subunit, matching up a sequence of its bases called the **Shine-Dalgarno sequence**[48] with a spot on the subunit called the **anti-Shine-Dalgarno sequence**. This process stabilizes the mRNA. At almost the same time, a formylmethionine-carrying tRNA (f-Met-tRNA) matches its anti-codon with the AUG codon on the mRNA, which appears shortly after the Shine-Delgarno sequence.[49]

---

[48] Technically, the Shine-Dalgarno sequence is found in prokaryotes. In eukaryotes, it's called the Kozak sequence. The function is essentially the same in both.
[49] The Shine-Dalgarno sequence is six bases long, usually AGGAGG, and typically appears eight bases before the AUG.

195

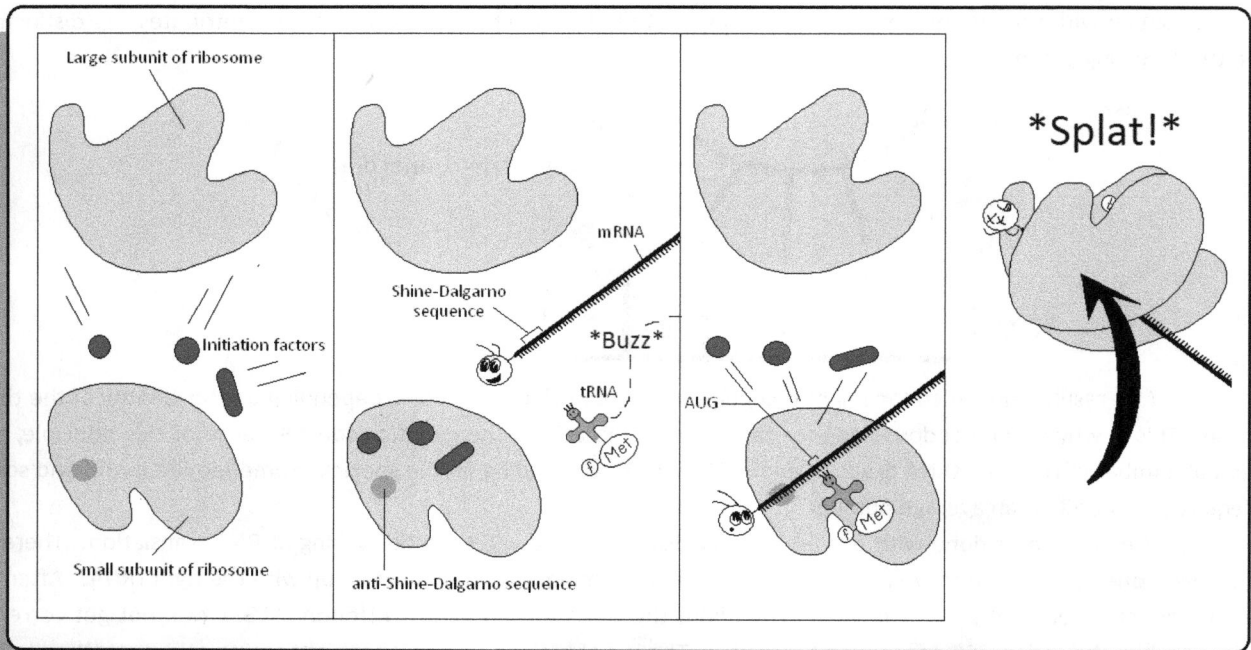

Once the f-Met-tRNA has bound to the starting AUG codon (it is Met because AUG codes for Met) and the messenger RNA is firmly attached to the Shine-Delgarno area of the small subunit, the large subunit attaches, much like two hands coming together to squish the fly and caterpillar in between. RNA translation is now ready to begin.

Notice, however, that "AUG" codes for both "start" and for methionine. This, naturally, begs an important question: Do all proteins have to start with methionine? The answer, surprisingly, is no. The methionine on this first tRNA has a small "formyl" chemical group attached to it. This formyl[50] group is a signal to the organism that this methionine is to be removed before the protein becomes mature. Only the first Met-tRNA is marked this way. The other Met-tRNAs are normal.

The Translation Cycle

Once the mRNA and first tRNA are trapped between the two subunits, a series of repeating steps commences. For reference, see the image on the following page. In that image, we pretend the small subunit is invisible so that we can observe the inner workings of this cycle, step by step. The mRNA is positioned horizontally across the large unit and is facing toward the left. The first tRNA is already in position, and in this case, we say it is resting in the **P-site**. At this point, the first thing to happen is that a tRNA matching the next codon enters and attaches to the mRNA at the **A-site**. Once the new tRNA is bound to the A site, the old tRNA transfers its amino acid onto the amino acid of the new tRNA. The old tRNA then departs. Once this happens, the ribosome moves along the mRNA so that the remaining tRNA is shifted into the P-site, and the A-site is freed for the next tRNA to enter. The

---

[50] A formyl group is just a HC=O attached via the C to another molecule.

energy molecule GTP is used to move the ribosome and fuel the cycle. The steps then repeat and will continue to do so until a termination codon is reached.

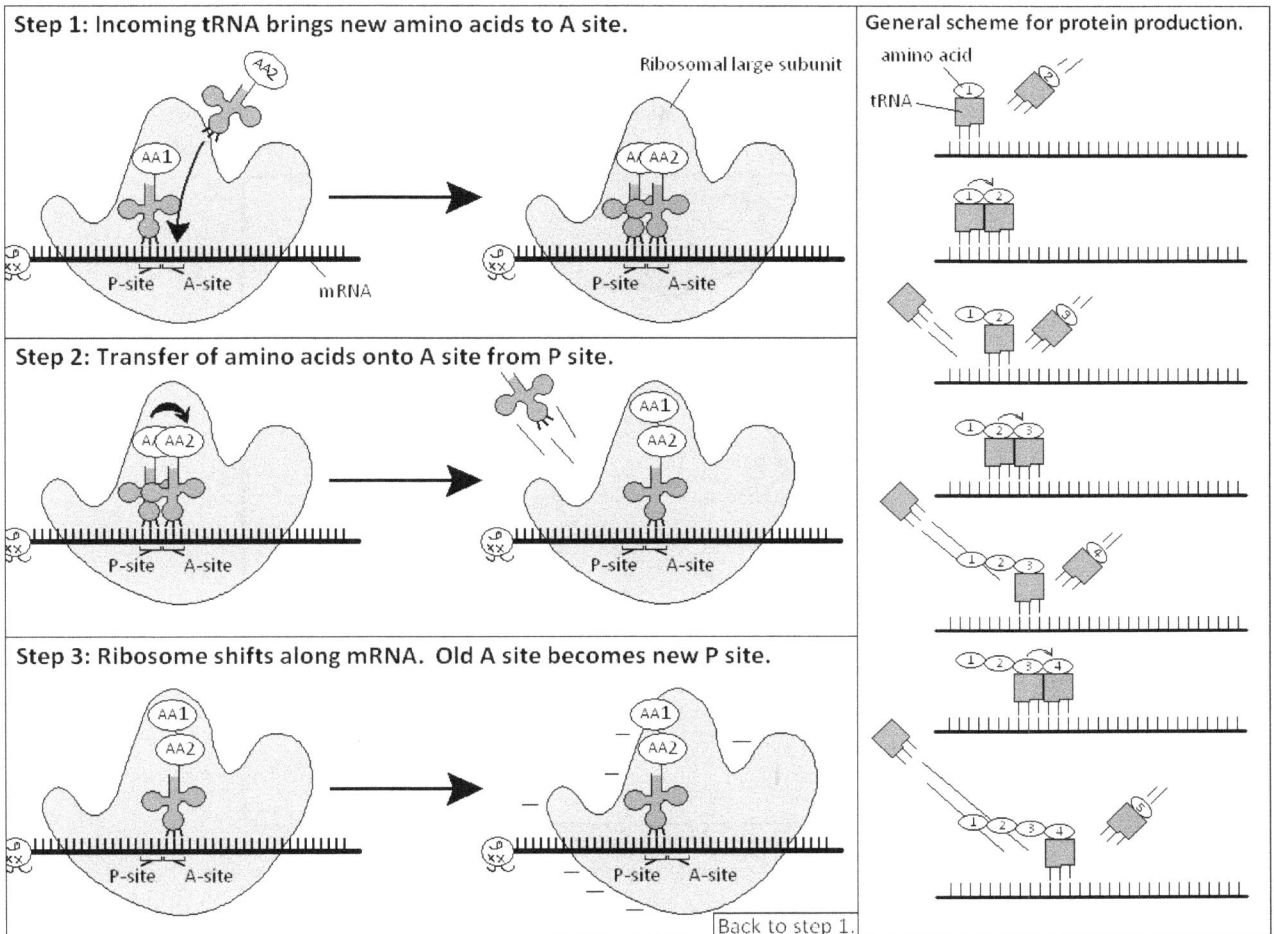

Step 1: Incoming tRNA brings new amino acids to A site.

Step 2: Transfer of amino acids onto A site from P site.

Step 3: Ribosome shifts along mRNA. Old A site becomes new P site.

General scheme for protein production.

To the right of the image is a schematic for the order in which amino acids enter into the reaction, and how these amino acids become linked into a chain. Basically, each new tRNA that enters adopts the existing chain of amino acids and allows the previous tRNA to leave. It then waits until the next tRNA enters, whereupon it gives the amino acid chain (plus one amino acid) to the new tRNA, and so forth.

When one of the three stop codons is encountered, a **release factor** protein binds to it and causes the entire structure to come apart; the subunits detach from the messenger RNA and the new protein is freed. It's sorta like my roommate; he shows up and all my guests bail.

If you'll recall, the large ribosomal subunit has a tunnel passing through it, like a hole in the palm of the hand. The growing chain of amino acids passes through this tunnel (not shown above) and emerges from the other side in the order in which the amino acids were added. Once they begin to emerge from this tunnel, their order determines the fate of the protein. Let's talk about that next.

197

Protein Destination

If you remember, prokaryotes (i.e. bacteria) have no nuclear membrane. Their mRNA also do not go through the post-translational changes, such as 5' capping, poly-A-tail addition and intron splicing. In fact, protein translation in prokaryotes can start before the mRNA is even done being transcribed from the DNA! In electron microscope images, one can sometimes see ribosomes attached to mRNA emerging from the DNA. The new protein being translated is too small to see in these structures, however, so the whole thing just looks like little blobs on strings.

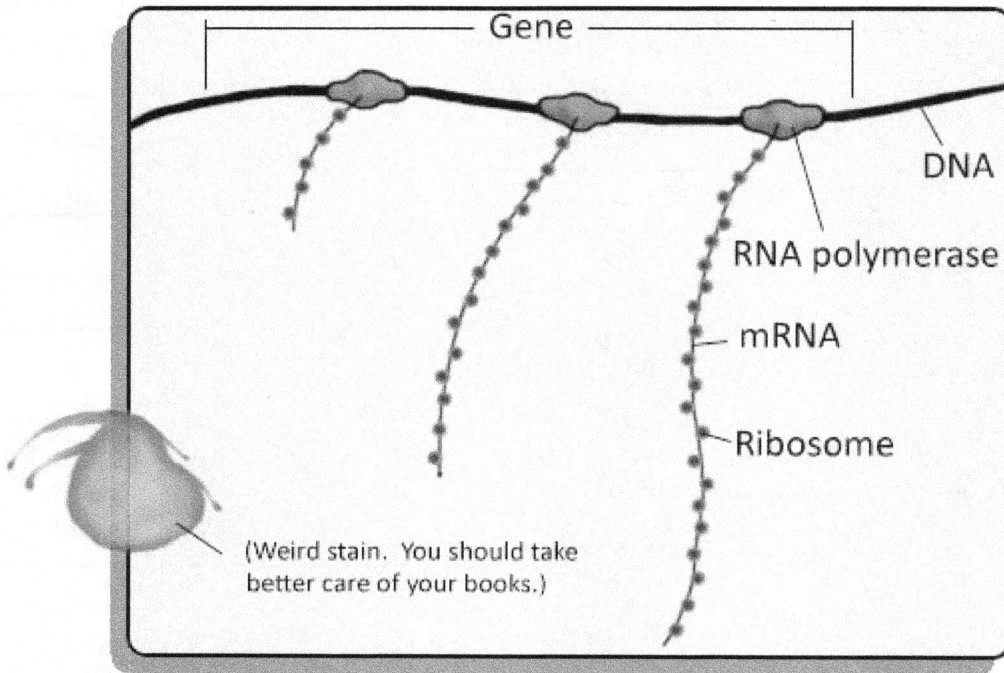

Gene

DNA

RNA polymerase

mRNA

Ribosome

(Weird stain. You should take better care of your books.)

In eukaryotes (such as you and me and possibly the odd thing growing in my sink), there is a cell nucleus that separates the events of transcription from translation. Once the pre-mRNA has gone through all the modifications and becomes a mature mRNA, it leaves the nucleus via a nuclear pore and is caught by a ribosome floating in the cytoplasm. For most mRNA, translation then proceeds (as described above) right there in the cytoplasm. The proteins are created, released and simply float away. For other mRNA, however, the process may be temporarily paused by a **signal recognition particle.**[51] Basically, the signal recognition particle "watches" for a specific sequence of amino acids emerging from the tunnel of the large subunit. If the first few amino acids match this sequence, the recognition particle binds to them and causes translation to temporarily stop. The ribosome/mRNA/recognition particle then attaches to proteins in the membrane of the endoplasmic reticulum, where translation resumes, only now the amino acid chain is threaded through a pore in the membrane. In this way, the protein is synthesized directly into the inside

---

[51] The signal recognition particle is not a particle like the protons, electrons, and neutrons. It's a protein/RNA complex. I normally refer to it as the signal recognition complex, but you'll see it written as the signal recognition particle in texts and publications, so I've presented it as such.

of the endoplasmic reticulum. Since the ribosome and mRNA are stuck to the wall of the endoplasmic reticulum throughout the translation process, this makes the endoplasmic reticulum look rough with little specks stuck to it.

The proteins that are threaded into the lumen of the endoplasmic reticulum in this way are destined for secretion from the cell, or will be incorporated as membrane proteins (such as those in the cell's membrane), or will be routed to organelles such as lysosomes to be used in the cell's digestion process. The signal sequence that caused all of this to happen is clipped before the protein is fully mature.

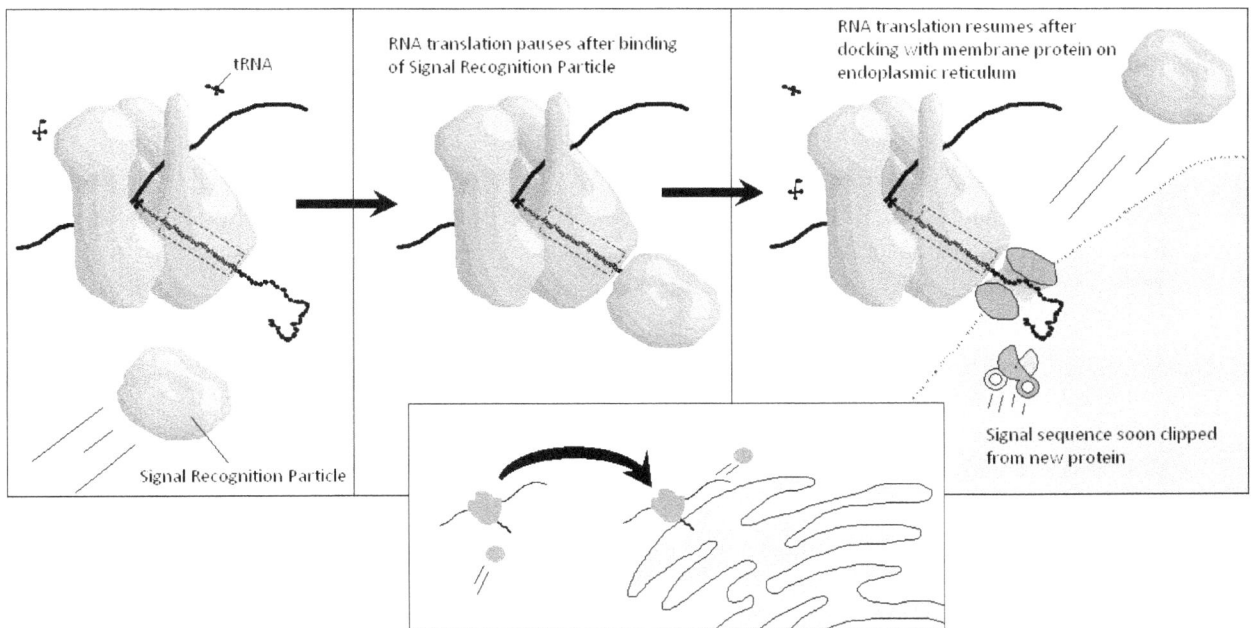

Translation Time

All of these impressive steps take place rather quickly. For example, ribosomes can add between 6-21 amino acids to a growing chain per second, depending on the amino acid and the species in question (eukaryotes being slower). That seems fast. After all, there isn't much *anyone* can do *that* many times a second. As it turns out, however, that isn't fast enough to sustain life. So how do organisms get around this? Well, there are several mechanisms. First, multiple copies of mRNA are typically being made at any given moment. So if each one is being translated, that increases the total rate of protein production. Secondly, two or more ribosomes can often bind to the same mRNA. While one is finishing up, another could be in the middle of making protein and still a third could just be starting! So, if the cell has 100 mRNA in circulation, protein production speeds up by a multiple of 100. If two or three ribosomes can fit onto the mRNA at once, protein production speeds up by another multiple of two or three. So how may mRNA do cells usually have floating around? Typically, they have on the order of *several hundred thousand*. So, if the cell has 100,000 mRNA and two to three ribosomes attached to each one, it means that amino acids are being added to growing protein chains at a rate of roughly three to four million amino acids per second. Now, *that* is fast, but those numbers are rather conservative. Usually many more than two or three ribosomes can fit on a messenger RNA, increasing the total rate of amino acid addition even more.

Summary

In this section, we discussed the basics of RNA translation.

- RNA translation is the second half of the Central Dogma of Biology, where the mRNA is used to determine the order of amino acids to add to a new protein.
- Nucleotides triplets (3 nucleotides), called codons, are each assigned to an amino acid. AAA goes to lysine, GGA goes to glycine, CCU goes to proline, and so forth. Often, many codons are assigned to the same amino acid. For example, GGU, GGC, GGA and GGG all are assigned to glycine. The complete list of codon assignments is called the genetic code.
- The order in which codons appear in the mRNA determines the order in which amino acids are added to a new protein.
- Transfer RNA, or tRNA, are 't' shaped single-stranded RNA blobs that have an anti-codon on one end and an amino acid on the other end. There are roughly 31 types of tRNA, on average, in most living systems. Each tRNA has its own anticodon, which pairs with a codon in the genetic code. Each tRNA is matched with the appropriate amino acid, a reaction that is controlled by enzymes called amino acyl tRNA synthetases.
- The Wobble Hypothesis predicts that some tRNA can fit onto more than one codon due to imprecise matching of the third base pair in the anticodon on the tRNA. This is due to the curved structure of the anticodon on the tRNA and is a big reason why multiple codons are assigned to single amino acids.
- Initiation of RNA translation starts when the small ribosomal subunit binds initiation factors. Once bound, mRNA and the first tRNA binds to the small subunit, the initiation factors leave and the large subunit attaches to the complex. The mRNA/ribosome is now ready to begin.
- RNA translation proceeds in a repeating cycle of steps.
- A new tRNA/amino acid enters and attaches at the A-site, as dictated by the codon that appears there.
  - 1) The amino acid chain is transferred to the new amino acid, and the old tRNA at the P-site leaves.
  - 2) The ribosomes moves along the mRNA so that the new tRNA is shifted from the A-site to the P-site. The process now repeats until a stop signal is found.
- When a termination codon appears at the A-site, a termination factor protein binds and causes the ribosome/mRNA/amino acid complex to come apart. The new protein is freed.
- During construction, the new protein is threaded through a tunnel that appears in the large subunit of the ribosome. As the new protein emerges from the other end of the tunnel, a signal recognition particle watches it for a specific sequence of amino acids. If the specific sequence is found, the signal recognition particle grabs the new protein end and causes production to pause while the entire thing moves and attaches to a channel protein in the membrane of the endoplasmic reticulum. When this happens, the new protein is

routed through the channel into the space inside the endoplasmic reticulum. New proteins routed to the endoplasmic reticulum are destined to be either 1) secreted from the cell, 2) embedded into one of the many membranes of the cell or 3) become part of the vesicle system, such as in lysosomes.

- Cells typically have hundreds of thousands of mRNA copies floating around, and multiple ribosomes can embark on RNA translation at the same time. This drastically increases the rate at which proteins can be produced.

Unit 25

Viruses and Transposons

In previous sections, we covered how DNA is reproduced. Then we switched tracks momentarily to talk about how the cell works hard to make proteins in the Central Dogma of Biology. Now, let's switch tracks again to talk about another key topic. For every hard worker, there is a moocher who lives off of him (or her). Politicians, bill-collectors, useless roommates – it appears there is no end. The same is true in biology. In the cellular world, there are entities whose sole purpose is to hijack cells and use their machinery to replicate and make their own protein. The most prevalent example is the **virus**. To this list, one might add a second entity known as the **transposon**. Let's discuss them one at a time.

Viruses

Viruses are packets of DNA or RNA encircled by proteins that invade a target cell and then take over the protein-making machinery of its cellular victim. In essence, they supply the genetic information while the cell supplies everything else. The virus' genetic information is replicated and their proteins are made. New viruses are then put together from these new parts, and once assembled they escape from the cell to repeat the cycle. Often, the cell that is invaded is killed when the viruses leave.

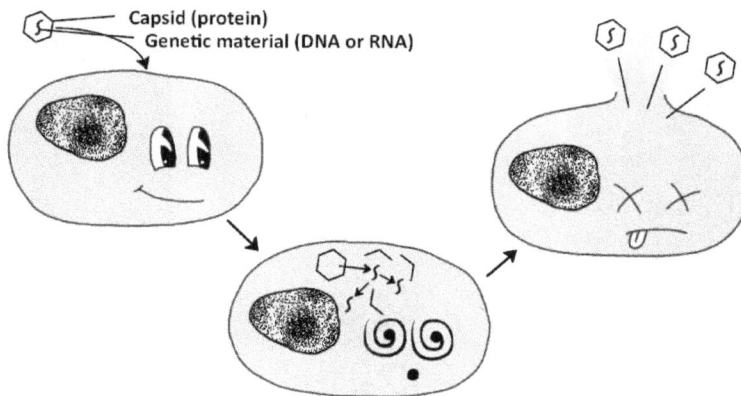

Capsid (protein)
Genetic material (DNA or RNA)

Viruses have relatively simple structures. Their shell, called a **capsid**, is generally made from repeating units of proteins. The capsid is usually one of two different shapes, tubular or icosahedral, although more complex shapes also occur. Inside the capsid is the genetic information, (which is either DNA or RNA). Around the outside of the capsid, some viruses have a lipid envelope. This envelope is the remnant of the lipid membrane from the last cell victim.

Basic virus shapes

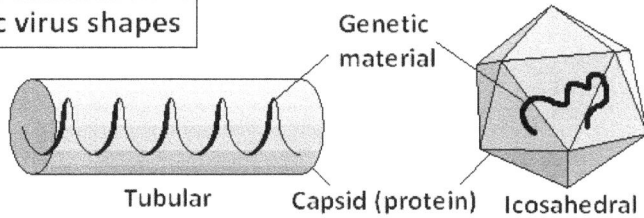

Genetic material

Tubular   Capsid (protein)   Icosahedral

Formation of virus envelope

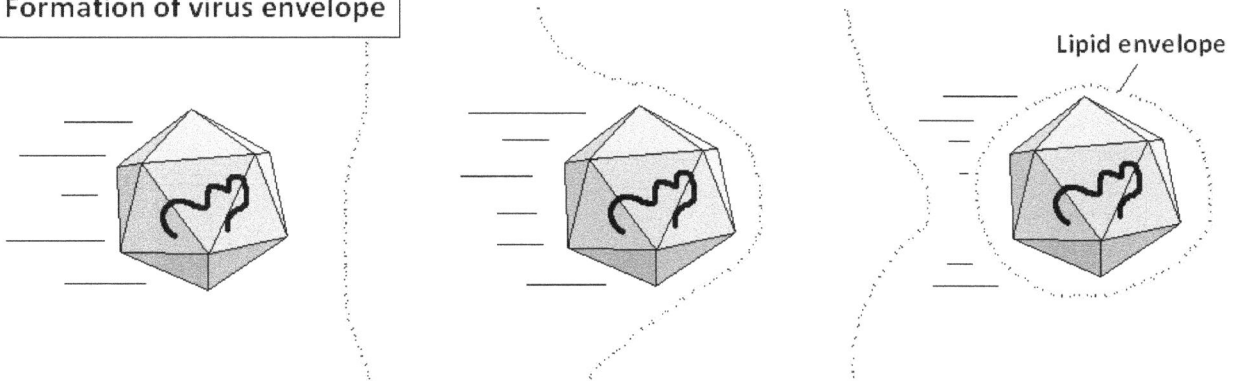

Lipid envelope

Apart from shape, there are two broad types of viruses: DNA viruses and RNA viruses. Each of these can be divided into single-stranded (ss) versus double-stranded (ds), referring to whether or not the genetic material comes packaged as just one strand or as a helix. The shorthand is to apply "ss" or "ds" in front of the virus to indicate the strandedness. For example, "ss-RNA virus" means a single-stranded virus that has RNA for its genetic information. As a rule of thumb, most DNA viruses are double-stranded, and most RNA viruses are single-stranded.

Virus by Nucleic Acid Type

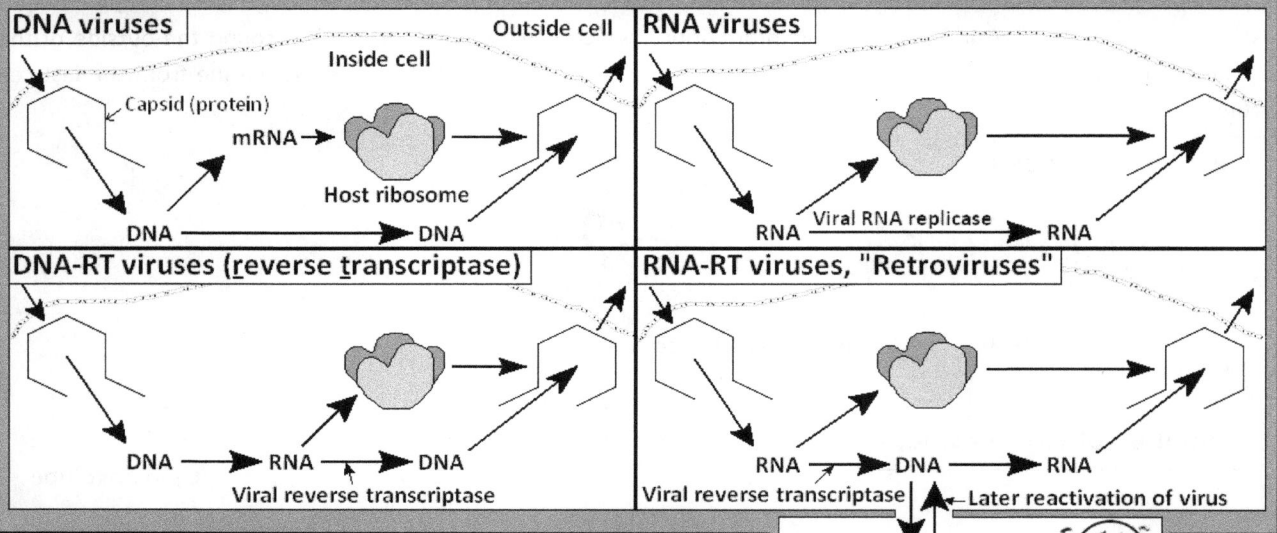

DNA viruses

Outside cell

Inside cell

Capsid (protein)

mRNA →

Host ribosome

DNA ————————→ DNA

RNA viruses

RNA ——Viral RNA replicase——→ RNA

DNA-RT viruses (reverse transcriptase)

DNA ——→ RNA ——→ DNA

Viral reverse transcriptase

RNA-RT viruses, "Retroviruses"

RNA ——→ DNA ——→ RNA

Viral reverse transcriptase | ←—Later reactivation of virus

Host genome integration

Single Strand Virus +/- Naming Convention

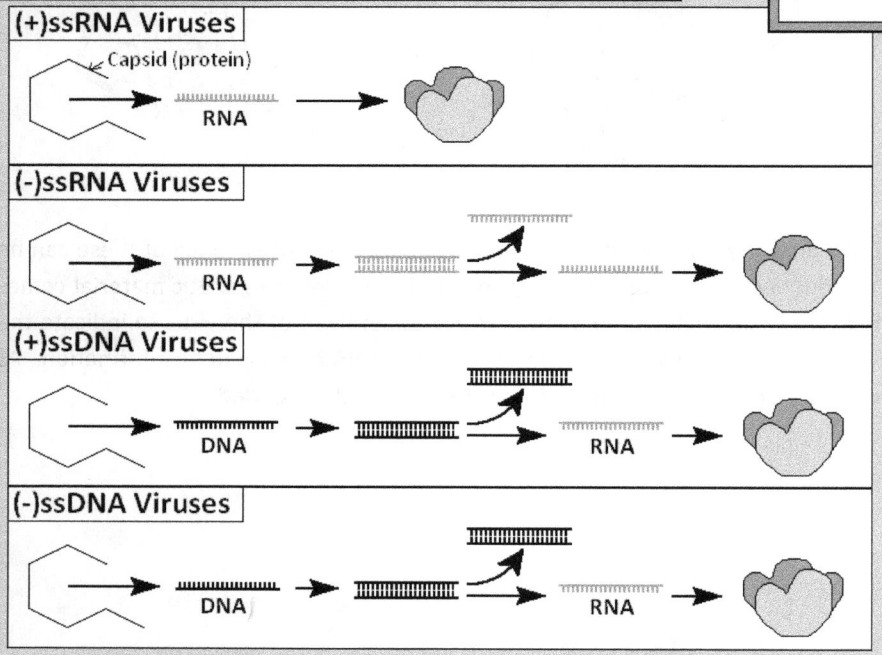

(+)ssRNA Viruses

Capsid (protein)

RNA

(-)ssRNA Viruses

RNA

(+)ssDNA Viruses

DNA → → RNA →

(-)ssDNA Viruses

DNA → → RNA →

Further, single-stranded viruses are given a (+) or (−) designation to signify whether or not they are similar to or complementary to the equivalent mRNA that makes its proteins. For example, if a single-stranded RNA virus busts into a cell, and its genetic strand heads directly to the nearest ribosome, we would say that the strand is equivalent; the virus is called a (+)ss-RNA virus. If, on the other hand, the virus first has to make a complementary strand using a

204

polymerase, and then that newly made strand goes to the ribosome, we would say the virus is (–)ss-RNA. The distinction between the (+) and (–) types is summarized in the diagram on the previous page.

Lastly, as if this were not complex enough, each type of virus (DNA vs RNA) has a subclass of viruses called reverse transcriptase viruses. These viruses make DNA from RNA (the reverse of transcription) as part of their replication cycle. Some of these kinds of viruses are further classified as retroviruses if that resulting DNA is then inserted into the host DNA. Their genetic information can then be reactivated sometime later to start the replication cycle all over again. These kinds of viruses are arguably the nastiest, because once their DNA has incorporated into yours, it's REALLY hard to get rid of them. HIV is known for this, which is why infection with it is – at least with our current technology – a life-sentence of illness.

It's not entirely clear where viruses originated. The current belief is that viruses have one of three possible origins. The first hypothesis is that they may have originally been primitive **bacterial-like parasites** that burrowed into cells. Over the millions and millions of years of life on Earth, these parasites became so dependent on their hosts that they lost their ability to make their own ribosomes and other organelles. They became simpler in design, until only a protein capsule and genetic information was needed in order to hop from one cell to another. Mitochondria and chloroplasts are examples of how this may have happened. These two organelles have a LOT of things in common with bacteria but have lost their independence because they can no longer make the things that their host cells supply to them. Perhaps in the distant future, mitochondria and chloroplasts will become simpler still, until one day they will hardly look like bacteria at all. Who knows?

The second hypothesis is that viruses coevolved with cells. Although we will be delving into this topic much more at length near the end of the book, I will discuss it here in passing. The most primitive form of life is believed to have used RNA as its genetic information (long before DNA). In those times, it is believed that viruses and cells probably didn't look much different from each other. Like my roommate, one lived off the other, and the relationship continues today. The viruses didn't go far in life, while the cellular organisms basically took off for great complexity.

The third hypothesis is that viruses started off similar to entities that we observe in labs every day. These entities are called **transposons**, or "jumping-genes." They have greater relevance than just their *hypothetical* start as viruses, however, so we'll devote an entire section to them.

Transposons

Transposons are sequences of DNA that have the ability to be cut out of one section of DNA and reinserted into another. It's like a person who cuts in line at random, then steps out of the line to pick a new spot somewhere else. These stretches of DNA are bordered by short sequences at both ends, and these ends are recognized by a protein called **transposase** that cuts the sequence out (putting the cut ends back together) and moves it somewhere else in the genome.

Many transposons contain the gene for transposase, making them self-actuating jumping genes. Other transposons do not, and they jump from one location to another simply because *other* transposons with the transposase are in the *same* genome. Still other transposons do not jump at all, but rather are *copied* and inserted into the new location. When the event is done the transposon will exist at both the original *and* the new location.

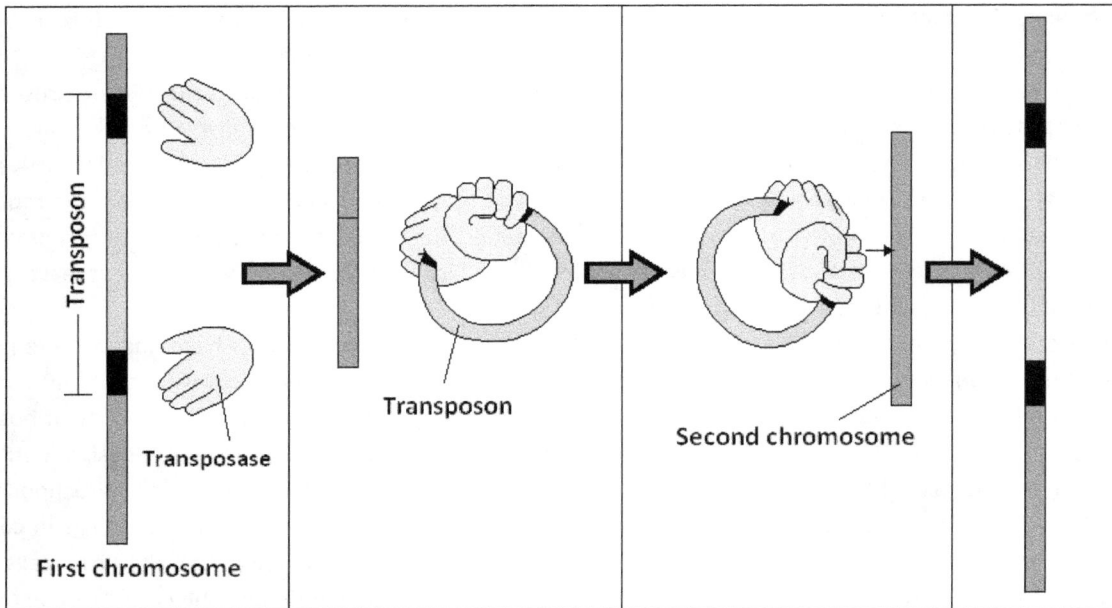

First chromosome

Transposon

Transposase

Transposon

Second chromosome

At any rate, the similarity between transposons and viruses is uncanny: they are both stretches of DNA that contain the information for proteins that allow them to hop from one host to another. The theory is that viruses slowly evolved from these entities. The first form of viruses was little more than a simple transposon. Over millions of years, transposons slowly became more adept at jumping from chromosome to chromosome until they finally developed the ability to go from one cell to another. It's a lot like mankind's progression to the stars. First we had to overcome our fear of gravity. Then we developed balloons. Then we developed planes, then jets, then space rockets, and someday (hopefully soon) we will have bases on Mars. It's all about baby steps. Primitive man was like transposons. We jumped from one place on Earth to another. Modern man is becoming a virus, packaging little parts of Earth together so we can infect another planet.

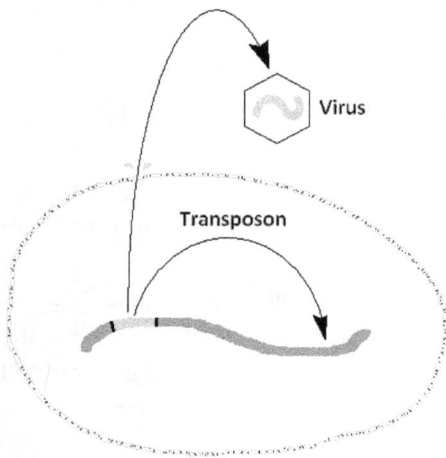

Virus

Transposon

Granted, at this stage it may not seem entirely clear just how a transposon gains the ability to go from jumping between chromosomes to jumping between cells. After all, where did these extra genes come from if they weren't made in a lab somewhere? Well, before we go any further, I must posit a warning: the belief that viruses were originally transposons is just a theory. That said – starting in the next section we are going to begin exploring where new genes come from, and hopefully then these ideas will be a little clearer.

Summary

In this section, we discussed viruses and transposons.

- Viruses are mobile packets of genetic information that invade cells and utilize their protein-making machinery to complete each viral replication cycle. They do not have ribosomes or any other organelles.

- Viruses are comprised of a protein capsule, known as a capsid, genetic material inside the capsid and sometimes a lipid envelope surrounding the entire thing. The genetic information is either DNA or RNA. The lipid envelope is the remnant of the last victim's plasma membrane.

- Both RNA and DNA viruses can be single stranded "ss-" or double stranded "ds-". Most RNA viruses are single stranded. Most DNA viruses are double stranded.

- Single stranded viruses have a designation of either (+) or (−). Positive stranded viruses are similar to the equivalent mRNA that codes for the protein. Negative stranded viruses are complementary to the mRNA that codes for the protein.

- Both DNA and RNA viruses have a subset of viruses called reverse transcriptase viruses, suffixed "-RT." These viruses use a protein called reverse transcriptase to make DNA from RNA. This is unusual in that it is the reverse of what cells do when they make protein. An even smaller subset of RNA-RT viruses are known as retroviruses, because the DNA they make gets inserted into the host's genome for later reactivation. HIV is an example of such a virus.

- Viruses may have one of three proposed origins. 1) Regressive evolution from bacteria-like parasites. Over time, they lost their ability to make their own proteins and have stripped down to the bare essentials in order to go from one cell to the next. 2) Co-evolution with cells since the first days of life. Cells evolved into cells, whereas viruses didn't advance much at all. 3) Viruses may have started off as transposons and gradually evolved to become more complex over time.

- Transposons are segments of DNA that are capable of being cut from one area of DNA and inserted into another. They usually are flanked by distinct regions of DNA recognized by an enzyme known as transposase that is usually encoded by the transposon's genetic information. Transposase cuts the transposon from where it is and moves it somewhere else. Some transposons work by being copied instead of cut, so that over time they increase in number and location.

Unit 26

Mutations

We have talked at great length about how the genetic code is used to make proteins, but a lingering question remains: how did the code get into that sequence? No, it did not happen because of a scientist in a lab. The scientist is only there because his genetic sequence allowed him to live and breathe and work and so forth. Therefore, we can't blame genetic sequence on mad scientists, because it puts the cart before the horse – so to speak. To help answer this question, we will now talk about the most important mechanism that drives the formation of genetic sequences: **mutations**.

Mutations

Mutations are changes in genetic sequence. For example, a change from ATTCG to ATCCG represents a type of mutation. In this case, the mutation is called a **point mutation**, because the change was only at one place in the sequence, here T to C. Other types of mutations exist as well. For example, an **insertion mutation** inserts one or more nucleotide bases into the sequence. So, ATTCG might become ATT[CGAGACCT]CG where the bracketed part represents the nucleotides inserted. Yet another type, **deletion mutations**, does the opposite: it omits X number of nucleotides from a sequence. ATTCG might become A[]G, where the brackets represent the missing nucleotides. Each of these types has numerous causes, some of which we will explore later in this chapter. For the moment, let's focus on the *consequences* of these changes.

If a mutation occurs inside a gene – or other area of DNA serving an auxiliary role – then a serious error could result. The protein that a gene makes might have a different amino acid at some point, or a completely different sequence altogether, causing the protein to behave differently. Almost always, the change causes the protein to behave less well, giving the organism a disadvantage in life. It may even cause the organism to die prematurely due to internal chemical problems. A whole host of genetic diseases fall under this category: cystic fibrosis, sickle cell anemia, Huntington's disease and so forth, to name a few.

Rarely, the mutation might cause the protein to behave more efficiently, in which case the organism has an advantage in life. Such an individual will tend to survive to mate more often than not. Each time it mates, the gene can be spread to more and more offspring. Over generations, the new beneficial gene becomes more frequent until the population at large has been "upgraded" with the good mutation. It needs to be double-stated for emphasis, however, that mutations rarely cause good things to happen. In short, you do *not* want a mutation. The odds are *very* unlikely that it will benefit you.

It may help to consider the following analogy: Imagine that you open a lengthy and well written report that is due in one hour. With eyes closed, you randomly click somewhere in the document and then randomly hit one or more letters on the keyboard. Finally you hit save. What are the odds that this behavior will result in a better report? Conversely, what are the odds that this change will just look like junk in the middle of a well written document? If you repeat this exercise a million times, ONE of those times may result in an improvement to the paper. Otherwise, you probably just dropped the quality of the report. Mutations work the same way.

That being said – let's zoom in on the actual mechanisms by which mutations cause problems. For point mutations, the alteration in genetic code usually results only in the change of a single amino acid, or an **amino acid substitution**, as it is called. This happens because the codons of the mRNA change, so when the ribosome reads the codon it matches a different amino acid than what was originally in that place. Sometimes, this change can insert a premature stop codon instead of an amino acid. In that case, the result is usually disastrous, because the protein is incomplete and totally useless. The image below shows these two possibilities. The first is an amino acid substitution that switches glutamine (gln) with glutamic acid (glu). The second introduces a premature stop codon.

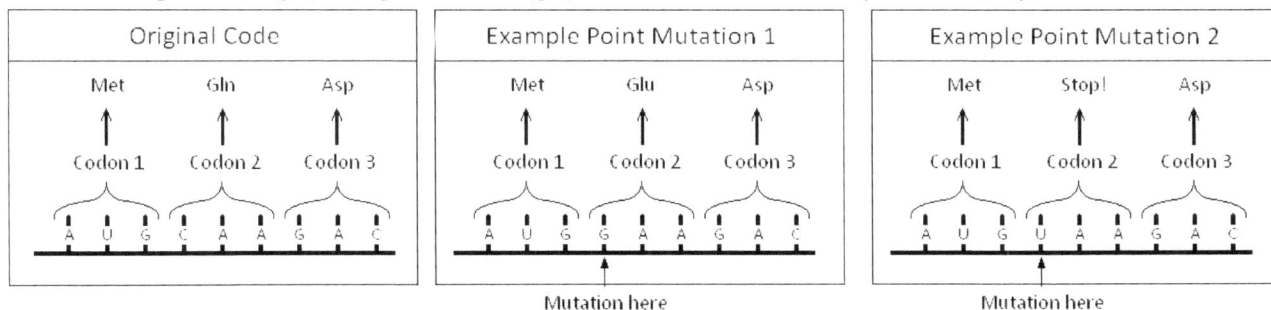

Aside from premature stop codons, point mutations are usually the least alarming of all the types of mutations. For example, if an amino acid substitution switches two amino acids that are of the same class (both acidic, or basic, or hydrophobic, etc.), the effect can often be negligible. If so, the mutation is considered silent, since it doesn't alter the behavior of the enzyme or cause any trouble to the organism. If the substitution occurs at some pivotally important part of the protein, however, the effect of the change could be disastrous. The mutation might shut the protein's activity down altogether, usually resulting in hardship for the organism having the mutation. A common example of this kind of mutation is the one leading to **sickle cell anemia**, where a protein in red blood cells called hemoglobin has a single mutation that switches one of its glutamic acids into valine. This causes the protein to change abnormally under certain physical conditions. When this happens, the red blood cells can get stuck in the small blood vessels of the body, causing that part of the body to begin to starve of oxygen. This simple mutation leads to a lifetime of incredibly deadly, painful crisis and frequent hospital visits. One last comment should be made about . point mutations before moving on to other mechanisms. If the point mutation occurs at the third nucleotide of a codon, the change can often have no effect on the amino acid sequence whatsoever. This is because of the mechanism described by the Wobble Hypothesis. For example, a codon change from CCC to CCA will have no effect, since both sequences code for proline.

Insertion and deletion mutations suffer from the same problems as point mutations in that they alter the codons and thereby introduce the possibility of amino acid substitutions and premature stops. They also introduce another particularly nasty problem called **frameshift errors**. A frameshift error is when a number of nucleotides indivisible by three are introduced into (or removed from) a sequence of codons. When this happens, the nucleotides shift over one or more places, so that the codons are no longer in the correct groups of three as they once were. From that point on, all the codons are read out of frame and each one produces a *completely* different sequence of amino acids. Here is an image to illustrate the concept:

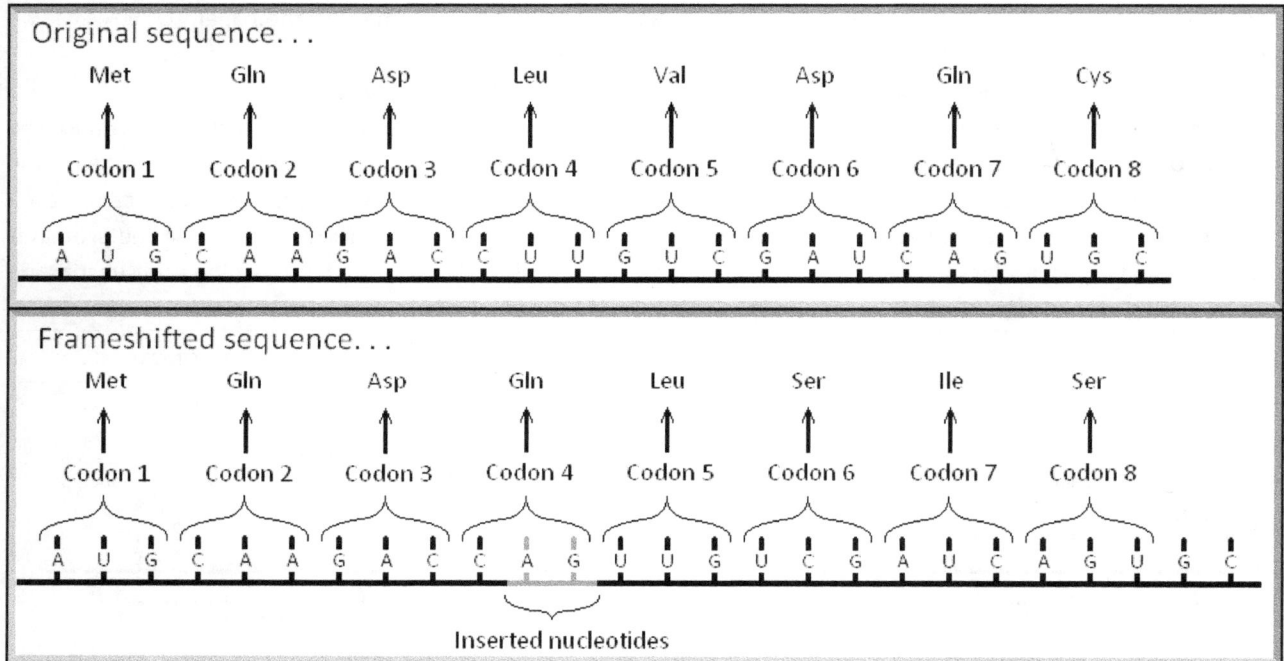

In this image, we see that the first three codons of each sequence code for met, gln and asp, in that order. After that, the frameshift mutation in the second sequence causes all subsequent codons to be read incorrectly, resulting in a completely different protein.

Finally, there is one last mechanism in which mutations can cause changes in an organism. As we saw earlier, not all of the genome is used to code for protein. Apart from the vast amounts of junk DNA, there are also sequences that are used to regulate the rate of transcription of various genes. In other words, these rate controlling sequences control how often some other proteins are made. Let's take an example. Suppose we have a rate controlling sequence that governs how quickly "gene X" is used to make its protein, "protein X." Suppose, for the sake of argument, that the sequence increases the frequency of gene X's transcription. If a mutation occurs in the rate controlling sequence, it will likely interfere with the sequence's job. As a result, gene X will be read less often, and protein X will drop in concentration. The delicate biological balance inside the cell will be destabilized. The cell will become sick and probably die. In the reverse scenario, let us suppose that the purpose of the rate controlling sequence is to slow down the frequency of gene X's transcription. In that case, a mutation in the rate controlling sequence will cause protein X to be made more often, and its concentration will increase. Like before, this can destabilize the cell and lead to sickness and/or death.

To summarize, mutations can lead to changes in a protein's structure or rate of production. In the vast majority of cases, these changes are hazardous to the health of the organism, because the organism depends on a delicate chemical balance in order to exist. Having investigated how mutations result in biological changes, let us now look at what makes mutations occur.

Causes of Mutation

One important source of mutations occurs during DNA replication. The DNA polymerase machinery is not perfect. Sometimes, for example, the DNA polymerase slips a little off the strand it is reading, and when it regains its footing it might not be exactly where it left off. This can lead to gaps or duplication in the new strand. Sometimes a stray H^+ will slip between the base pairs, changing the way they pair up so that DNA polymerase inserts the wrong partner on the new strand. These are just two possibilities. Though there is some proof-reading capability in the polymerase enzymes, the rate of error has been estimated to be somewhere in the vicinity of one in a million base pairs. Human cells have just over 3 billion base pairs, which means each time one of our cells replicates, it experiences about 3,000 mutations! The rapidly dividing organs, such as the lining of the GI tract and blood cells, experience the most mutations, simply because they divide the most. Yep. Just by sitting there, you're mutating. Over the years, these mutations slowly accumulate, and sooner or later they occur in a place that screws up the control of the cell cycle. Once this happens, the cell will begin replicating out of control, becoming a cancer.

Another source of mutations occurs from mutagenic chemicals. While there are huge varieties of such chemicals, several general patterns can be observed. For instance, some are carbon-ring structures that hide in between the ladder steps of the DNA helix, (a process called **intercalation**). By doing so, they interfere with normal binding or appear as just another DNA base. When replication occurs, the polymerase misreads this invading chemical as a base and tries matching it with the "appropriate" base pair. This naturally changes the base sequence.

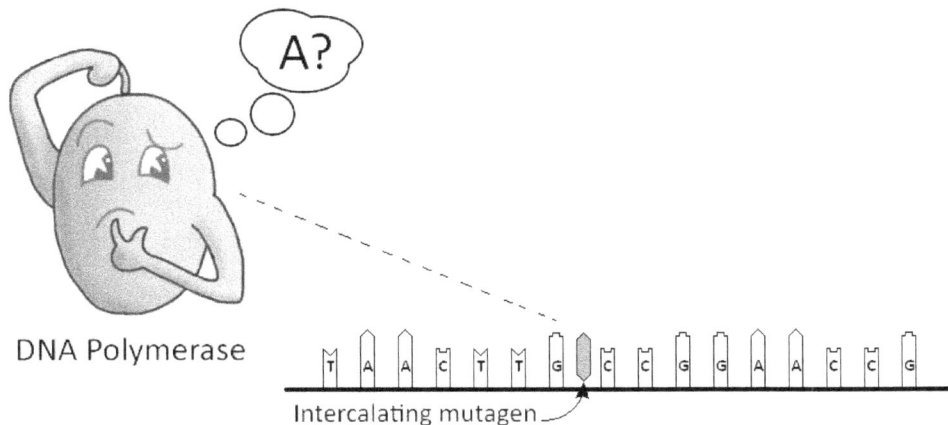

DNA Polymerase

Intercalating mutagen

Alternately, some mutagenic compounds are structurally close enough to nucleotides that they are inserted into DNA at the time of replication. Later, these nucleotides shift their structure so that they look like a *different* base. Once the DNA replicates again, it is matched with the wrong base, causing a mutation. One example is 5-bromouracil, shown here:

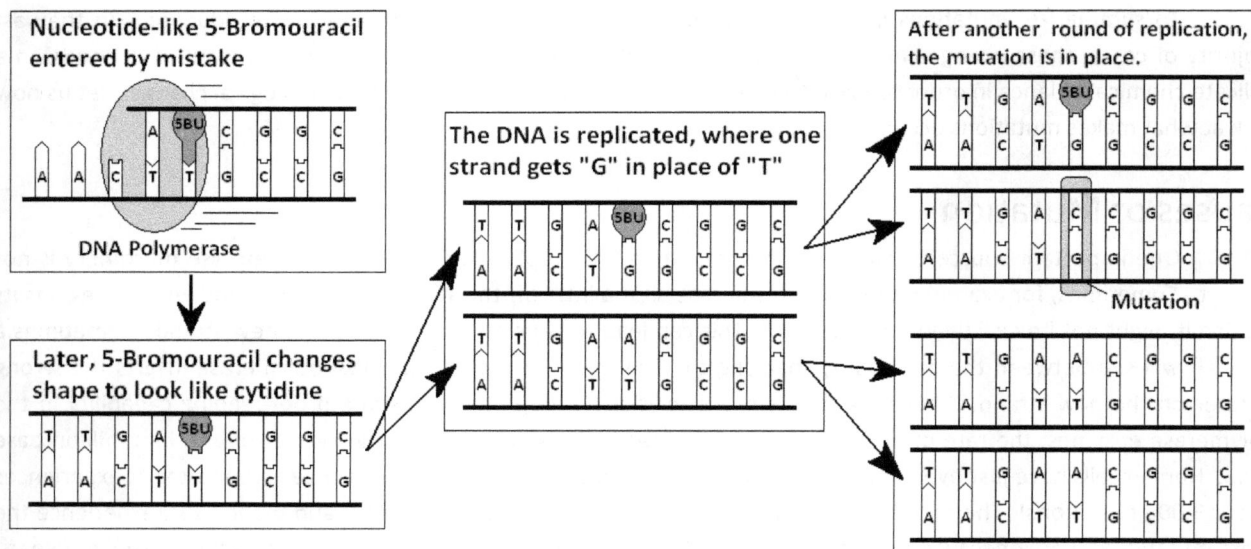

Nucleotide-like 5-Bromouracil entered by mistake

DNA Polymerase

Later, 5-Bromouracil changes shape to look like cytidine

The DNA is replicated, where one strand gets "G" in place of "T"

After another round of replication, the mutation is in place.

Mutation

Other culprits of mutation that you'll likely hear about in popular media are **oxygen radicals**, which cause **oxidative damage** to DNA (and other parts of the cell). The more oxygen that is around, the more it tries to interrupt what DNA is doing (see Unit 19). It reacts with it, changing its structure. This, of course, causes all kinds of chaos for the cell. Cells do their best to repair the damage, but once in a while the damage causes a permanent mutation. Examples of common species capable of producing oxygen radicals include molecular oxygen (O_2), hydrogen peroxide (H_2O_2), superoxide anions (O_2^{2-}), and organic peroxide[52] (ROOH, where R is some arbitrary carbon structure). Some of these radical sources are found in commercial products. Others are just floating around in the air looking for trouble. The first example, O_2, gives us a love-hate relationship with this molecule. Without it, we die. With *too much* of it, or with uncontrolled presence in the body, we get sick. This is a big reason why plants do not horde O_2 after they make it.

Having explored the topic of mutagens, let's move on to the other culprits in the mutation-causing business. The next on the list is radiation, which can take many forms. Radiation is a general term that refers to both electromagnetic radiation and nuclear decay. *Electromagnetic* radiation is the large spectrum of light particles traveling in waves. We discussed these briefly in the sections on photosynthesis. The light that we see is part of that spectrum. *Nuclear decay* was discussed briefly near the beginning of the book when we talked about atoms. It refers to the various pieces that spring out of the unstable nuclei that are breaking down. Think of nuclear decay as a stream

_____

[52] Regular peroxide, HOOH, will also cause oxidative damage.

of "heavy" particles like protons, neutrons, and (oddly enough) electrons,[53] whereas electromagnetic radiation is a stream of light particles.

Either way, both types of radiation can be harmful to DNA. Ultraviolet light, for example, is a form of electromagnetic radiation that causes the nucleotides to rearrange internally, forming a localized abnormality in the strand. Sometimes the cell's repair mechanisms can recognize this damage, but sometimes not. When it goes unrepaired, (or *incorrectly* repaired), a mutation results.

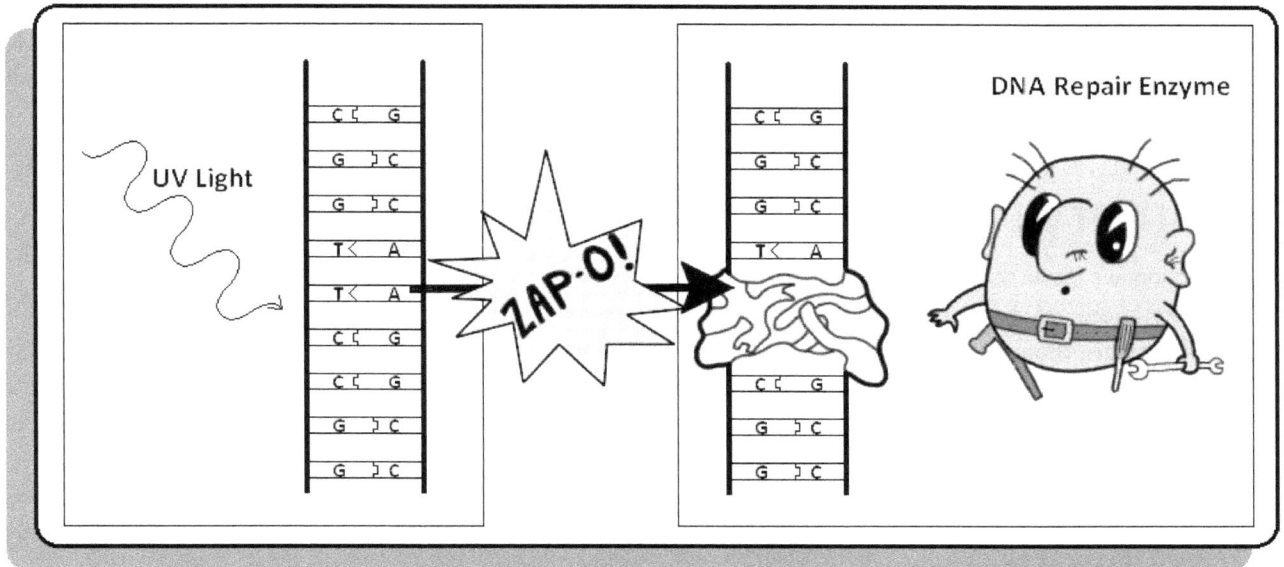

Stronger forms of radiation, like X-rays, gamma rays and nuclear decay, will often splinter the DNA backbone or break off nucleotides completely. Think of these as bullets shot at a rope ladder, tearing it all to pieces. As with UV damage, cells have repair proteins that search for characteristic damage caused by these kinds of radiation, but the repairs aren't always perfect. Once in a while a mutation sneaks in.

As if mutagens and radiation weren't enough, there are also biological causes of mutations as well. The first we'll discuss is due to transposons – those "jumping genes" we discussed in the last chapter. When these genetic drifters hop into DNA, they occasionally jump right into the middle of a gene, interfering with its transcription. Naturally, this changes the protein that is encoded there. Sometimes they insert into the sequence that control the rate of a gene's transcription, speeding up or slowing down the creation of the respective protein as we discussed above.

Another biological cause of mutation is due to viruses, or more specifically *retroviruses*. If you'll recall, retroviruses are RNA viruses that create a DNA template of their information and insert it into the genome of the host. When they do this, it can cause the very same havoc as transposons; they can plop their DNA right in the middle of your important DNA, completely messing up the code that is there, having the same consequences discussed previously.

---

[53] It may seem odd that electrons can fly out of the nucleus, especially since they don't hang out there to begin with. In nuclear decay, however, sometimes neutrons will split in half, producing the positively charge proton and the negatively charged electron. This electron zips off into the distance. Also, protons, neutron and electrons are not the only entities generated by unstable nuclei. Some electromagnetic radiation is emitted from the nuclei. Indeed, this is the reason the sun emits so much light; its atoms are undergoing nuclear chemistry.

Ending it all!

If exposed to too many mutations, regardless of the cause, the cell may begin to function abnormally. In response, these damaged cells may often undergo a deliberate process of suicide, called **apoptosis**. While this may sound counterproductive, the reason for the strategy is quite good. If a cell sustains too many mutations, the chances of uncontrolled cell division (i.e. cancer) increase. Apoptosis, therefore, is a survival mechanism put in place to protect the organism from the cancer-causing effects of mutation.[54] Unfortunately, this defense mechanism is not perfect either, because apoptosis is a process that requires proteins. If a mutation disrupts proteins needed for apoptosis, then the cell will be unable to perform suicide. In other words, if a cancer forms in that cell, it will not be able to kill itself, and the organism is in major trouble.

Summary

In this section, we discussed mutations.
- Mutations are changes in the sequence of nucleotides in an organism's genetic material (usually DNA).
- There are three broad categories of mutations: 1) point mutations, 2) insertion mutations, and 3) deletion mutations. Point mutations are changes in a single nucleotide. Insertion mutations are when one or more nucleotides are introduced into a sequence. Deletion mutations are when one or more nucleotides are cut from a sequence.
- Mutations alter the functioning of a living system by changing the codon of the messenger RNA or altering the rate of protein transcription. If a mutation occurs at the 3rd position of a codon, it may not always result in a change of amino acid type.
- Mutations are considered silent if they result in no change of biological function. Mutations occurring in junk DNA or point mutations in the 3rd location of codons are usually silent.
- If a mutation occurs within a stretch of DNA that controls the rate of transcription of some gene, the mutation may cause the gene to be transcribed at a new rate. It may speed the rate of transcription, resulting in more proteins being made, or it may slow transcription, resulting in less proteins over time.
- The causes of mutations include internal errors of DNA replication and repair, mutagens, radiation, and biological entities such as viruses and transposons.
- Mutagens work by one of several mechanisms: They can 1) intercalate between neighboring base pairs (thereby looking like a base), 2) be structurally similar enough to a nucleotide that they are inserted at the time of replication (and later change shape to look like a different base), or 3) they can directly react with a base to change its structure.
- Radiation is a general term that refers to electromagnetic waves (the light we see being an example) or the particles that are emitted from unstable atomic nuclei. Both types interact with DNA, changing its structure or breaking its backbone. Repair mechanisms are not always successful at reversing the damage, and in those cases a mutation remains.

---

[54] There are other reasons and instances of apoptosis, however, but this is an important one.

- Retroviruses and transposons can insert themselves in DNA. This behavior can result in mutations. If insertion takes place within a gene or rate controlling sequence, the chemical function of the cell will change.
- Mutations are almost always bad. On rare occasion, mutations can result in an improvement to the existing protein design. The rare improvements, likely, are rarely groundbreaking. Over the course of millions and millions of years, the result is a gradual trend toward steadily more efficient genetic designs.

Unit 27

The Cell Cycle

The Cell's Day

In previous sections, we covered how DNA is reproduced. Then we switched tracks momentarily to talk about how protein is manufactured in the Central Dogma of Biology. We also saw how mutations creep into that scene and how viruses take advantage of the entire thing. Now let's have a look at a cell's typical "day." The cell spends its time alternating between two different states: **interphase** and **mitosis**. During interphase, the cell conducts normal business; it makes various proteins and spends at least part of its time replicating its DNA. During mitosis, the replicated DNA molecules are separated from each other, usually into two new cells. At the end of mitosis, interphase resumes and the process repeats. This entire cycle is known as the **cell cycle**, shown to the right.

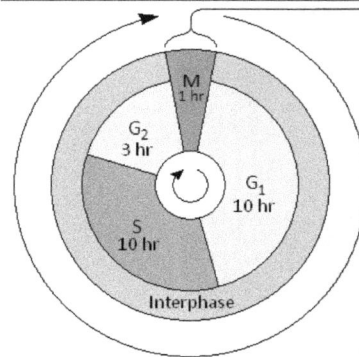

The Cell Cycle with Sample Hours

Prophase
Metaphase
Anaphase
Telophase

M = Mitosis
G_1 = Gap stage 1
S = Synthesis stage
   --> DNA replicated
G_2 = Gap stage 2

As you can see in the image on the previous page, interphase is divided into three distinct stages: G_1, S and G_2. Although the image assigns a certain amount of time to these individual stages, the hours given – though typical – are only samples. The actual time spent in the individual stages is variable and depends on the cell in question.

In the **G_1 stage**, (the "G" standing for "gap"), the cell begins replicating various organelles such as mitochondria, chloroplasts and ribosomes in preparation for mitosis. The length of time that a cell spends in G_1 is extremely variable. At one end of the spectrum are cells that are incredibly efficient at cell division, such as embryonic cells; they spend so little time in G_1 stage, that they seem to steamroll right past into S phase. At the other end of the spectrum are cells such as muscle and nerves that *never* leave G_1. They never divide again. This means that the nerve cells you have at birth are the same you carry the rest of your life. They don't divide to make more, so take care of them or you'll end up like my housemate. It's a terrifying fate, because I've never met anyone else who could come up with an infinite supply of incredibly terrible ideas. It's like the time he convinced me to write a biology book. I'm still paying for that one.

At any rate, after the cell has replicated its organelles, it moves into the S stage, ("S" standing for "synthesis"). This is the period in which the DNA is replicated using their origins of replication that were discussed in Unit 21. Like the G_1 stage, the length of time that a cell spends in the S stage can be variable, although it is not nearly as extreme. By the end of the S phase, the genome of the cell has been copied. Two genomes are now present.

The cell is now ready to move forward into the next stage, G_2, also standing for "gap." In this stage, the cell makes last minute preparations for the much-anticipated eviction of the other genome, bringing us to mitosis.

Mitosis/Division

By the time the cell enters mitosis, the DNA has been replicated; each chromosome has been copied. Here is an image of a eukaryotic chromosome being replicated.

After the DNA is copied, each chromosome has a clone attached to it via a central region called the **centromere**.[55] The centromere is a complex structure of proteins that forms around the same time as DNA

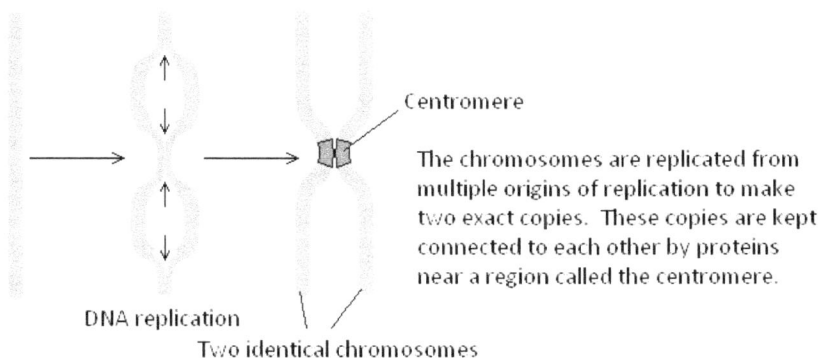

Centromere

The chromosomes are replicated from multiple origins of replication to make two exact copies. These copies are kept connected to each other by proteins near a region called the centromere.

DNA replication

Two identical chromosomes

replication. Its purpose is to keep track of the two copies; soon they will be going their separate ways, so the cell needs to keep close record of them. Let's see how the separation occurs. . .

---

[55] Though this forms an "X" like pair of chromosomes, do not confuse this with the "X chromosome." It is not the same thing. We will encounter the X chromosome in later sections.

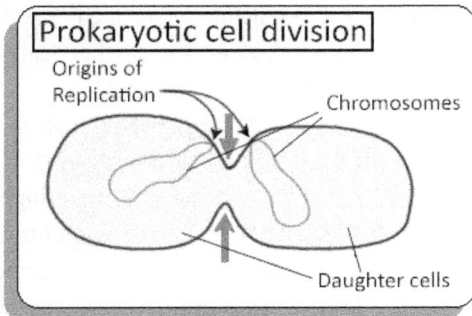

Prokaryotic cell division

Origins of Replication

Chromosomes

Daughter cells

For prokaryotes, this is a simple process: the bacteria have their origin of replication machinery attached to the cell membrane near where the cell will split in half. The two new chromosomes will go their separate ways as the two new cells partition off their membranes (see picture to the right).[56] No centromere is necessary. Interestingly, this is also how mitochondria and chloroplasts segregate their chromosomes, lending more evidence to the theory that they were once free-living prokaryotes.

In eukaryotes, the process of DNA segregation is a little more complex and is broken down into several stages: **prophase**, **metaphase**, **anaphase** and **telophase**. Some books break these stages down into intermediate stages as well, but the general pattern still holds.

During prophase, two **centrosomes** appear near the nucleus and migrate to opposite sides of the cell. The centrosomes are connected to a large network of microtubules that span across the cell, and the star-like structures that they form are called asters. This network begins to align itself based on where the cell

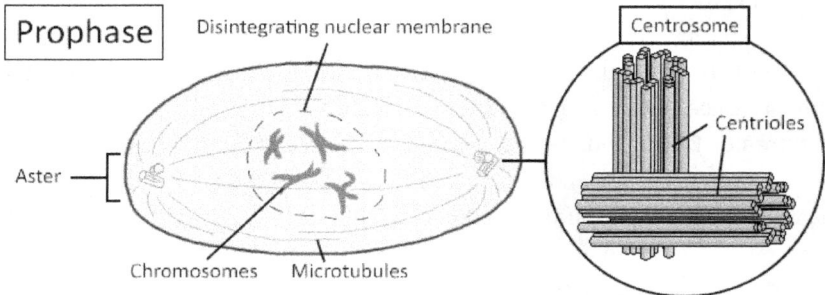

Prophase

Disintegrating nuclear membrane

Centrosome

Centrioles

Aster

Chromosomes   Microtubules

will soon divide. At this point, the nucleus slowly begins to disappear, and the replicated DNA has been packed up very tightly, tight enough that it appears as discrete chromosomes under a microscope. There are twice as many chromosomes as usual, each one copied and attached to its clone via the centromere to form a number of little X figures. The image to the top-right is a simplification.

Near the end of prophase, little protein complexes called **kinetochores** form around the centromeres, (see image accompanying "anaphase" on next page). Once the kinetochores have attached to the centromeres, microtubules from the aster networks begin to connect to the kinetochores. After this connection has been established, the microtubule machinery is now ready to begin the next phase.

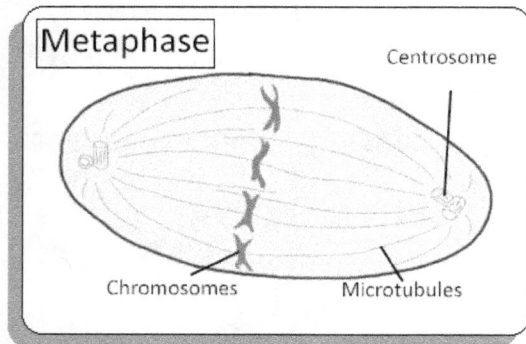

Metaphase

Centrosome

Chromosomes   Microtubules

Next is **metaphase**. During this part of mitosis, the microtubule system pulls the chromosomes onto the line that will become the future boundary between the two new cells.

Next, **anaphase** starts when the chromosomes begin to separate from each other. During anaphase, the microtubule aster systems pull their respective chromosomes in opposite directions. The X-shaped chromosomes pull apart; the individual chromosomes go their separate ways. The next image shows kinetochores grasping onto the chromosomes as they are being pulled. It also shows a simple visual pneumonic for remembering anaphase.

---

[56] I should point out that cell division in prokaryotes is not called mitosis. It is just called cell division. Mitosis is a eukaryotic phenomenon.

218

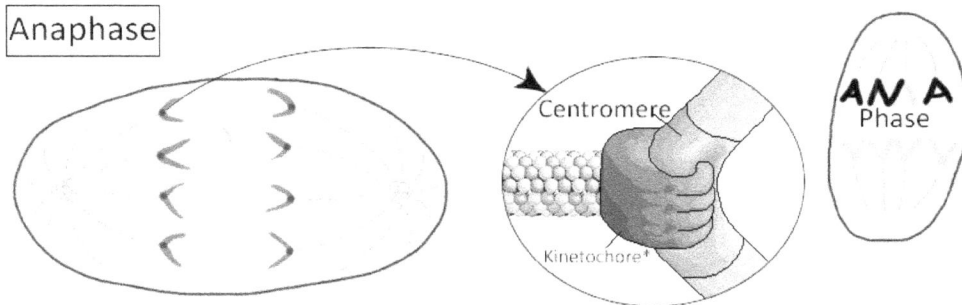

Anaphase

Centromere

Kinetochore*

AN A
Phase

* Kinetochores don't really look like hands; I've drawn them like that to facilitate memory. Their actual structure is quite complex.

Last in the process of mitosis is **telophase**. In this stage, the microtubule aster systems begin to degrade and the cell (usually) starts to divide via **cytokinesis**. The cell nuclear membrane also begins to reemerge.

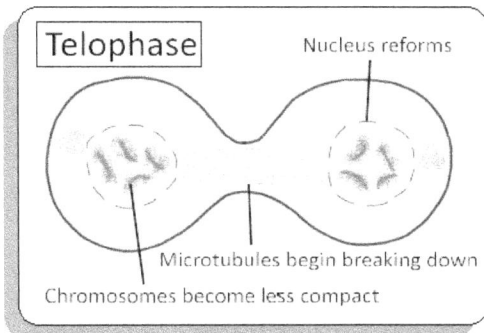

Telophase

Nucleus reforms

Microtubules begin breaking down

Chromosomes become less compact

As it turns out, the process of cytokinesis is quite variable, depending on the cell type and species (see image on next page). In most animal cells, the cell membrane begins to pinch inward, slowly cutting the cytoplasm in half, roughly 50% for each daughter cell. In plants, the cells are surrounded by tough walls, and the division takes place by the gradual creation of a wall between the two cells. This wall begins as small islands along the new boundary, which slowly connect until the two daughter cells are finally cut off. In other species, such as algae, the cell wall is slowly built inward, analogous to the way the cell membrane splits off daughter cells in animals. In some protists and fungae the cytoplasm doesn't divide at all, in which case mitosis is referred to as **endomitosis** (mitosis without cytoplasm division). This results in a cell with multiple nuclei, called a **polynucleated cell** or **coenocytic cell**. Examples of this can also be found in some human cells, such as the megakaryocytes, (which make blood platelets), and skeletal muscle cells.

Various forms of Cytokinesis			
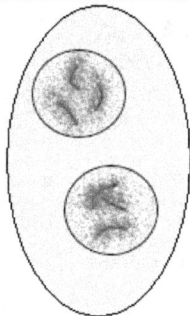	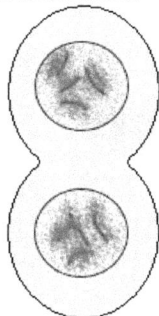	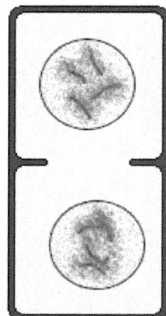	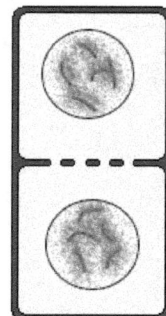
No cytokinesis. Results in polynucleation, seen in some protists and fungae. It is also common in a few types of animal cells. (i.e. skeletal muscle cells).	Membrane pinches in two. Common mechanism for most animal cells and bacteria.	Cell wall pinches inward. Common mechanism for most fungae.	Cell wall pinches inward and also forms as islands along the dividing line. These coalesce into a final dividing wall. Common mechanism in plants.

Regardless of the presence of cytokinesis, at the end of telophase the cell reenters interphase and returns to business as usual.

The Cellular Clock

At this point, a question might be occurring to you: How do the cells control the transition between G_1, S, G_2 or mitosis (M)? This question is an active area of research, primarily because cells that forget how to time these events can begin dividing rapidly, without restrictions. These we call cancer cells. In normal cells, the cell cycle appears to be governed by a protein called **cell-division-cycle** protein, or cdc. This particular protein guards a switch that moves the cell from one stage to the next. Cdc, however, has a baseline state of being inactive. Cdc doesn't lift a finger to do anything unless it has to, so the cell has a mechanism in place that periodically forces cdc to be active. As a cell progresses through the two G stages, it produces little motivator proteins called **cyclins**. There are two types of cyclins: **S-cyclin** and **M-cyclin**. Throughout the G_1 stage, the concentration of S-cyclin slowly builds up until they force cdc from its lazy form to active form. This throws the switch in favor of DNA replication and causes a chain reaction that bumps the cell from the G_1 stage to the S stage. The S-cyclin is subsequently broken down, allowing cdc to be lazy again. At the beginning of G_2 stage, M-cyclin begins to build up. When M-cyclin reaches a critical concentration, they cause cdc to become active again, pushing the cell into mitosis. The M-cyclins are then broken down during the course of mitosis and cdc once again becomes lazy. That in a nutshell is how the cell cycles through its divisions, by slowly making S-cyclins and then M-cyclins that keep cdc on its toes.

In this cartoon, we see two S-cyclin proteins grumbling because they cannot get cdc to throw the cell into DNA replication (S stage). Near the end of G_1, the number of S-cyclins has become so high that the cdc is alerted to throw the switch. A similar cartoon could be drawn for the transition into mitosis. M-cyclins would slowly build up until they become numerous enough to wake cdc, which would throw the switch to enter mitosis.

Summary

In this section, we discussed the cell cycle, including mitosis and interphase.

- Normal eukaryotic cells alternate between interphase and mitosis. The amount of time that is spent in the individual stages varies. Some eukaryotic cells, such as muscle and nerve cells, stay in interphase indefinitely.
- Prokaryotic cells also alternate between interphase and a mitosis-like stage, called (quite simply) cell division. Their division is typically much simpler than eukaryotic mitosis.
- Interphase is divided into G_1, S and G_2 stages.
- During G_1, or "gap 1" stage, the cell replicates various organelles and conducts normal business of protein production.
- During S stage, or "synthesis" stage, the cell replicates its DNA in preparation for mitosis.
- During G_2 stage, or "gap 2" stage, the cell completes final errands necessary to enter mitosis.
- Mitosis is divided into four substages: prophase, metaphase, anaphase and telophase.
- During prophase:
  - 1) two centrosomes (each containing two centrioles) migrate to opposite sides of the cell and serve as centers for the formation of a cell-wide microtubule system. These star-like microtubule centers are called asters.
  - 2) The nuclear membrane disappears.
  - 3) The DNA begins to condense into discrete chromosomes. Each chromosome is attached to its duplicate via proteins connected to their centromere region.
  - 4) Kinetochores form near each centromere, and the microtubules radiating out from the asters attach to the kinetochores.
- During metaphase, the chromosomes are lined up along the middle line.
- During anaphase, the chromosomes detach from their duplicates and are pulled by the microtubule/kinetochore structures toward opposite asters.
- During telophase:
  - The chromosomes begin to unpack, making them seem less discreet.
  - The microtubule aster system breaks down.
  - The nuclear membranes of the two new nuclei begin to form.
  - Cytokinesis (if present) divides the cell.
- The cell alternates from G_1 to S and from G_2 to M by making little proteins called cyclins. These cyclins build in concentration until they cause a protein complex called cdc to start a chain reaction that changes the cell's stage in the cycle. S-cyclins cause cdc to switch the cell into S stage. M-cyclins cause cdc to switch the cell into M stage. Once these stages are accomplished, the respective cyclins are destroyed, and the cell reverts to the next G stage.

Unit 28

The Cellular Basics of Sexual Reproduction

In the last unit, we had a look at a cell's typical day. It spends part of the time doing business as usual and part of the time dividing in a process called mitosis. For any purely **asexually** reproducing eukaryotic species, mitosis is essentially the end of the story for cell division. For **sexually** reproducing species, such as humans like you and me, there is a little more to the saga. As it turns out, there is a small population of cells within your body whose business it is to divide in a very *different* way. These particular cells are involved in the production of **gametes**, the sex cells, and the way they divide is called **meiosis** (pronounced "my oh sis."). This particular process shares many features with mitosis but has a few important differences as well. A study of these differences will lead us to a vastly higher understanding of biology in general, because the process of meiosis is common to a great many organisms on this planet, (not just humans), and is an important part of the stage upon which we study heredity and evolution. To tackle the learning of these topics, we'll need to back up a little for a quick discussion of the genome and how it's organized.

The Homologous Chromosome

For all practical purposes, eukaryotic cells can be divided into two groups, **haploids** and **diploids**.[57] Diploids have a special feature built into their genome: **homologous chromosomes**. These chromosomes are paired, each

---

[57] Technically there are also triploids, quadruploids, and others, but these are *way* outside the scope of this book and are infrequent enough to be viewed as the exceptions to a general rule. Regarding these exceptions now would only serve to guarantee confusion.

having similar gene layout and function. Let's learn by example. Suppose we are examining a diploid genome, and one particular chromosome in it has a gene starting at nucleotide 10,355 and is 1200 nucleotides long. Further suppose that the protein made from this gene affects the color of the organism's eyes, say making them blue. The homologous partner chromosome would *also* have a gene positioned at nucleotide 10,355 that is 1200 nucleotides long, and the gene at this location would make a protein that also affects the color of the eyes, although the color does not have to be the same. For example, the eye color from the homologous chromosome might be brown. Naturally, you may wonder which chromosome gets to color the eyes, which is an excellent question. We'll be covering that topic in the upcoming chapters on heredity. For now, the important thing to understand is that in diploids, there are two genes for every function and that at every point where there is a gene on one chromosome, there will be a gene at the equivalent point on the partner chromosome. Haploid cells do not have the "extra" genes.

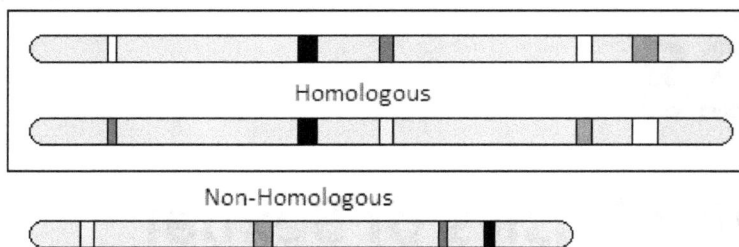

Homologous

Non-Homologous

Here are sample genomes from generic diploid and haploid organisms.

Diploid Nucleus

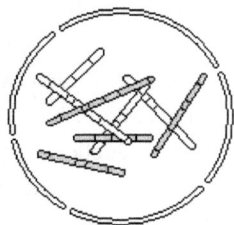

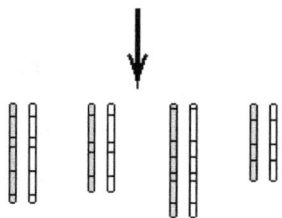

Haploid Nucleus

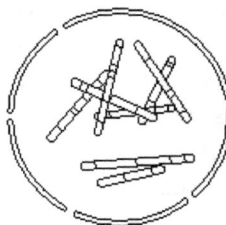

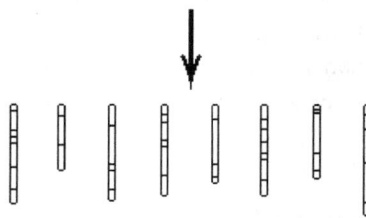

Paired chromosomes

Nonpaired chromosomes

(The coloring is purely for illustrative purposes. Chromosomes do not have color.)

So far the punch-line is simple: in diploid organisms each chromosome has a partner that is strikingly similar. This little feature lies at the heart of the differences between meiosis and mitosis. If you'll recall, during mitosis the replicated chromosomes remain attached to each other by their centromere regions, forming little X-structures. The purpose is to ensure that both chromosomes go their separate ways into the two new nuclei. In meiosis, as we will see, four new cells are created through two division cycles, and a similar strategy of keeping chromosomes attached to each other is employed to ensure that each of the replicated homologous chromosomes go their separate ways.

First look at Meiotic Divisions

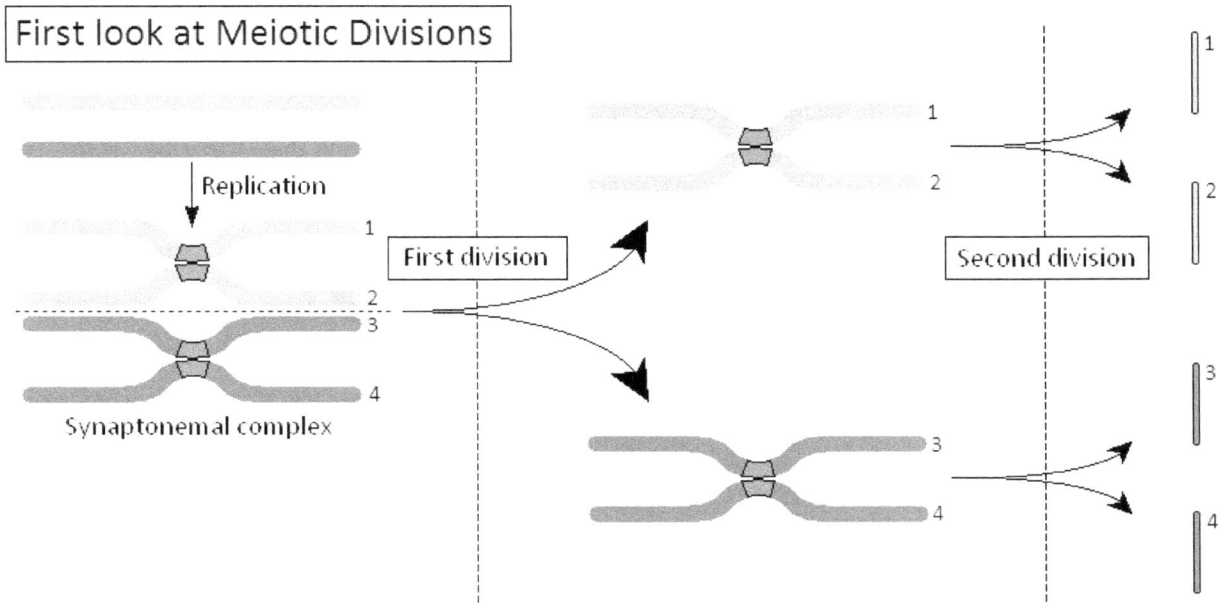

To facilitate this goal, the replicated chromosomes in the X-structures[58] are saddled up next to each other to form a **synaptonemal complex**,[59] as shown above. Meiosis proceeds with two mitosis-like rounds of division. During the first division, the synaptonemal complex is split down the middle, so that each X-structure goes its separate way. During the second round of division, the individual X-structures are split, just as they are in mitosis, so that the individual chromosomes go independently. When all is said and done, the little trick assures that each homologous chromosome has now been sent to a different new cell.

As it turns out, there is more to the story . . . a very important part, in fact. During the construction of the synaptonemal complex, the homologous chromosomes will often switch parts of their DNA with each other. In this process, the DNA of the two chromosomes line up perfectly, then switch parts as illustrated by the following graphic.

---

[58] Once again, please do not confuse this with the "X" chromosome. The X chromosome is unrelated to this particular phenomenon. We will learn about the X chromosome later.

[59] There is a great deal of protein involved in the formation of the synaptonemal complex. These are not shown in the images to maintain simplicity.

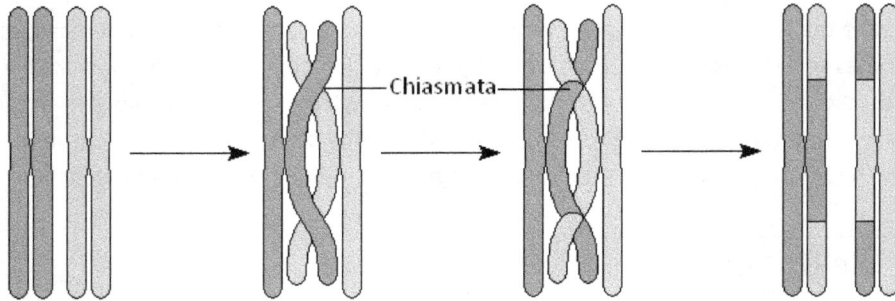

This is called **crossing over** or **recombination**, and the actual spots where the chromosomes cross are called **chiasmata**. Generally, chiasmata usually occur between gene locations and not within them. In other words, DNA typically does not cross over right in the middle of a gene, but rather in a junk region.

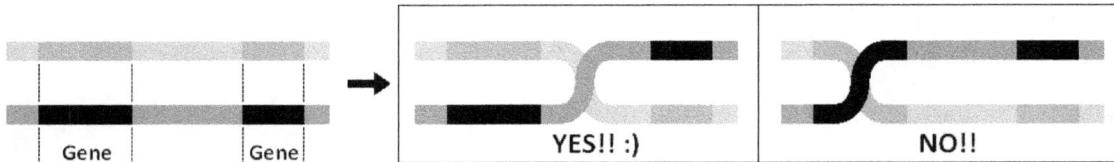

Taking recombination into consideration, an updated view of the meiotic division process would look like this:

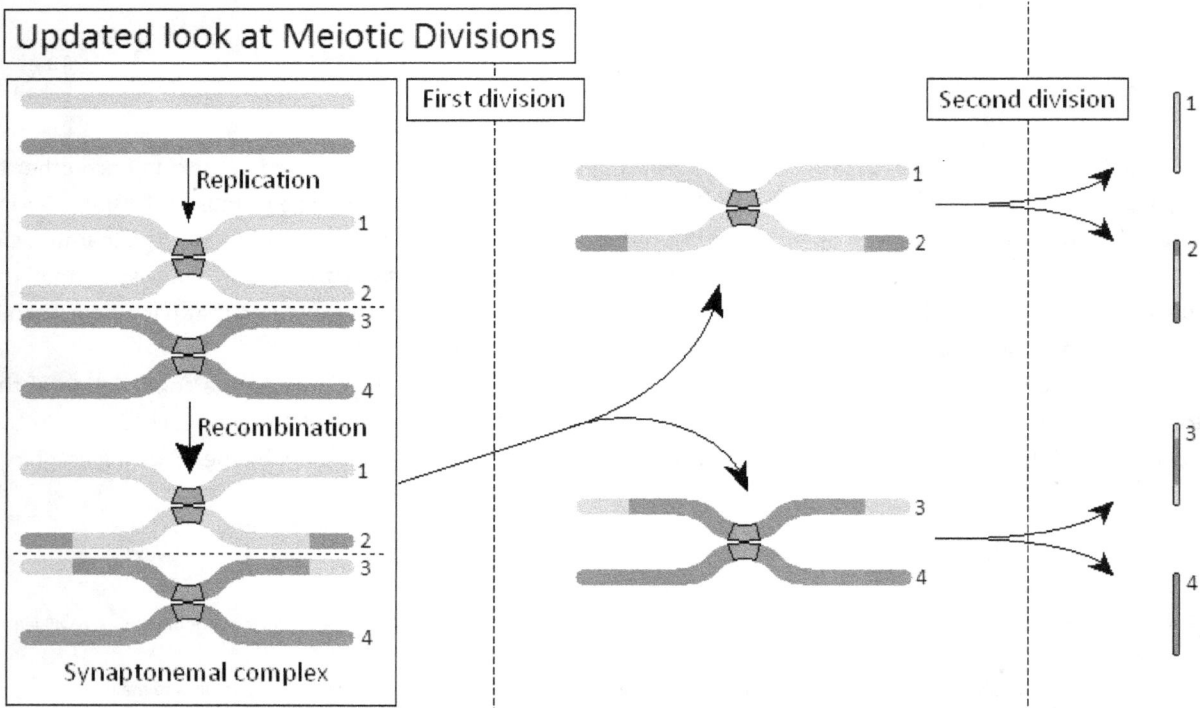

The above images show only the nearest two homologous chromosomes exchanging the tips of their chromosomes. The light gray and dark gray chromosomes on the outside of the synaptonemal complex can recombine with each other as well. Taking all this into account, the number and positions of the recombination events is variable and essentially impossible to predict. As such, the above image represents only one possibility among the nearly limitless number. Here are a few more possible outcomes:

Examples of other possible recombinations. . .

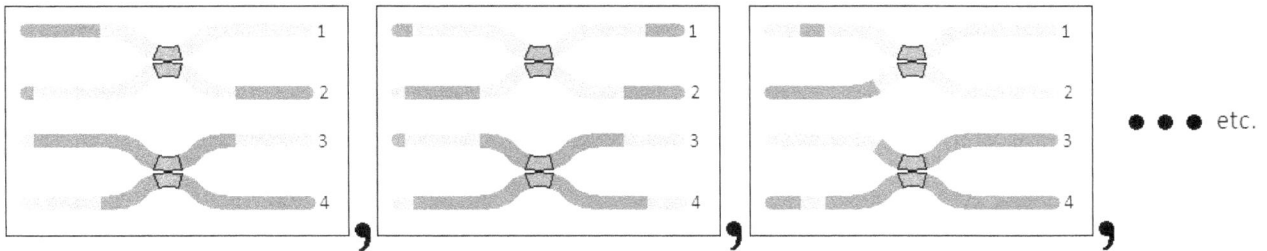

●●● etc.

As you can see, there are many ways to duplicate, recombine and separate homologous chromosomes, and as we'll discuss more near the end of this unit, it leads to an unbelievable spectrum of diversity amongst a population of individuals. Before we pick that discussion back up, however, let's have a closer look at meiosis itself.

Meiosis

In the last chapter, when we studied mitosis, we saw that it involves four broadly defined stages: prophase, metaphase, anaphase and telophase. Meiosis follows a similar pattern, but division happens *twice*. It starts with one "parent" cell that has duplicated its chromosomes and undergoes *two* mitosis-like divisions to create four "child" cells. The business of homologous chromosome recombination takes place during the first division. The combined process is shown on the following page.

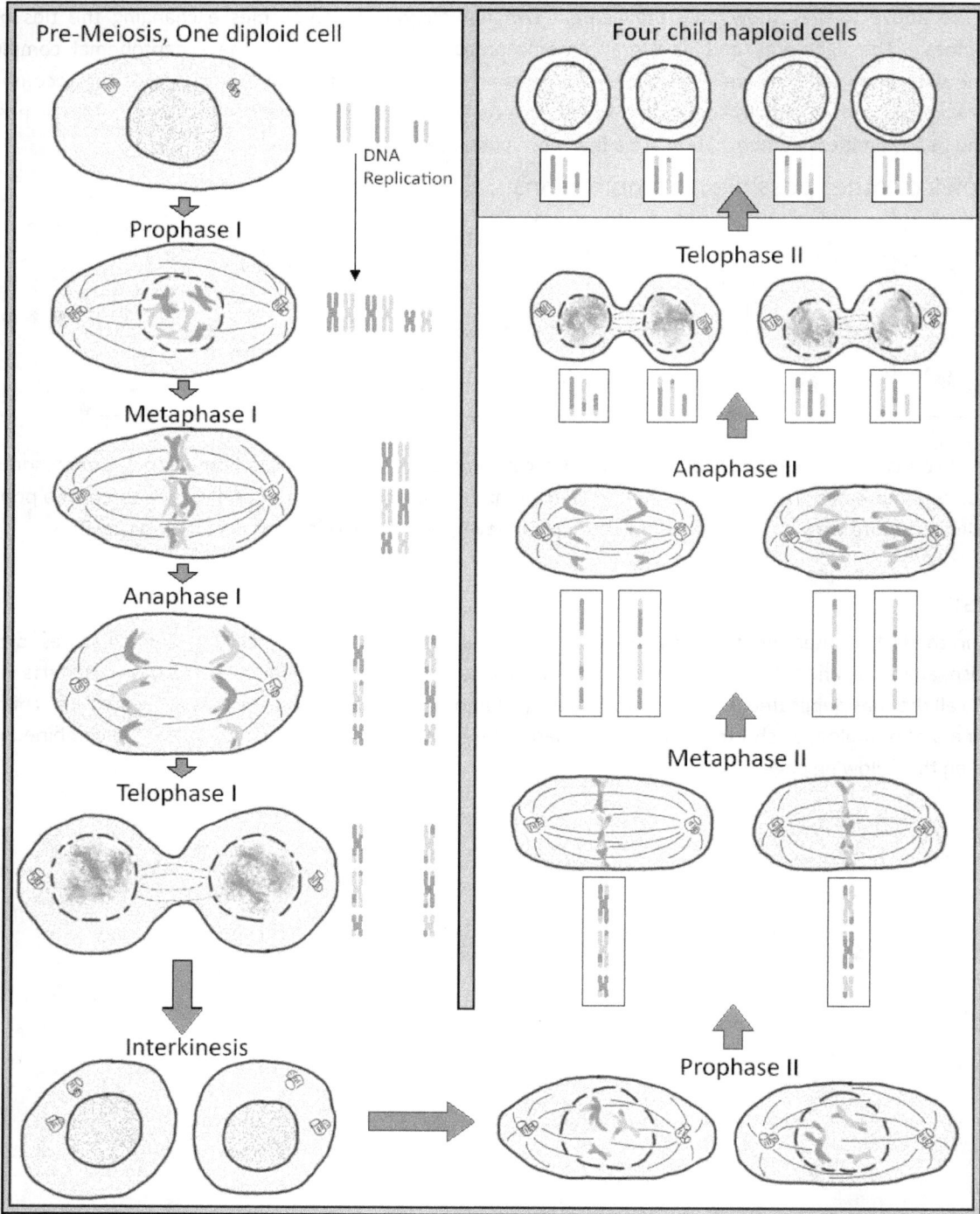

Pre-Meiosis, One diploid cell

DNA Replication

Prophase I

Metaphase I

Anaphase I

Telophase I

Interkinesis

Four child haploid cells

Telophase II

Anaphase II

Metaphase II

Prophase II

As the previous image shows, the process starts with a diploid cell that has replicated its chromosomes. These chromosomes undergo recombination around the time of metaphase I. After the completion of telophase I, the cell has split into two smaller cells, which are now both haploids. It can be difficult to understand why they are considered haploids, however, so look at the arrangement of chromosomes that arises at this stage. Though some duplication and recombination has taken place, both cells basically have only one representative from each of the original homologous pairs. Each exists in their duplicated state and tied to their duplicate via their centromere. After a brief intermission-like **interkinesis** stage, (during which the centrioles and other organelles replicate), the second division proceeds and these duplicated chromosomes go their separate way. By the end, the process results in four child cells, which then go on to further modification to become gametes.

Diversity of Gametes

As we saw earlier, each individual homologous pair can recombine a vast number of ways, creating a nearly endless spectrum of possible outcomes. We considered only a few of these possibilities before we looked at the picture of meiosis. Diploid cells typically have many more than just one homologous pair, however, and each pair is recombining independent of the others. That means the number of possibilities increases many-fold. Suppose, for example, that one homologous pair can recombine 1,000,000,000 different ways (a ridiculously low number compared to the *actual* number of ways they can recombine – but no matter). If a second pair can recombine 2,000,000,000 ways, the two pairs can recombine a total of 1,000,000,000 x 2,000,000,000 = 2,000,000,000,000,000,000 ways. That is only with two homologous pairs! Many diploid cells have ten or more homologous pairs. Humans have 23 homologous pairs! At this point, we have a massive number of ways we can recombine DNA. It gets worse though. . .

If you look back at anaphase I, you'll notice that each homologous pair splits apart independent of each other. For example, the dark gray chromosome from the first pair goes left and the right gray one goes right. In the second pair, the dark gray chromosome goes right and the light gray one goes left. The number of ways this **independent assortment**[60] of chromosomes can occur is 2^X, where X is the number of homologous pairs the diploid cell has. If the diploid cell were from a human, the number of ways would be 2^{23}. The statistics technically get more complex as we go, since many of these "new" combinations are already covered with our first set of numbers, so we'll leave it to the mathematicians. When all is said and done, however, the point is that you can make an essentially limitless number of possible gametes. The odds are so staggering that in all likelihood NO two gametes will be the same.

When reproduction occurs, one gamete from a male organism (usually called a **sperm**) will join with one gamete from a female organism (usually called an **egg**). If both the male *and* female can each produce a practically limitless combination of gametes, the total possible number of ways to combine the two gametes is squared. At this point, the argument has hopefully been driven straight home: No two siblings are identical.[61] This is the advantage of

---

[60] The fact that each chromosome pair separates independent of each other is called the Principle of Independent Assortment

[61] What about twins? Well, twins, triplets, quadruplets, etc., result from abnormal divisions of the fertilized egg. This takes place *after* the sperm has already united with the egg, which means that each duplicate individual receives the same combination of DNA. This is the reason all such siblings are "identical." Otherwise, for standard siblings, the odds of two being the same are so remote that it is pointless to even consider. On the other hand, I feel inclined to point out that twins, triplets, quadruplets, etc. are not technically identical either – only their DNA is. If you take microscopic samples of any tissue from their bodies and compare them, you'll find vast differences. On the macroscopic level, they look practically identical, but that is because our vision is not very discriminating.

sexual reproduction. The number of possible outcomes is vast. If organisms only reproduced by mitosis, they would only replicate the same old boring genome each time. What happens if the environment changes and that particular genome is no longer able to keep up? The entire line of organisms might die off.

That is not to say that some organisms don't survive by asexual reproduction. Bacteria are notorious for reproducing asexually. They just divide in half and do it so rapidly that they flood the environment with themselves. Take your hand and touch practically any object near you; the odds are good you just touched a LOT of bacteria. Bacteria populate areas very easily, but die just as easily too. Take an antibiotic and bomb them with it, and more often than not the entire colony will get wiped out. Rapid clone-producing division grants the advantage of rapid environment invasion, but it also leads to the disadvantage of being poorly adapted when the environment changes.

Let's take a moment to toss the effects of mutation into this argument. As we saw above, meiosis usually does not swap DNA in the middle of genes. As a result, meiosis shuffles around existing genes, but usually does not create new ones. Mutations introduce the changes required to make new genes. Now, if we add in the basic rate of mutation that we saw several units ago, (the one in a million nucleotides per division), we increase the genetic variability even more. Therefore, sexual creatures achieve incredible variability from meiosis and mutations, whereas organisms such as bacteria rely mostly on mutations for their variability, (although in the next chapter we'll look at a few other mechanisms they use to get around this limitation). For now, let's look at an example of meiosis.

Example of Meiosis: Mammals

It may be instructive to consider the most obvious example of meiosis, that which occurs in human beings. The process in humans is sufficiently similar to many animals (especially mammals) that it can be widely generalized. In males, meiosis occurs in the **seminiferous tubules** of the **testis** (i.e., **testicles**), and the result of this meiotic process is the **sperm** cell. The generation of sperm cells is called **spermatogenesis** and begins with a diploid **primary spermatocyte**. The first round of division, (meiosis I), produces **secondary spermatocytes**. Meiosis II produces four haploid **spermatids**, each with a completely unique DNA combination. These spermatid cells mature into **spermatozoa**. They have a long flagella tail, a single mitochondria wrapped around the junction of the tail, and the cell body with a special cap called the **acrosome**. This cap is full of enzymes that help the sperm to penetrate the egg. The process of spermatogenesis occurs from the onset of puberty and continues until death.

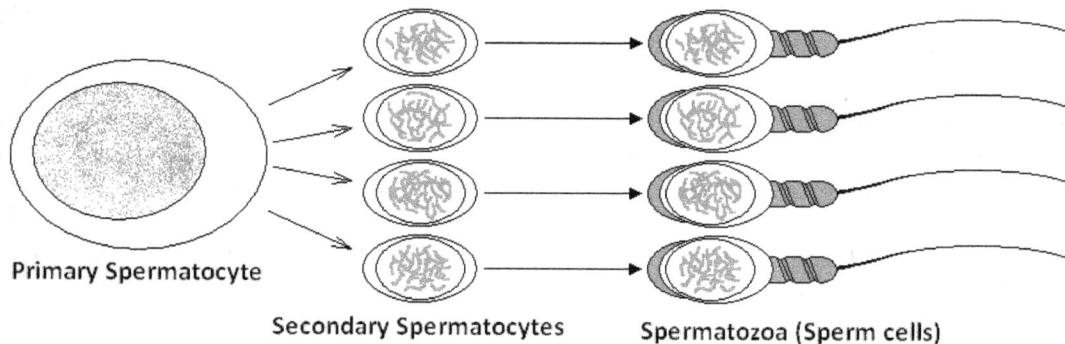

Primary Spermatocyte

Secondary Spermatocytes

Spermatozoa (Sperm cells)

In females, meiosis occurs in the ovaries, and begins shortly after their formation in utero (i.e., while still a fetus in the womb). The initial cells that start the process are called the **primary oocytes**, which are diploid. They pause just after completion of the prophase I and remain at that step until the onset of puberty. At that time, each month one primary oocyte within the two ovaries is chosen alone to complete meiosis. The chosen cell then resumes meiosis I, but unlike spermatogenesis, the resulting daughter cells are not the same. One cell is called the **first polar body**, while the other is the **secondary oocyte**. The secondary oocyte continues through meiosis II, once again producing two uneven daughter cells. One is called the **ootid**, which ultimately matures into the **ovum** (egg). The other cell becomes a **second polar body**. The first polar body sometimes completes meiosis II as well, becoming two secondary polar bodies.

This entire process of ovum production is called **oogenesis**, and is outlined here:

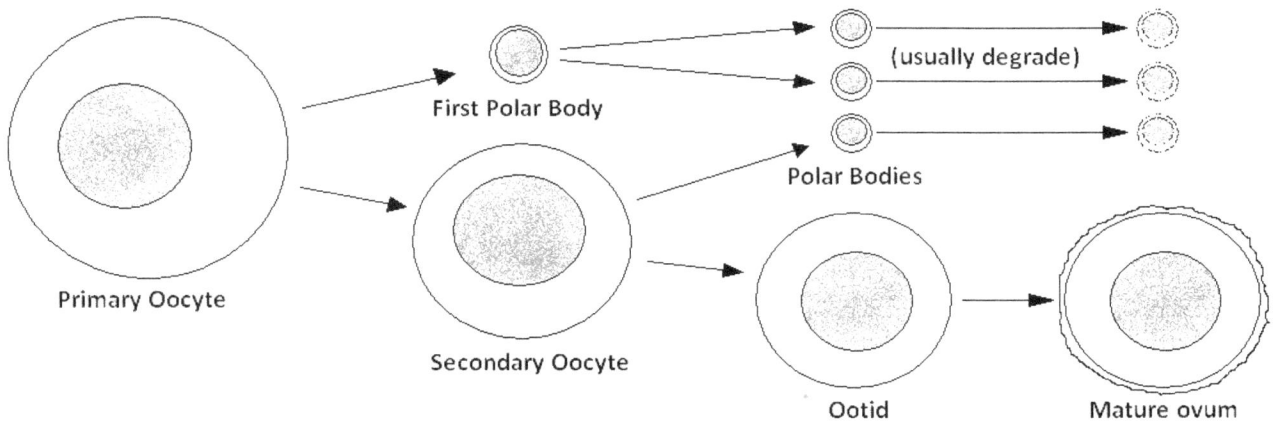

As you can see from the diagram, one primary oocyte produces one ovum and three secondary polar bodies. Why doesn't it just make four oocytes? The answer to that is a practical one. Fertilization occurs when the sperm enters the egg, and the first few events that follow take place within the volume of the egg. Eggs, therefore, need to have as much volume and material as possible to maximize the chance that fertilization will succeed. So, one egg ends up hogging all the stuff from its runt sister cells, who each become tiny secondary polar bodies. There doesn't appear to be much purpose of polar bodies, and they usually disintegrate after a short time. That's life, I guess. The favorite sibling gets all the inheritance.

Fertilization and Life Cycles

When the sperm and egg unite, their haploid nuclei combine to create a new diploid nucleus, marking the start of an organism with genetic information that differs from that contained in the cells of either parent.

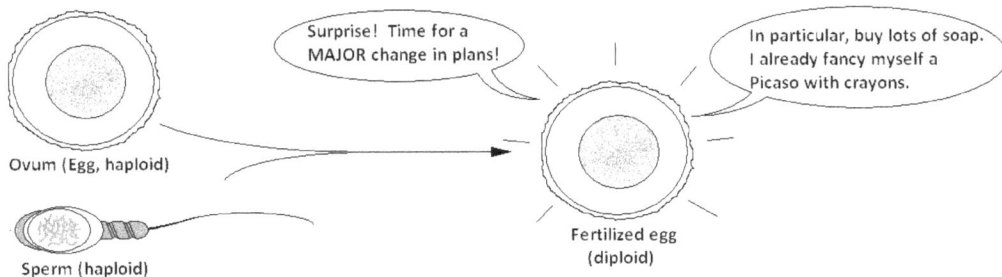

231

We are now prepared to examine a more global understanding of life by realizing that it is essentially the continuation of a cycle. Life within the human species – like most animals – is spent primarily in the diploid "individual" form. A small part of the time is spent in haploid form as gametes. In many advanced plants, the pattern is the same. Comparing animals and advanced plants with progressively more primitive plant forms, we could draw the life cycles like this:

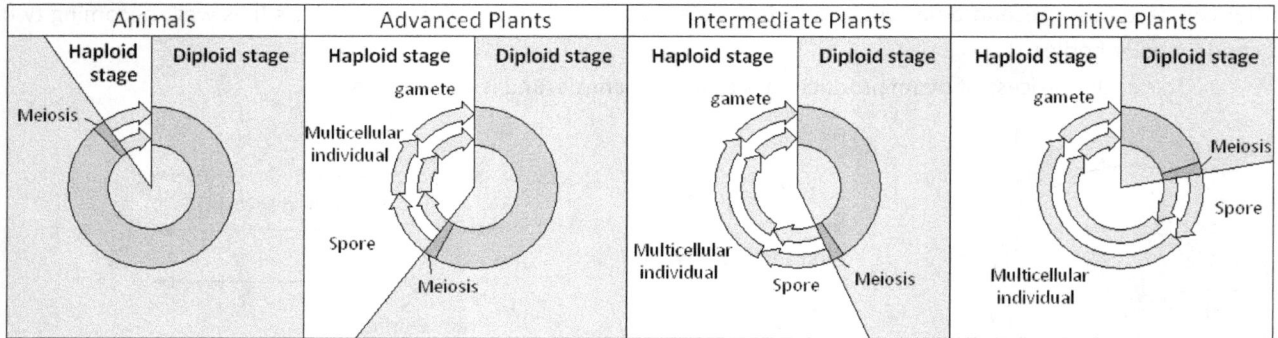

The trend for plants is rather clear: the further you go back, the more dependence there is on the simpler haploid form of existence. How do we know this, and does the trend exist for animals? Well, we know about intermediate and primitive plants because of similarity with fossils. For example, ferns are not much different than their fossil forms; they've been around with little variation for almost 400 million years. The modern day ferns live about half their lives as small multicellular haploids, called **gametophytes**. These create gametes, which then join to make the diploid fern. The fern then creates spores through meiosis, and the spores reform the gametophytes. At that point, the cycle begins again. The same thing is true for primitive plants. A few representatives are still around today, and they spend very little time as diploid individuals. So, what about animals? Well, practically all animals have diploid life cycles. They partake in a brief haploid stage for the purpose of reproduction and that is all. Even the reptiles and fishes, arguably similar to the fossil representatives, live their lives mostly in diploid form. On the other hand, if you go back far enough – before the dinosaurs, before things walked on land, before fish in the sea, you see fossil forms that get progressively smaller and smaller. The trail of plants and animals begins to converge into simpler paths that are not unlike each other. If plants were mostly haploids back in those days, one can guess that animals began with similar life cycles as well. So, in a global sense, the pattern appears to hold. Life started simple and slowly got more complex.

In line with the evidence, the theory goes that life started haploid and later became diploid, but why would life need to make the switch? The answer is not entirely certain, but one thought is that diploid organisms have a more secure genome; they have two versions of each chromosome. If one copy breaks down for some reason, they have another copy to rely on. Haploids do not. If their copy breaks down, they are done.

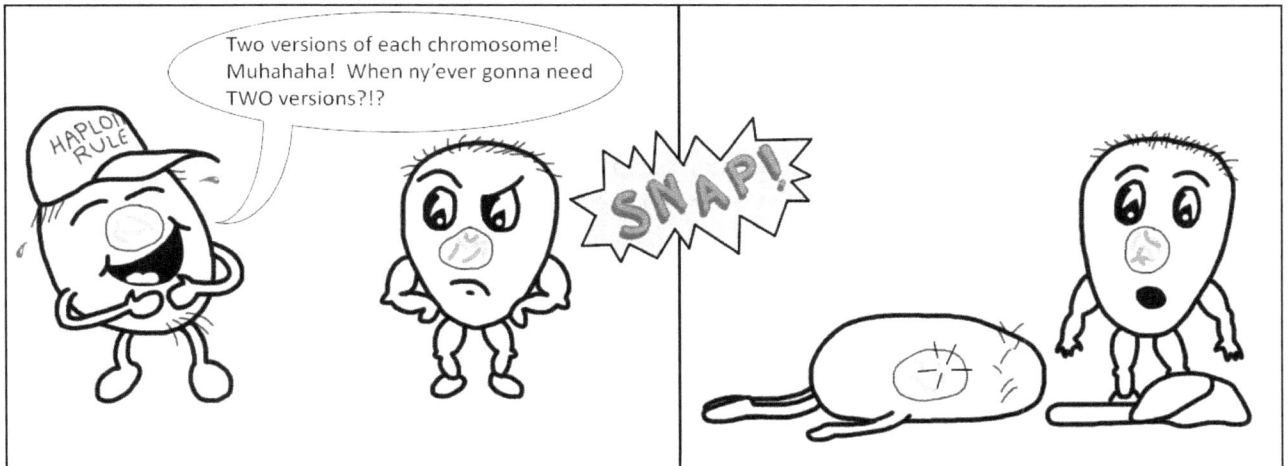

Plants have an easier form of life than animals do. After all, they do not need to be concerned with all the thermodynamic and engineering difficulties associated with movement. They don't need heightened senses in order to detect resources and avoid getting chased down by some nasty critter. Plants just sit there, soaking minerals from the ground and catching rays from the Sun. If an animal eats some of their leaves, it's no big deal; they just make more leaves. In that sense, it isn't hard to see why some forms of plants are still around that can make it as haploids. With animals, however, the slightest genetic problem can create all kinds of havoc. It definitely helps to have an extra version of your chromosomes hanging around.

Summary

In this section, we discussed meiosis.

- A homologous chromosome is one that has a partner chromosome that has a similar genetic layout and length. The gene appearing in a particular position will have the same general purpose, (i.e. eye color, hair texture, whatever), in its partner chromosome. While the homologous genes have the same general purpose, they do not always result in the same effect. For example, one might influence the eyes to be blue. The other might influence them to be green. How color is decided is the topic of future chapters.
- Non-homologous chromosomes have completely different genetic plans. Their genes do entirely different things and are positioned at different places.
- Organisms whose chromosomes all have homologous partners are called diploids. By comparison, organisms with no homologous chromosomes are called haploids. Diploid genomes are believed to be more stable because they have two "versions" of each chromosome.
- Meiosis is a cell division process that results in four unique daughter cells. It takes place in two distinct mitosis-like divisions.
- During the first division of meiosis, specifically around metaphase I, the homologous chromosomes pair up in synaptonemal complexes and swap portions of their DNA. This is called crossing over or recombination. By

the end of telophase I, the homologous chromosomes are each separated into two new cells, which are haploid.

- During meiosis 2, the duplicated DNA chromosomes are segregated into four total haploid child cells. In the male, these four cells mature into sperm cells. In the female, only one of these four matures into an egg. This egg inherits the majority of the original cytoplasm and materials. The other three sister cells, called polar bodies, inherit very little cytoplasm. The polar bodies eventually die off, with little known purpose.

- The number of ways that DNA can be recombined during meiosis is vast. This contributes to the wide variety of genetic combinations found in sexually reproducing species. Asexual organisms do not benefit from this genetic variation and depend on mutations and a few other minor mechanisms to increase their genetic variation.

- After the sperm and egg unite, the resulting cell becomes a new diploid individual, and for the initial stages that follow, this new individual stays the same size as the egg, utilizing the materials inherited by the egg.

- Most modern plants and animals spend the majority of their life cycles as diploid organisms, with very little time spent in the haploid form. The more primitive forms of plants spend a larger amount of time as haploids, so that they can appear as two entirely different physical forms. Ferns are an example, in which their haploid stage is referred to as a gametophyte. Primitive animals are also believed to have spent most of their life cycle in the haploid form.

- Diploid organisms are believed to have a more robust genome because each chromosome has a homologous version. If one gene experiences a mutation (i.e. breaks), the other version of the gene located on the homologous chromosome is still there to carry on the function. The theory is that life gradually shifted from a haploid form to diploid to take advantage of this extra protection.

Unit 29

Additional Mechanisms for Genetic Variability

In the last unit, we touched on the amazing variability that arises from sexual reproduction. In this unit, we'll begin to look at how variability is accomplished in bacteria. Specifically, we will turn our attention to three bacterial processes: **transformation**, **conjugation** and **transduction**.

Transformation

Many species of bacteria are capable of taking up fragments of DNA that are floating around. After acquiring the new DNA pieces, the bacteria incorporate them into their own genome, and if the pieces contain genes, the bacteria gain the benefits of the new genes. This is called transformation, and the most obvious example, (and possibly the most relevant to us as humans), is when they pick up genes that give them immunity to an antibiotic. This process is especially obvious when you add the dead remains of antibiotic-resistant bacteria to a culture of living bacteria that are *not* resistant. Some portion of the non-resistant bacteria will pick up the DNA pieces from the dead bacteria and become resistant to the antibiotic! To illustrate how remarkable this is, that would be like a person catching the DNA from a feather and gaining the ability to fly, or my neighbors catching one of my skin cells and then walking upright. Anyway, this is one of many reasons why antibiotic resistance arises.

Note that not all bacteria can pick up and incorporate external DNA, but many can be induced to be able to do so. Recall that many bacteria have cell walls that protect them from the environment. In order for these armored bacteria to pick up stray DNA, you need to treat them with chemicals that break down the cell wall first. This, however, is a bit of a trick, and these kinds of bacteria are much less likely to be transformed in natural settings without our laboratory tinkering.

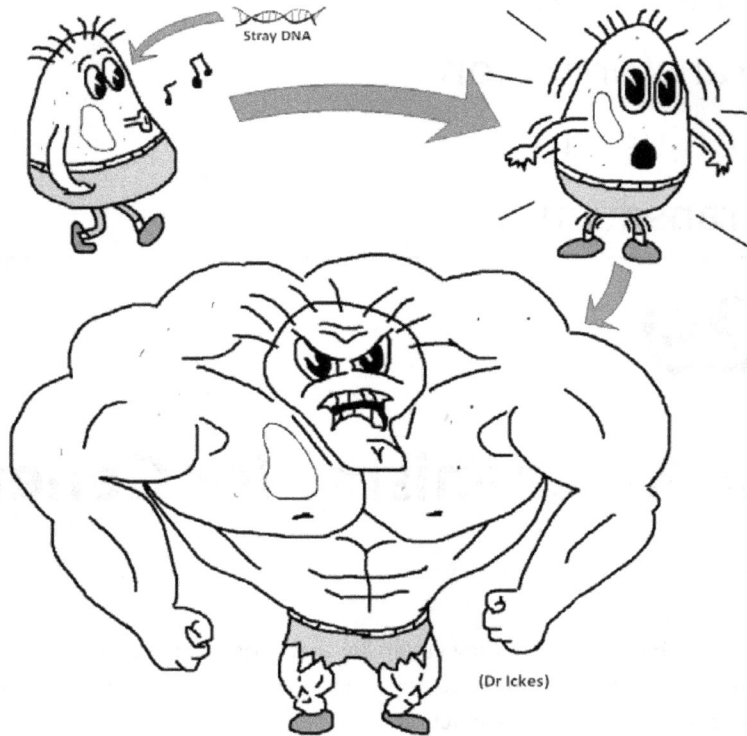

Bacterial Transformation

Stray DNA

(Dr Ickes)

Conjugation

Bacterial cells do not have sex. Well, at least they don't have sex in the form that we recognize. As it turns out, some bacteria undergo a process called conjugation, the closest thing to sex happening in their little world. During this process, one bacterium pulls up next to another and extends a bridge of cytoplasm (called a **pilus**), like a pirate ship broadsiding another ship, or one cell sticking a straw in another. Once the pilus is established, bits of DNA are transferred from the pilus-forming bacteria. The odd thing is, it's usually not the bacterial chromosome that gets transferred but rather little loops of DNA called **plasmids**.

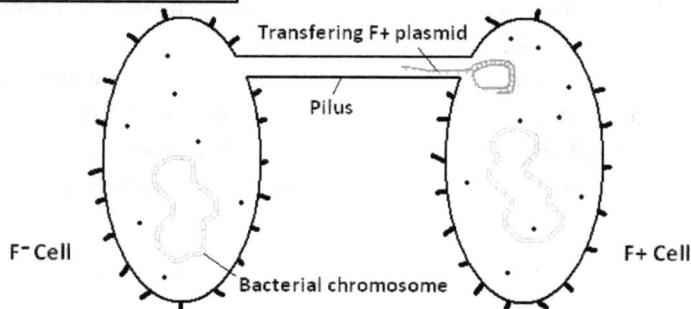

Bacterial Conjugation

Transfering F+ plasmid

Pilus

F⁻ Cell

F+ Cell

Bacterial chromosome

236

If you'll recall from our discussion of cell anatomy in Unit 17, plasmids are small rings of DNA that are very common amongst bacteria. Quite often, they carry genes that give the bacteria chemical upgrades. For example, one plasmid might confer the ability to breakdown a normally-harmful chemical in the environment, giving the host bacteria an extra boost in an otherwise poisonous world. Another plasmid might contain the genes for synthesizing chemicals that are vital, whereas the bacteria would otherwise just have to rely on these chemicals being premade in the environment.

In short, plasmids are tiny upgrades, and conjugation is one way the bacterial world passes them around. There is an important hurdle to this free-trade of genes, though: only bacteria having the genes for making the **pilus** can initiate conjugation. This particular ability comes encoded in what is called the **sex factor** (or **F factor**, for "fertility"), which is the set of genes that work to make the pilus and all other proteins necessary for the genetic transfer. Bacteria that have these genes are called F+ cells and are considered "male." Those without are designated F- cells and considered to be female.

After pilus formation, the DNA that contains the sex factor genes are copied and passed to the F- cell. At that point, the F- cell also becomes F+ and can initiate conjugation with other F- cells. Naturally, one might think that before long all bacteria would become F+, but this isn't the case. As it turns out, it takes long enough to efficiently make a pilus that F- cells have plenty of time to divide, so that in many cases the proportion of F+ to F- cells in a population can actually drop. So many F- cells, so little time. . .

There is a theory that proposes that the sex factor genes are a kind of virus, and that bacteria do not conjugate deliberately. Rather, the sex factor uses the bacteria to form the pilus, which acts as a stationary "viral" capsid of sorts. Once the sex factor has invaded a new cell, it prepares itself for another cycle, invading more and more bacteria over time. In other words, the sex factor genes exist purely because they can and for little other reason than that.

The important consequence to this process is not just the replication and propagation of genes that think only with their pilus, but that *other* genes can sometimes be transferred along with the sex factor genes. Through conjugation, bacteria can spread new DNA that confers unique immunities to antibiotics or new abilities to synthesize chemicals. So no matter how inconvenienced the bacteria might feel about the whole thing, conjugation carries the added benefit of transferring new genetic information.

Transduction

There is one more major way in which bacterial cells acquire new genetic information: through viral infections, a process called transduction. Recall from several units earlier that when a virus invades a cell, it uses the cell's protein-making machinery to manufacture new capsids. These capsids act as escape pods for the virus, taking up the viral genetic information and heading off to a new cell. Sometimes, however, other DNA, (like the little circular plasmids, for example), are taken up by the virus by mistake. When this happens, the viral capsid takes off for a new cell and inserts this "bystander" DNA into the new cell. In this manner, viruses help to spread genetic information throughout a bacterial colony.

When it comes to bacteria, there is one particular virus that scientists use a lot to deliberately insert genetic information into bacteria cells. This particular virus is called a bacteriophage, and it looks like a little spider. Basically, scientists have developed techniques to insert various plasmids into the capsids of these viruses, then add the plasmid-containing capsids to a population of bacteria. The capsids then inject the new genetic material into the bacteria. Depending on the plasmid used, the recipient bacteria may gain a new ability or resistance, and scientists can use that to study the bacteria or the genes they inserted.

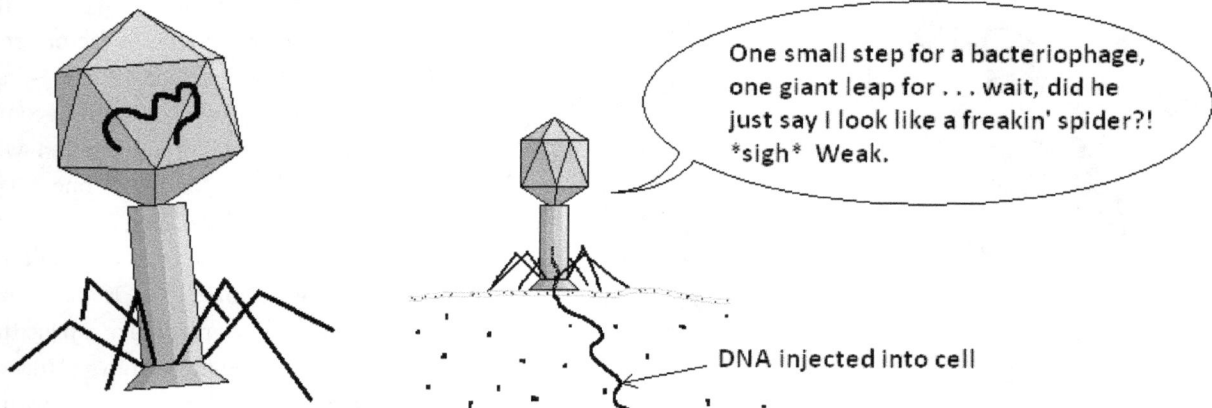

One small step for a bacteriophage, one giant leap for . . . wait, did he just say I look like a freakin' spider?! *sigh* Weak.

DNA injected into cell

Summary

In this chapter, we discussed three major ways in which bacteria acquire new DNA.

- The process of bacterial transformation occurs when bacteria pick up stray DNA that is floating around. If this DNA contains genes for useful proteins, the bacteria gains whatever abilities the proteins grant. Transformation can make bacteria newly immune to an antibiotic, or give the bacteria the ability to synthesize chemicals that it could not previously make.

- Bacterial conjugation is the closest thing to sex that bacteria have. It begins with an F+ cell (a cell having the sex factor genes). This cell creates a straw-like pilus that bridges the gap between it and another cell. The other cell is called the F- cell. Once the pilus attaches, DNA (usually containing the sex factor genes, as well as other stuff) is copied and transferred to the F- cell. After the F- cell acquires the sex factor genes, it becomes an F+ cell and can initiate conjugation with other bacteria cells. Like transformation, this process can make bacteria newly immune to an antibiotic or be able to make new chemicals.

- The DNA that is transferred in conjugation is usually a plasmid.

- Transduction is the process in which a viral capsid accidentally injects non-viral DNA into target bacteria. This DNA can consist of gene fragments or an entire plasmid. Either way, like transformation and conjugation, it can result in bacteria gaining new abilities.

- A common bacterial virus often used in labs is the bacteriophage, which looks like a little spider.

Unit 30

Basic Inheritance

We are now in a position to start studying **inheritance**.[62] This is a rough road ultimately leading us to the central bastion of modern biology: evolution. We'll take this road one step at a time and learn the concepts carefully. Let's briefly review what we've covered to this point. DNA is a collection of 4 different kinds of base pairs. The order of these pairs is used as a code to hold information. Discrete stretches of base pairs that code for protein are called genes. Genes are surrounded by genetic "punctuation marks" that tell the cell's machinery when to start and stop reading. Messenger RNA is transcribed from the DNA in genes, and this mRNA is then taken to ribosomes and translated using the genetic code to make protein, where the order of the codons in the messenger RNA determines the order of the amino acids in the protein. Proteins have many different roles in the cell; some help control chemical reactions (such as those in glycolysis and the citric acid cycle[63]), while others have structural roles, and so forth. Therefore, the order of the DNA in genes has a direct influence on the overall function of the protein, and through that function the DNA influences the performance of the cell and overall organism. This connection can be obvious, such as a gene that makes someone's eyes blue, or makes someone tall, or gives someone blond hair, and so forth, or

---

[62] Here, inheritance is defined as the passage of genetic material from parent to offspring at the time of fertilization.

[63] Recall that the citric acid cycle is also known as the Krebs Cycle or Tricarboxylic Acid Cycle. The name used in your class will be the one favored by a particular textbook or instructor.

the genes can have a subtle influence, such as making someone more or less susceptible to emotional problems, making someone vulnerable to some types of infections, etc. In these latter cases, it becomes difficult to say what effects are due to the genes and what are due to the environment. This is what the whole "Nature vs. Nurture" hubbub is all about.

Anyway, in genetics we say that an organism's genetic information (the order of its DNA) is its **genotype**, while the effect of that genetic information (eye color, hair color, etc.) is its **phenotype**. The phenotype that appears most commonly in the wild (i.e., natural settings) is called the **wildtype**. For example, suppose we monkeyed around with a mouse's DNA (put in a new genotype) so that we could get it to glow purple (a new phenotype). This new phenotype might certainly be quite fascinating, but it's far from the typical wildtype characteristics or normal dull grey fur.

Now that we've reviewed and covered these basic points, let's talk a little more on what homologous chromosomes are. We've mentioned them throughout the course of the previous sections, but their significance has not been particularly clear until now.

Homologous Chromosomes

For diploid organisms, every chromosome has a homologous copy. Now, if you'll recall, homologous chromosomes are pairs of chromosomes that share a similar genetic pattern. If one chromosome has a gene affecting hair color positioned at nucleotide 40,005, then its homologous chromosome will also have a hair-coloring gene at nucleotide 40,005. The two genes aren't necessarily the same gene though. One might color the hair yellow, while the other might color it brown; they are two *versions* of the hair-coloring gene.

In genetics, we call different versions of the gene **alleles**. We call the position of the gene its **locus**. To recap: a gene is a physical stretch of nucleotides that codes for a protein. A locus is the *location* of the gene on the chromosome. An allele is the *version* of the gene that is present at the locus. So for example, the hair-color locus we discussed before contains the hair-color gene, and two alleles are possible – one is the yellow-hair version and one is the brown-hair version. Within the population as a whole, many more than two alleles could be floating around of course. For instance, another individual might have an allele for green hair, or burgundy, or chartreuse, or razzmatazz. Who knows? There are lots of versions of genes floating in the gene pool, but within one individual, you find a max of two for any given locus.

Homozygous   Heterozygous

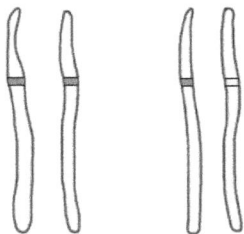

Suppose the two alleles at a location are identical, (both coding for blue hair, say). In that case we say the organism is **homozygous** at that locus. If the alleles are not identical, (one for blue hair and one for orange hair), then we say the organism is **heterozygous** at the hair color locus.

Let's adopt a sample critter to illustrate. Take the fictional creature *Mungabeastius biggus*. Let's say one particular mungabeast has two alleles for hair color: light gray and white. When the mungabeast makes gametes, each homologous chromosome goes its separate way during meiosis. Fifty percent of the gametes will get the allele for light gray hair and the other 50% will get the allele for white hair. That means that 50% of the mungabeast's offspring will inherit the light gray allele while the other will inherit the white allele. This is called the **Principle of Segregation**, which is a fancy term that basically means that the two alleles of a locus will get segregated away from each other when it comes time to make gametes. Both the mother and father do this, and when mating occurs the homologous pair is reformed when each parent donates his or her version of that chromosome.

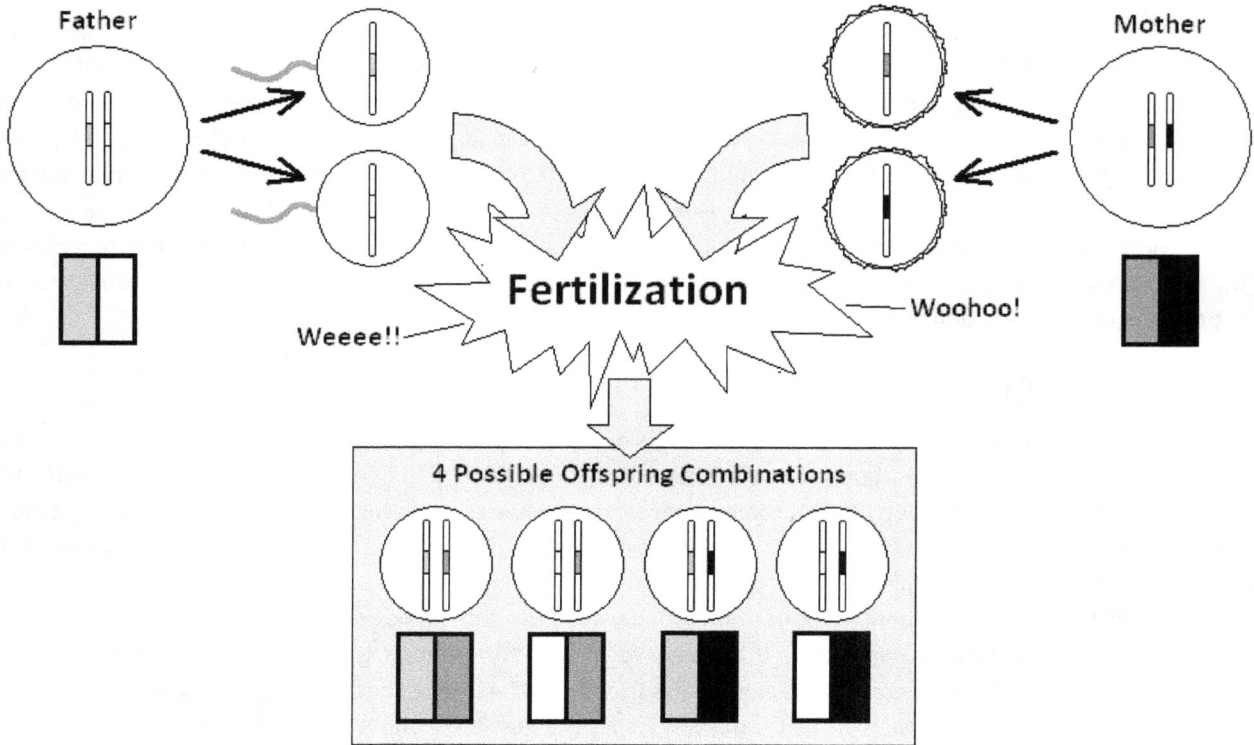

The trick is that when the gametes come together at fertilization there are a number of combinations that can result. To see how this is true, consider the image above. Suppose that dad has the aforementioned light grey and white alleles, whereas mom's two alleles at the same locus are one dark grey and one black. Now, when mom and dad contribute gametes during fertilization we find that there are four possible outcomes of this jumble. The first is light gray and dark gray. The second is white and dark gray. The third is light gray and black. The last is white and black. This genetic shuffle is at the heart of inheritance, and we'll see it over and over and over . . .

But let's back up to cover an interesting question about what happens during the time that the two alleles are within the same cell. Suppose an organism is heterozygous for hair color, like our mungabeasts. In other words, let's say it has one allele for black hair and one for white hair. Which version of the gene gets **expressed**? What color of hair will the organism actually *have*? To examine the answer, let's look at some basic patterns of inheritance that result as a consequence of the Principle of Segregation.

Blending Expression (or Partial Dominance)

The simplest resolution to which allele would lead to the organism's properties is that *sometimes* there is a stalemate between the two. Expression in this fashion is a blend of the two traits. For example, suppose that both mom and dad have the same alleles, say alleles that lead to blue or yellow pigment production. If both alleles created proteins that worked equivalently well, the combined color would end up being green (blue + yellow = green). Both mom and dad would have green hair. This is a simple blending, or **partial dominance**, as we say in biology, because both alleles assert themselves, each winning partially.

However, what will happen if they have offspring? Both parents could produce two types of gametes: those packaged with the blue allele and those packaged with the yellow allele. Then, when you put the gametes together, you have four different ways of doing so. One offspring could get both blue alleles, two offsprings could get one blue and one yellow, and one offspring could get both yellow alleles.

This event is typically represented by drawing a **Punnett square**.

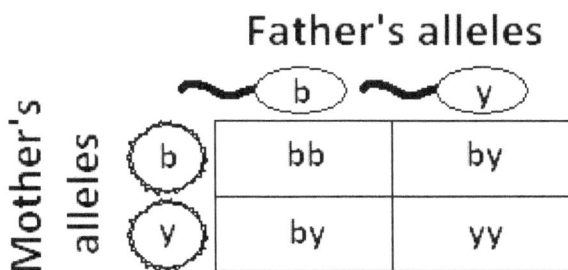

Father's alleles

Mother's alleles

	b	y
b	bb	by
y	by	yy

If you examine the Punnett square, the first thing to note is that three different phenotypes emerge. The blue/blue (bb) genotype leads to a blue haired phenotype. The blue/yellow (by) genotype leads to a green haired phenotype, and finally the yellow/yellow (yy) genotype leads to a yellow haired phenotype. The distribution of these phenotypes is in the ratio of 1:2:1, meaning if the parents had four offspring, the odds are that one would have blue hair, one would have yellow hair, and two offspring would have green hair. This is the typical pattern in partial dominance inheritance.

Dominant/Recessive Expression

There is another possible outcome to the struggle between two alleles, an outcome called **dominant and recessive expression**. The basis of this kind of allelic interaction is as follows: the function of the protein of one allele is almost unnoticeable compared to the function of the other. In other words, it seems as though one allele overpowers the other. For example, one allele might code for a protein that is simply defective; it has no function at all. If the competing allele makes a protein that has *any* function at all, it wins by default. Another possibility might be that one allele's protein completely negates the effects of the other allele. Regardless of the reasons behind it, the expression you see is very distinctive: anytime the two alleles are put together into an organism, one allele gets its way over the other. The phenotype that results is determined by the dominant allele. The other allele is silent.

In the image that follows, we see a mungabeast that is heterozygous for both hair and eye color genes. In the picture, the hair alleles behave with partial dominance so that the black and white alleles mix to make the grey hair phenotype. With eye color, the genes for the mungabeast behave with dominant/recessive nature so that only one of the alleles wins (in this case the grey one[64]). Note that the picture doesn't necessarily represent the way hair color

---

[64] Hey, what do you expect? It's *grayscale* after all!

and eye colors work in other organisms. It's just meant to show how blending and dominant/recessive expressions work.

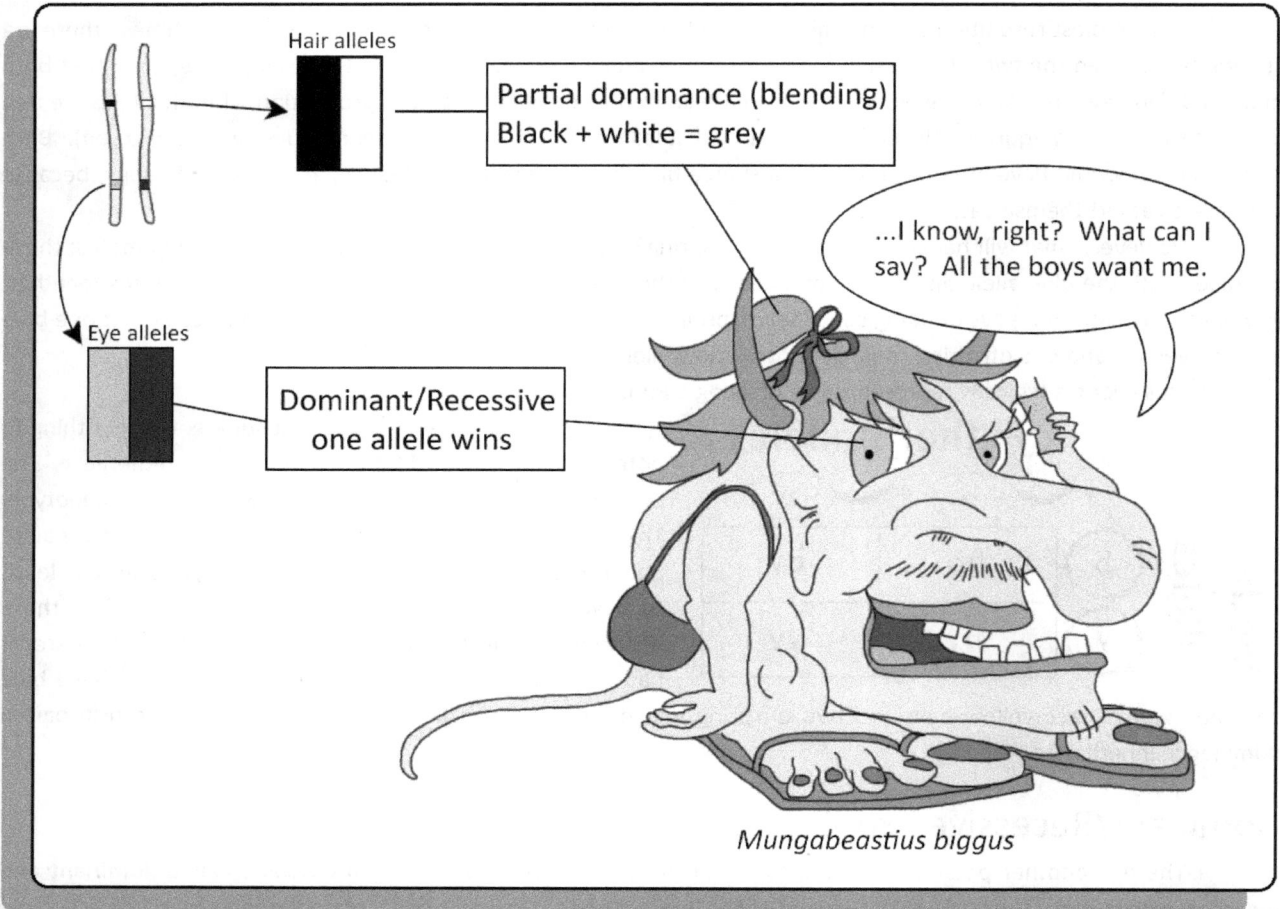

Mungabeastius biggus

As it turns out, dominant/recessive expression has a great deal of historical importance in the study of inheritance, because it was the careful experiments to investigate this phenomenon that put the field of modern genetics on the map. Particularly significant were the experiments by **Gregor Mendel**, an Austrian monk living in the mid-nineteenth century, who used careful cross-mating of pea plants to draw important conclusions.

Gregor Mendel started with a pure strain of pea plants that always gave red flowers, and another pure strain that always produced white flowers. Based on our first experience with expression patterns, one would guess that by crossing them you would get a pink flowered pea plant. What Mendel found was that all the first generation offspring had red flowers. The white trait was nowhere to be seen. Strange! Moreover, when he crossed all these first offspring generation flowers with each other, three out of four made red flowers, and one out of four made white flowers. The white flower trait had somehow survived hidden in the first child generation of all red flowers!

We can decipher the mystery by drawing another Punnett square using the first generation as the cross (see below). When we do this, we get a second generation that all have a genotype of Rw. These are all red plants. If we now cross these plants with each other, and draw the outcome in another Punnett square, we see an assortment of three possible genotypes: RR, Rw and ww. The first two genotypes are red plants, and only the ww genotype is white. Those two Punnett squares illustrate the nature of dominant and recessive alleles. In the presence of a dominant allele (here the R allele), recessive alleles (here the w allele) do not contribute to the overall phenotype.

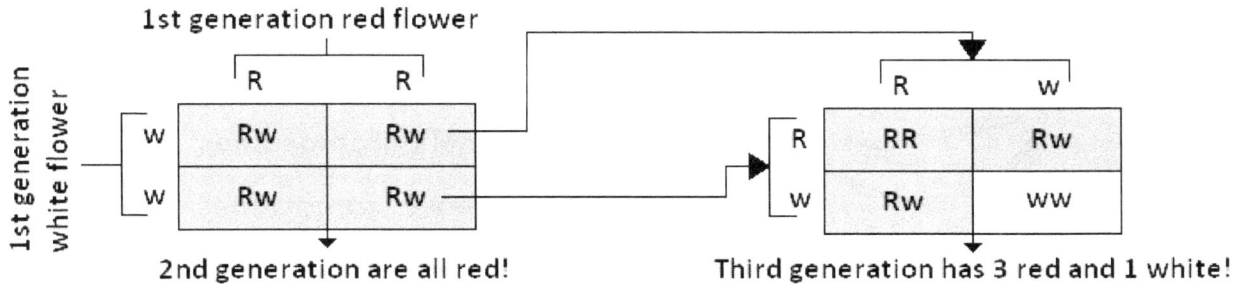

In conclusion, the pattern of inheritance that emerges from dominant / recessive gene expression is a 3:1 ratio.

Middle-of-the-Road Expression

On one end, you have the perfectly blended expression, and on the other end you have the dominant/recessive expression. Think of these as two extremes. One could easily imagine a situation that is midway between the two: one allele is highly outspoken over the other, but the other allele is not entirely silent. Suppose Mendel's pea plants had been like that. Instead of finding the ratio of three red and one white plant, you might find one red, two slightly-less-than-red and one white. Without careful observation and measurement, you might simply conclude there were three red and one white and jot down the result as dominant and recessive.

The purpose of pointing out this middle-of-the-road pattern is that inheritance can be a complicated game to play, and deciphering what each gene is doing can be a monumental task. In any event, Punnett squares can help you see the underlying phenomenon. For any given pair of alleles, four types of offspring are possible, and depending on the degree to which the two alleles compete with each other, you can get a clear 3:1 ratio, or a 1:2:1 ratio, or a ratio that is deceptively close to 3:1. A sharp eye and careful measurement are crucial in biology!

Other Expression Patterns: Sex-Linked and Maternal Inheritance

Up to now we have only discussed homologous chromosomes. Technically, not all of a diploid's chromosomes are homologous. There are usually two chromosomes in the entire bunch that do not have a second copy. These are the **sex chromosomes**, designated "X" and "Y." All the others, (those we've talked about previously), are referred to as the **autosomal chromosomes**. A diploid organism will either have one **X chromosome** and one **Y chromosome**, or two X chromosomes. For most species, if an organism has two X chromosomes, it is female. If it has one X and one Y, it is a male.

Diploid Nucleus

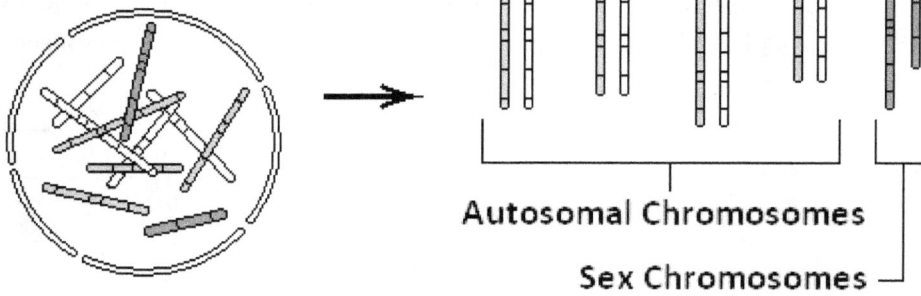

Autosomal Chromosomes

Sex Chromosomes

A unique phenomenon called **sex-linked inheritance** occurs when an allele pair is located on the sex chromosomes. Many of the gender-specific genes responsible for making us male or female are found on the X and Y chromosomes, and as a result, large parts of their respective DNA differ dramatically. Still, they are paired, but a locus on one may not have a homologous locus on the other. This latter fact means that some sex-chromosome alleles will not be paired with an equivalent partner like the alleles on the autosomal chromosomes. So, if a recessive gene appears on an X chromosome, and there is no equivalent locus on the Y chromosome to affect it, it will automatically be expressed in a male (since the recessive gene has no competition from an allele on the Y).

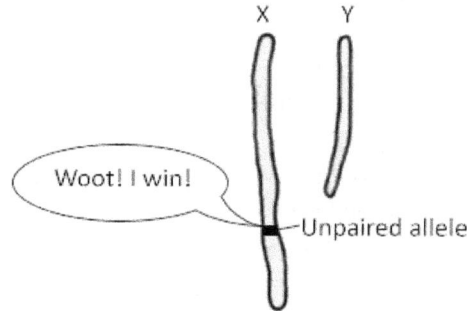

If, however, another X chromosome is present, such as in a female, the usual dominant / recessive or blending patterns will be seen.

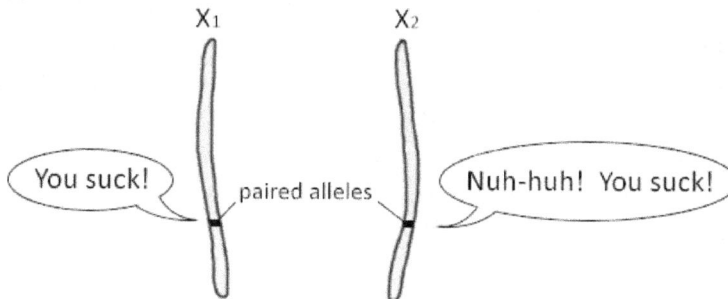

The important point is that the inheritance will be different for males and females. More often than not, males get the raw deal. Since they only get one copy of the X chromosome, any bad allele on that chromosome gets expressed. This harkens back to the argument that haploid organisms have less stable genomes. They only have one copy of each allele, and if that copy is not good, they're in for trouble. Males in that sense are partially haploid, you might say,[65] at least in so far as their X chromosome is concerned, since many of the genes on that chromosome have no partner.

On the other hand, if there are alleles on the Y chromosome that are not present on the X chromosome, males once again have only the one copy to use. In other words, since males have one Y chromosome and one X chromosome, whatever bad genes appear on the Y chromosome will generally be expressed.

The punch-line for sex-linked chromosomes is that these kinds of traits are expressed mostly in males, primarily because they only have one copy of the allele in question, either on the X chromosome or on the Y chromosome.

Lastly, there is one more pattern of inheritance that differs from the usual autosomal patterns we discussed at the beginning. This last pattern is **maternal inheritance**. As you may recall, the mitochondria (the powerhouses of the cell) have their own DNA, which may be left over from when they lived as independent bacteria-like organisms. Mitochondria are typically inherited from the mother (stored in the egg). This means that if a mitochondrion has a gene that gives rise to a very specific trait, all the offspring of the affected mother will inherit the mitochondrial gene, regardless of the father's genetics. This form of inheritance is generally easy to follow, because the trait will follow along the maternal lines of a family tree. Mothers will have children that are all affected. Affected daughters will have children who are all affected, and so forth. In contrast, the males of the family will never pass the trait to their offspring. For us males, this is a bit of a reprieve, since we normally get the brunt of the genetic attack.

---

[65] It is technically not correct to refer to males as haploids. I only point out this slight parallel just to illustrate how diploid genomes can be more stable than haploids.

Maternal Inheritance

Those are the basic inheritance patterns. In the net unit, we'll take some time to discuss pedigree charts, which help us to show things about inheritance using pictures. After we talk about pedigree charts, we'll take a look at what happens when we examine the patterns that result from *more* than one locus (hair color and eye color, for example).

Summary

In this section, we discussed the basic patterns of inheritance.

- A gene is a sequence of genetic information that codes for a protein. For any given gene, there are usually a number of different versions of it floating around in the population. Each version is called an allele. The location where the gene appears is called its locus.
- A creature's genotype is the set of alleles the creature has. Its phenotype is how those alleles have been expressed. The phenotypes that appear in the wild are called the wildtype.
- Homologous chromosomes that have the same allele (i.e. same version of a gene) at some specific locus are said to be homozygous at that locus. Homologous chromosomes that have different alleles for a given locus are said to be heterozygous at that locus.
- When a male and female make gametes, partner alleles on homologous chromosomes are sorted independently into different gametes. This is called the Principle of Segregation.
- If a male and female each have two different alleles for a given locus, then there are a maximum of four ways in which those alleles can be partnered in the offspring.

- In individuals that are heterozygous (two different alleles), a blending effect can sometimes be seen as the two alleles compete to be expressed. This type of inheritance pattern is called partial dominance, because each allele expresses some dominance. Offspring of heterozygous individuals fall into ratios of 1:2:1 based on their phenotype.

- When one allele is completely dominant over another, the dominant allele will determine the phenotype of the organism. This type of inheritance pattern is called dominant/recessive. This is a common pattern and is historically important. Offspring of heterozygous individuals of dominant/recessive alleles fall into ratios of 3:1.

- Some inheritance patterns fall in between the partial dominance and dominant/recessive patterns, so that the offspring appear to have a 3:1 ratio, when in reality a subtle difference leads to the 1:2:1 ratio.

- Most diploid organisms have two sex chromosomes, designated X and Y. Typically, the male are XY, and the females are XX. The non-sex-chromosomes are referred to as the autosomal chromosomes. The Y chromosome typically differs radically from the X chromosome.

- Sex-linked inheritance patterns occur when the allele producing a particular trait appears on one of the two sex chromosomes. In males, this often leads to an allele that has no partner, so that it always expresses its phenotype. In females, the allele in question will have a partner allele (since there are two X chromosomes) so that the phenotype is determined by the same mechanics as alleles on autosomal chromosomes.

- Females in most species provide the mitochondria for their offspring. The mitochondrion in the sperm of the father is broken down and not passed to offspring. Due to this, a special form of inheritance called maternal inheritance can result, where all the children of the mother inherit the mitochondrial trait in question, without exception.

Unit 31

Pedigree Charts

When we talk about inheritance, we often draw little diagrams that show how a trait travels from one generation to the next. These little diagrams are called **pedigree charts**, and a sample of one is as follows:

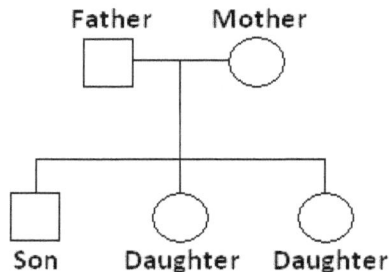

In this pedigree chart, we have two parents and three children. The first thing to notice is that males are designated as squares and females are designated as circles. When a pedigree chart is used to show how a trait travels in a family, the squares and circles are darkened in. For example, let's say we want to see which individuals in the family have freckles (a supposedly dominant/recessive trait in humans). In this case, we might see:

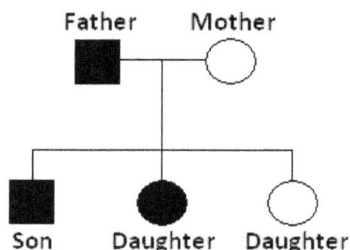

This chart means that the father and two of the three children have freckles. Dominant traits can usually be spotted in this manner. If a parent has a dominant trait, then half or more of the children will often have the trait as well (it can surprise you though). If one parent is homozygous for the trait (say homozygous for freckles, f/f), then all the children should end up with freckles.[66] Since less than all the children have freckles, we can conclude that, (unless something else is involved to complicate things), the father is not homozygous dominant.

Pedigree charts are especially good at helping us figure out if a gene is dominant or recessive. Let's look at another chart and discuss the possibilities:

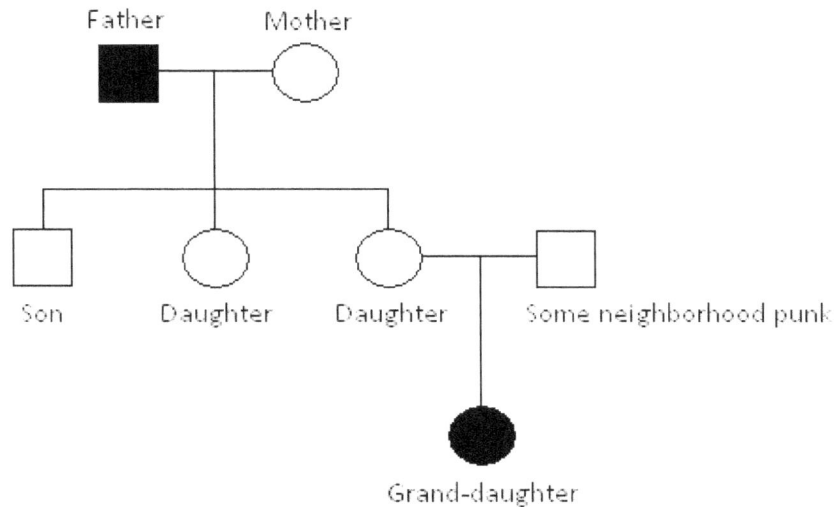

In this chart, we see that the father has some kind of trait. Say he has the inheritable disease "Goofadoofosis," and we see that the granddaughter has it as well. We can conclude from this chart that – in the absence of complicating factors – Goofadoofosis is not inherited in a dominant fashion. If it were, the second daughter would have had it as well. Since the second daughter does not have it, it must be inherited in some other fashion, possibly as a recessive trait. At any rate, this is commonly what a pedigree chart will look like for recessive traits. The trait will appear periodically, then vanish and skip generations. Since it is recessive, it requires both parents of any given union to have the recessive allele and if the allele is particularly rare in the population, the disease will skip many generations before appearing again. In contrast, as we saw earlier, dominant traits tend to have a clear path; every generation is affected.

---

[66] That is, of course, unless other genes are involved, in which case it gets a little more complex. We'll investigate this a little later.

The pedigree charts for sex-linked inheritance are very recognizable as well:

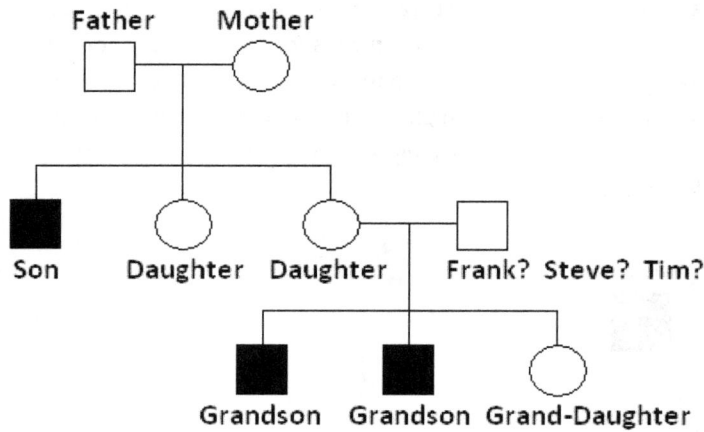

Father   Mother

Son   Daughter   Daughter   Frank? Steve? Tim?

Grandson   Grandson   Grand-Daughter

 As discussed in the previous section, sex-linked inheritance tends to zap male offspring much more often than females, so the pedigrees for these kinds of traits will show an unusual concentration of affected males and little to no affected females. The most likely explanation for this particular pedigree chart is that the grandmother had the allele in question on one of her X chromosomes. The son inherited that X chromosome, as did the daughter on the far right. That daughter then gave the X chromosome to both her sons.

 The pedigrees for maternal inheritance will typically show all affected individuals linked through the mothers:

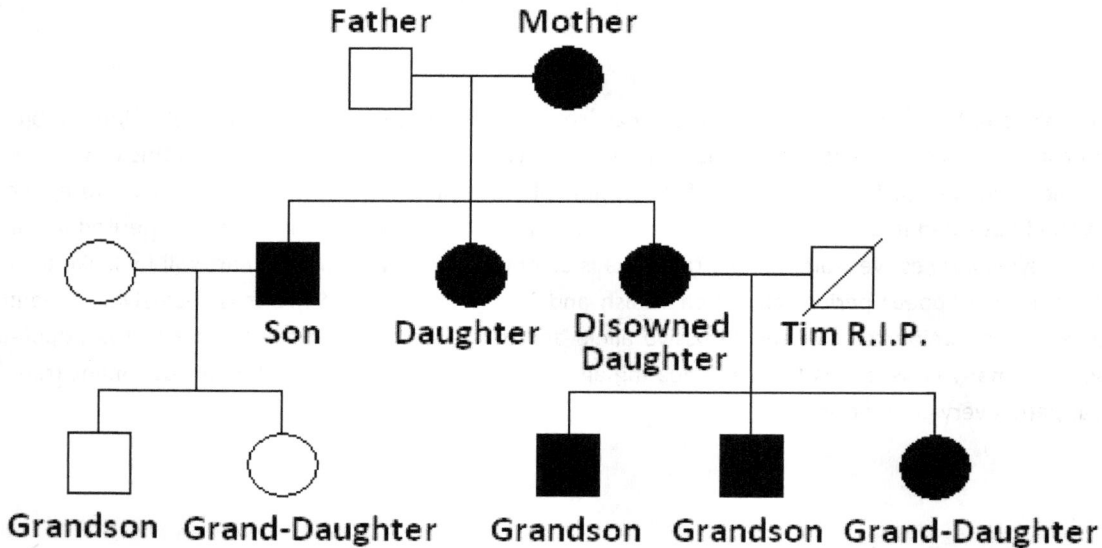

Father   Mother

Son   Daughter   Disowned Daughter   Tim R.I.P.

Grandson   Grand-Daughter   Grandson   Grandson   Grand-Daughter

 Note how the male on the left does *not* transmit the trait to the offspring, but all the females do. This is very common for the maternally inherited traits.

 Lastly, individuals that are deceased are usually drawn with a slash through them (such as Tim in the above diagram).

Summary

In this section, we discussed basic pedigree charts.

- In pedigree charts, squares are used to designate males. Circles are used to designate females. Individuals having the trait in question are usually drawn with their symbol darkened in.
- In autosomal dominant inheritance, half of the children of an affected adult will tend to show the trait.
- In autosomal recessive inheritance, the trait will skip multiple generations and ther unexpectedly show up again.
- In sex-linked inheritance, there will be an unusual number of affected males and very few – if any – females affected.
- In maternal inheritance, the pedigree chart will show all children of the affected mother as having the trait in question.
- Deceased members of a pedigree chart are often drawn with a slash through their symbol.

Unit 32

Two Loci Inheritance

We are now in position to examine what happens with the inheritance pattern of two simultaneous loci. That is to say that we will be watching the inheritance pattern of two separate pairs of alleles at the same time.

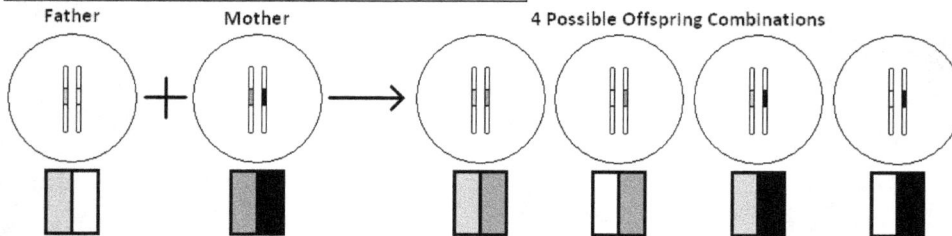

When examining only 1 pair of alleles . . .

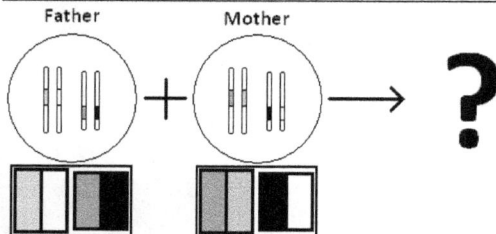

When examining 2 pairs of alleles at a time . . .

254

Why might studying this be important? Well, biological organisms contain more than one allele pair; they typically contain thousands upon thousands. Further, most traits that we see around us, (such as height, facial structures, diseases, and a slew of other things), are the result of a combination of many alleles working together as a group. The ultimate goal is to understand how these traits are passed from generation to generation, and the first step in that goal is to see how two pairs of alleles that each governs their own trait are inherited. In the next unit, we'll tackle how two pairs of alleles working together to make a single trait are inherited. Throughout this unit and the next, we'll be using a **three-step strategy**: 1) determine the types of gametes a genotype can produce, 2) put these together into a Punnett square and 3) group the offspring genotypes based on their expression. The best way to understand this is to see it in action, so let's work through several examples.

Two Loci, Two Traits, Two Chromosome Pairs

When each allele pair is on a different pair of homologous chromosomes (shown below) each allele will segregate from the other. Let's consider our fictional mungabeast and examine two different traits: hair color and eye shape. Let's say that for hair color, it has one blue allele and one yellow allele, and for eye shape it has one allele for circular eyes, and one allele for triangular eyes.

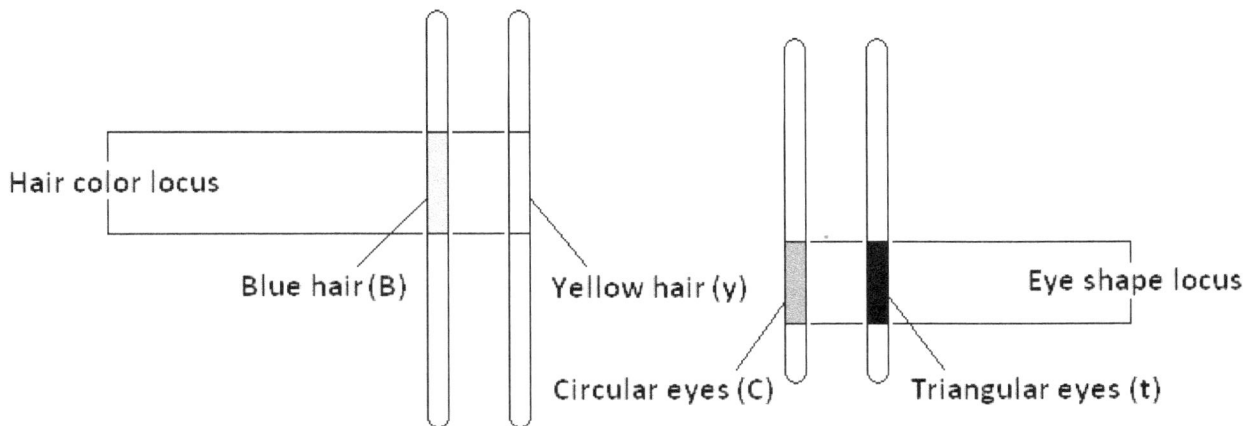

For shorthand, the blue allele will be "B" and the yellow allele will be "y." The circular allele will be "C" and the triangle allele will be "t." So, given this information, we know that the overall genotype of the organism is By/Ct. The first step in our strategy is to determine the types of gametes the organism can produce. When the mungabeast makes sex-cells, 25% of the gametes will have a blue allele and a circular allele (B/C), 25% the blue allele and triangle allele (B/t), 25% the yellow allele and circular allele (y/C), and the last 25% the yellow allele and triangle allele (y/t), as shown in the next image.

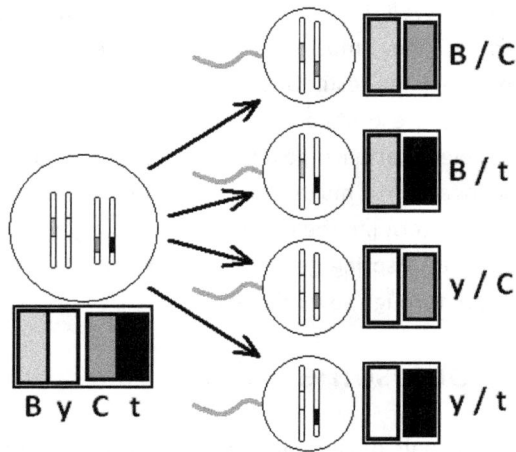

B / C

B / t

y / C

y / t

B y C t

Now suppose this mungabeast mates with another mungabeast that has the same genotype (By/Ct). To determine the possible combinations in the offspring, we form the top and side of a double Punnett square with the possible gamete combinations. When we cross the gametes, we get a complex pattern with sixteen possible squares, each square being 6.25% likely to appear in the offspring. The resulting phenotypes (the color of hair and shapes of the eyes, etc.) depend on the individual expression patterns of the independent allele pairs.

	B/C	B/t	y/C	y/t
B/C	BB/CC	BB/Ct	yB/CC	yB/tC
B/t	BB/Ct	BB/tt	yB/Ct	yB/tt
y/C	By/CC	By/tC	yy/CC	yy/tC
y/t	By/Ct	By/tt	yy/Ct	yy/tt

→ 16 Possible offspring

↓

So, each is 1/16 likely

↓

6.25% each

The last step is to see what phenotypes emerge. Let's say that both allele pairs behave with perfect blending (like the first pattern we looked at in the previous sections). What will the possible offspring look like? Let's look at just the hair color phenotype first and see. Twenty five percent of the offspring have two blue alleles, so they have blue hair. Twenty five percent have two yellow alleles, so they have yellow hair. Fifty percent have one blue and one yellow allele, so they have green hair. Looking at hair color alone, you see the same 1:2:1 pattern we saw in the previous section.

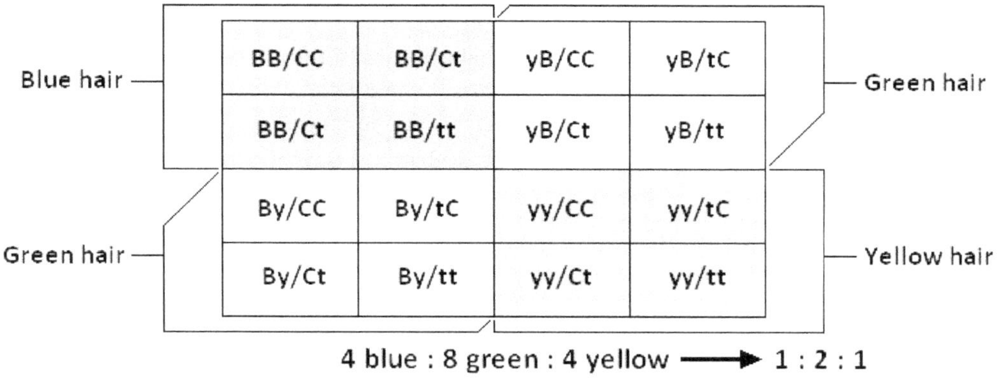
Blue hair
Green hair
Green hair
Yellow hair
BB/CC BB/Ct yB/CC yB/tC
BB/Ct BB/tt yB/Ct yB/tt
By/CC By/tC yy/CC yy/tC
By/Ct By/tt yy/Ct yy/tt
4 blue : 8 green : 4 yellow → 1 : 2 : 1

A similar examination will reveal the same pattern for eye-shape.

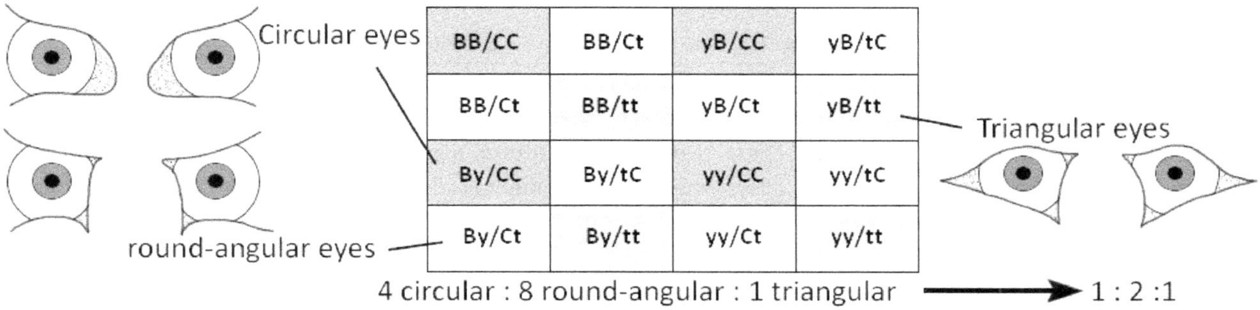
Circular eyes
Triangular eyes
round-angular eyes
BB/CC BB/Ct yB/CC yB/tC
BB/Ct BB/tt yB/Ct yB/tt
By/CC By/tC yy/CC yy/tC
By/Ct By/tt yy/Ct yy/tt
4 circular : 8 round-angular : 1 triangular → 1 : 2 :1

Let's use these results to see how many have, say, green hair and circular eyes. In order to have green hair, the mungabeast must have the By genotype, and in order to have circular eyes, it must have CC. So, to have green hair and circular eyes, it must have a genotype of By/CC. If you look at the double Punnett square, you'll see that two individuals meet that criterion. Since each of the two represents 6.25% probability, the total probability is 2 × 6.25% = 12.5%. Therefore, 12.5% of all the offspring will have *both* green hair *and* circular eyes. If we follow this approach for all the phenotypes, we arrive at the following table.

Phenotype	Probability
1 Blue, Circle (BB/CC)	6.25%
1 Yellow, Triangular (yy/tt)	6.25%
2 Green, Circle (By/CC)	12.5%
1 Yellow, Circle (yy/CC)	6.25%
2 Green, Triangular (By/tt)	12.5%
1 Blue, Triangular (BB/tt)	6.25%
2 Yellow, Round-angular (yy/Ct)	12.5%
2 Blue, Round-angular (BB/Ct)	12.5%
4 Green, Round-angular (By/Ct)	25%

From this list, you can see that when you have two pairs of alleles that each behave with blending expression, the total number of phenotypes is quite diverse and spread out.

Suppose, however, the two allele pairs behave closer to the dominant/recessive pattern of expression. Let's assume that the uppercase letters represent dominant traits; so blue (B) is dominant over yellow (y), and circle (C) is dominant over triangle (t). In that case, all the green haired individuals will have blue hair instead, and the round-angular eyed individuals will have circular eyes instead. An examination of the double Punnett square will then yield the following phenotype distribution.

Phenotype	Probability
9 Blue, Circle (B-/C-)	56.25%
3 Yellow, Circle (yy/C-)	18.75%
3 Blue, Triangle (B-/tt)	18.75%
1 Yellow, Triangle (yy/tt)	6.25%

The dashes in the table above indicate that it doesn't matter what the other allele is; the dominant allele gets its way regardless. To get an idea for how these numbers were found, consider the first entry, B-/C-. Nine of the sixteen will have at least one B *and* one C allele; these will be blue-haired *and* circular-eyed. To find the percent probability, we just divide nine by sixteen, which equals 56.25%. That means 56.25% of the offspring will have blue hair *and* circular eyes.

In that example, both parents had the identical genotype (By/Ct), but it would be easy to determine the outcome of any combination of parental genotypes. For example, if the father has, say, BB/Ct and the mother has By/tt, simply 1) list all the possible gametes that BB/Ct can produce and put them at the top of a double Punnett square and put all the possible gametes that By/tt can produce on the side. Then 2) fill in all the possible offspring in the sixteen given squares. Once that is done, 3) group the offspring genotypes based on how they compete for expression, be it blending, dominant/recessive, or anything in the middle range.

That said, let's look and see what happens if both allele pairs are located on the same homologous chromosome pair. Our overall three-step strategy will stay the same, but as we'll see, the major difference that will occur will be a large shift in the probabilities away from the typical 6.25% for each square.

Two Loci, Two Traits, One Chromosome Pair

When the loci were on *different* chromosome pairs, each individual allele went its own way so that 25% of the gametes had alleles B/C, 25% had B/t, 25% had y/C, and 25% had y/t. Now, we look at two loci stuck on one chromosome pair.

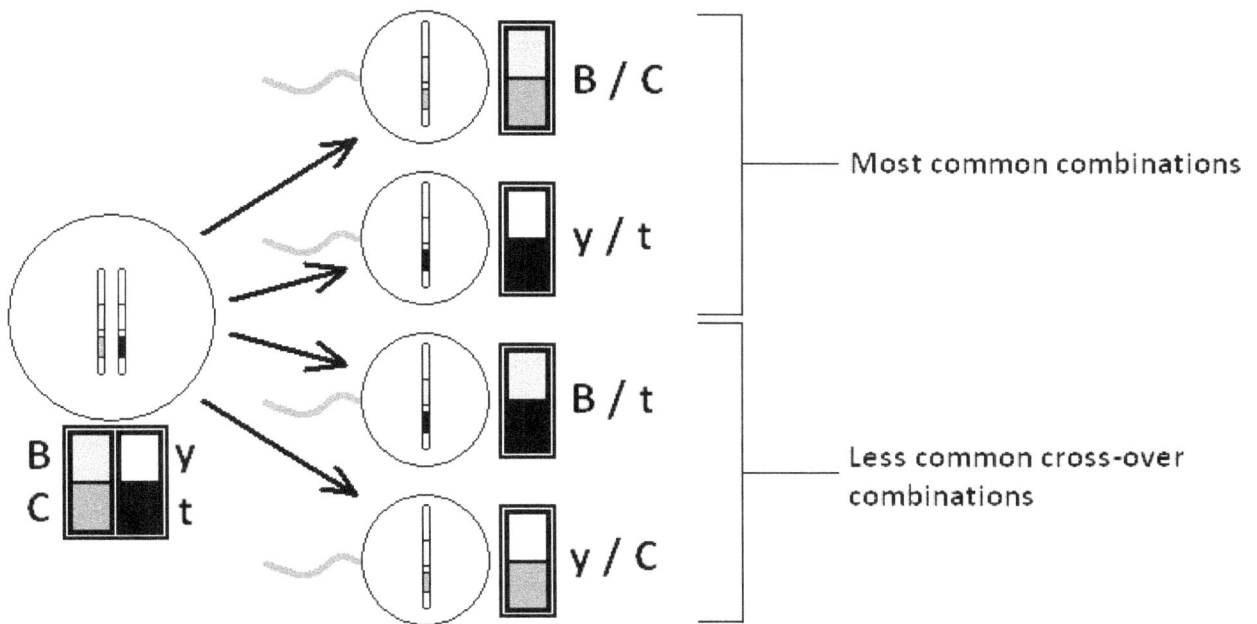

When the two loci are on the same chromosome, the alleles tend to segregate together into the same gametes. Most of the gametes that result are of one of two types: B/C and y/t. A small percent switch partners during cross over to result in a small population of B/t and y/C gametes. Let's ignore this small fraction for the moment.

Suppose the mungabeast having this genotype meets another mungabeast with the same genetic makeup. To determine the possible offspring outcomes, we once again make a Punnett square. On the top we put the two most common combinations, B/C and y/t. On the side we put the same thing (since we've said that both parents have the same genetic makeup). As is usual with a simple Punnett square, we have four possible outcomes, as shown here.

Father's alleles

		B/C	y/t
Mother's alleles	B/C	BB/CC	yB/tC
	y/t	By/Ct	yy/tt

The yB/tC and the By/Ct are the same thing,[67] so if we compare the ratios of the offspring types, we get a 1:2:1 genotype pattern: one BB/CC, two By/Ct, one yy/tt. What will the phenotypes be? To determine that, we

---

[67] The order in which we write the letters makes no difference.

examine how each individual allele competes for expression. If they both behave with blending expression, you'll have a 1:2:1 phenotype distribution. The BB/CC will be blue-haired and circular-eyed. The two By/Ct will be green-haired with round-angular eyes, and the yy/tt will be yellow-haired with triangular eyes.

If, on the other hand, both allele pairs use dominant/recessive inheritance, then the By's will look the same as the BB's, and the Ct's will look the same as the CC's. You'll end up with the 3:1 pattern that is typical of the dominant/recessive type of inheritance: three blue-haired, circular-eyed individuals, and one yellow-haired, triangular-eyed individual (the one with the yy/tt).

A similar argument could be made if one pair behaves with dominant/recessive expression and the other with blending expression. Simply add up each phenotype and compare the ratios.

So far that's not unlike the genetics of a single allele pair, but what about the tiny fraction of gametes that resulted from crossing-over? Well, to be thorough, once you consider the possibility of crossing-over, then you are once again back to 4 different types of gametes that can be produced: B/C, B/t, y/C and y/t (just like in the first half of this unit!). The difference is in the *probabilities* of each gamete combination. Previously, each gamete had a probability of 25%; each was equally likely. Now we are facing the situation of *two* loci on a *single* pair of homologous chromosomes, and the probability of the gametes having the cross-over products (B/t and y/C) will *not* be the same as the probability of having B/C and y/t. In fact, the chance of having a cross-over will depend on the distance between the two loci. The farther apart the two loci are, the higher the probability that cross-over will occur between them, and the less rare the B/t and y/C will be.

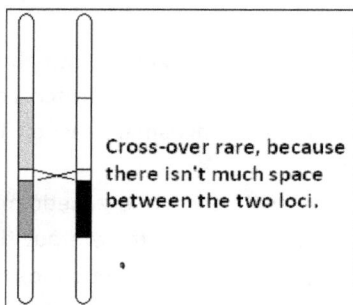

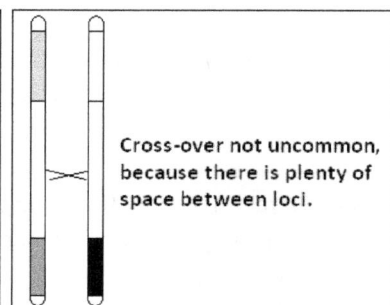

Cross-over rare, because there isn't much space between the two loci.

Cross-over not uncommon, because there is plenty of space between loci.

CONCLUSION:
The closer two genes are together, the more likely they will be inherited together.
The farther apart they are, the greater the chance they will segregate independently into different sex-cells.

Suppose the two loci are closely spaced to one another. In that case, it would be rare for cross-over to occur; B will almost always be packaged into the same gamete as C and y will almost always be packaged with t. Only a small percentage of the gametes will have B packaged with t and y packaged with C. As such, you might see the following probabilities for the gametes: 47.5% B/C, 47.5% y/t, 2.5% B/t, and 2.5% y/C.[68] Alternately, if the two loci are widely separated from each other, it might not be so rare that B crosses over to travel with t, or that y crosses over to travel with C. In that case, you might see a probability spread like this: 30% B/C, 30% y/t, 20% B/t, 20% y/C.

Regardless, this difference in gamete probabilities is one way in which geneticists can get an estimate of how far apart two loci are from each other. The farther apart they are, the more independently they are segregated into gametes, and the more independently they are inherited in offspring. If two alleles seem to travel together most of the time, we can conclude that they are closely spaced on the same chromosome.

---

[68] These are just hypothetical numbers. They were not found using any formula.

How does the possibility of cross over affect the overall distribution of genotypes and phenotypes? Well, since we technically have 4 types of gametes, we need a double Punnett square. That, of course, means sixteen different offspring genotypes are possible. However, since the four different gamete types are not evenly distributed, the sixteen different genotypes are not evenly distributed either! When the four different gametes were each 25% likely, the sixteen resulting genotypes were each 6.25% probable. Now, the gamete distribution is skewed, so you end up with heavily skewed offspring genotype probabilities. For the sake of clarity, I'll draw out the table of probabilities that would result from a probability of 45% B/C, 45% y/t, 5% B/t and 5% y/C. Also, to minimize complexity, let's just assume that the two parents have the same genotype.

	45% B/C	45% y/t	5% B/t	5% y/C
45% B/C	20.25% BB/CC	20.25% yB/tC	2.25% BB/tC	2.25% yB/CC
45% y/t	20.25% By/Ct	20.25% yy/tt	2.25% By/tt	2.25% yy/Ct
5% B/t	2.25% BB/Ct	2.25% yB/tt	0.25% BB/tt	0.25% yB/Ct
5% y/C	2.25% By/CC	2.25% yy/tC	0.25% By/tC	0.25% yy/CC

This double Punnett square lists the probabilities of each individual genotype. How did I find these numbers? Well, to determine the combined probability of two independent probabilities, you multiply them together. So, the combined probability of 45% and 45% is:

0.45 x 0.45 = 0.2025 (20.25%)

Why did I write 0.45 and not 45% in the equation above? The reason is that 0.45 is another way of writing 45%. They are equivalent values, just different notation.

Several of these squares have the same genotype in them. For example, yB/tC, yB/Ct, By/Ct and By/tC are the same thing, so together their combined probability is 20.25% + 0.25% + 20.25% + 0.25%= 41.00%. Analyzing the rest of the genotypes in this manner we arrive at the following table:

Genotype	Percent	Out of 10,000 offspring
BB/CC	20.25%	2,025
By/Ct	41.00%	4,100
yy/tt	20.25%	2,025
BB/Ct	4.50%	450
By/tt	4.50%	450
By/CC	4.50%	450
yy/Ct	4.50%	450
BB/tt	0.25%	25
yy/CC	0.25%	25

The last column represents how many offspring out of 10,000 offspring one might theoretically find to have the given genotype. So, if you had a brood of 10,000 children, like several people I grew up with, then you could expect roughly 2025 to have the BB/CC genotype. Just as before, the next step is to break this down into phenotype distribution, which requires analyzing how each individual combination gets expressed. Let's suppose that the blue/yellow alleles behave in dominant/recessive pattern, whereas the circular/triangular alleles behave with blending pattern. That means that all the By genotypes will look the same as BB genotypes. The CC's, the Ct's and the tt's will each reflect their own phenotype. The table we just made will collapse into the following:

Phenotype*	Percent	Out of 10,000 offspring
B-/CC (blue hair, round eyes)	24.75%	2,475
B-/Ct (blue hair, round-angular eyes)	45.50%	4,550
yy/tt (yellow hair, triangle eyes)	20.25%	2,025
B-/tt (blue hair, triangle eyes)	4.75%	475
yy/Ct (yellow hair, round-angular eyes)	4.50%	450
yy/CC (yellow hair, triangle eyes)	0.25%	25
* B is assumed to be dominant over y		

Once again the dash means the other allele is irrelevant. This table also gives us an idea of how wildly disproportionate some of the phenotypes can be. For example, the odds are that to find a yellow-haired, triangle-eyed offspring, the parents would have to have several thousand children before even *one* is born.

Let's take one more example of the above genotype frequency and consider what the phenotype frequencies would look like if BOTH allele pairs behaved in dominant/recessive pattern. In that case, all Ct's would look like CC's, so the phenotype frequencies would collapse into:

Genotype	Percent	Out of 10,000 offspring
B-/C-	70.25%	7,025
B-/tt	4.75%	475
yy/C-	4.75%	475
yy/tt	20.25%	2,025

Notice how close the pattern comes to the 3:1 pattern (B-/C- close to 75% and yy/tt close to 25%). That is the effect of having the two allele pairs on the same chromosome pair. The closer they are on the chromosome, the smaller their chance of being separated from each other into different gametes. As their chance of separation gets smaller, the 4.75% gets smaller as well. Suppose the two loci are close enough that the phenotype probability of B-/tt and yy/C- are each, say, 0.001%. In that case, you would have a table like this:

Two alleles positioned really close on one chromosome		
Genotype	Percent	Out of 100,000 offspring
B-/C-	74.999%	74,999
B-/tt	0.001%	1
yy/C-	0.001%	1
yy/tt	24.999%	24,999

From this table, you can hopefully see that the closer two alleles get, the more they begin to behave as though they were a single locus. In this particular example, you'd need 100,000 offspring before the statistical outcome of one cross-over even happens to show the true nature of the individual loci. For some reason, I always seem to live next to the family that tries to do this.

So as the alleles get closer together, they begin to behave more and more like a single allele pair, but a similar argument could be made about their behavior when they are positioned farther and farther away. In the latter case, the more distance there is between them, the more independent they are. The more independent they are, the more they behave like two loci on two *different* chromosome pairs. It's a spectrum of inheritance patterns.

Single allele pair <---> two closely spaced allele pairs <---> two widely spaced pairs <---> two allele pairs on two chromosomes pairs

Summary

In this section, we discussed the inheritance patterns of two loci.

- When two loci are located on two separate homologous chromosome pairs, four possible gametes can result, each with 25% probability. When two such individuals mate, these 4 gamete types can lead to sixteen possible offspring, each 6.25% likely. The combined result is the same as a statistical superposition of two independent allele pairs.

- The phenotype distribution that results from sixteen different genotypes depends on the expression of the individual allele pairs. If both allele pairs behave in dominant/recessive fashion then the phenotype distribution will collapse into a 9 : 3 : 3 : 1 spread. If both allele pairs behave in a blending fashion, then there will be a large spread of intermediate phenotypes.

- When two loci are located on a single homologous chromosome pair, the individual alleles will tend to be inherited together. This results in two common gametes, along with two fewer common cross-over gametes. When two such individuals mate, the combined genotype distribution will be skewed away from the usual 6.25% probability per square. As before, the phenotype distribution will depend on the degree of dominant/recessive and blending tendencies.

- The closer two different loci are to each other on the same chromosome, the less likely will be a cross-over event. Two closely spaced loci will tend to produce an inheritance pattern similar to a single locus, whereas two widely spaced loci will tend to produce a pattern more like that of two loci on two different chromosomes. The degree to which two loci behave similar to a single locus gives us an idea of how far apart they are on the chromosome.

Unit 33

Polygenic and Complementary Traits

In the last unit, we explored the way in which genotype and phenotype distributions occur in the inheritance of two allele pairs, each one governing a different trait. During that exploration, we used the three-step strategy: 1) determine gamete probabilities, 2) fill in the Punnett square and 3) group the genotypes by expression. Now, let's wade a little deeper into the gene pool by studying the situation in which *two* allele pairs govern a *single* trait. A trait governed by more than one allele pair is called a **polygenic trait** because it requires more than one gene for its expression. Most human features are controlled by multiple genes . . . facial features, height, etc.

At any rate, in this unit we'll look at the length of the noses of the mungabeasts and pretend it is controlled by two different sets of alleles: locus A and locus B. In addition, we'll take a look at a unique distribution that arises when the trait requires at least one dominant gene at each locus.

Inheritance of Polygenic Traits

Let's look at an example of how these traits pass from one generation to the next. If the loci that control the single trait are located on different chromosomes, then the two pairs separate independently into different gametes, creating four possible gametes of equal likelihood, just as we saw in previous situations.

As before, if two such individuals mate then sixteen possible genotypes can result in the offspring. Since each gamete is 25% likely, each one of the sixteen offspring is 6.25% likely (0.25 × 0.25 = 0.0625).

	A/B	A/b	a/B	a/b
A/B	AA/BB	AA/bB	Aa/BB	Aa/bB
A/b	AA/Bb	AA/bb	Aa/Bb	Aa/bb
a/B	Aa/BB	Aa/bB	aa/BB	aa/bB
a/b	Aa/Bb	Aa/bb	aa/Bb	aa/bb

AA/bB is the same as AA/Bb, so we collect them together into a combined likelihood of 6.25% + 6.25% = 12.5%. Collecting the other genotypes in a similar manner we get the following list of genotype probabilities:

Genotype	Probability
1 AA/BB	6.25%
2 AA/Bb	12.5%
1 AA/bb	6.25%
2 Aa/BB	12.5%
4 Aa/Bb	25%
2 Aa/bb	12.5%
1 aa/BB	6.25%
2 aa/Bb	12.5%
1 aa/bb	6.25%

As with all the other examples, the phenotype distribution that results depends on the pattern of expression that the individual alleles follow. Suppose that both allele pairs follow a blending expression. For instance, let us say that AA contributes three inches to the nose length, Aa contributes two inches and aa contributes one inch. Suppose locus B contributes similarly (BB contributes three inches to the nose length, Bb contributes two inches, and bb contributes one inch). Since both allele pairs work together to determine the nose length trait, the combined effect of genotype AA/BB would be a nose of length 3 + 3 = 6 inches. The genotype Aa/bb would have a total nose length of 2 + 1 = 3 inches. Genotype aa/bb would have a total nose length of only 1 + 1 = 2 inches. These are only a few examples, but if we examine all the possible genotypes, we would find that many of them result in the same nose length, so we can group them as follows:

Genotype	Percent	Nose Length (in)
AA/BB	6.25%	6
AA/Bb or Aa/BB	25%	5
Aa/Bb or AA/bb or aa/BB	37.50%	4
Aa/bb or aa/Bb	25%	3
aa/bb	6.25%	2

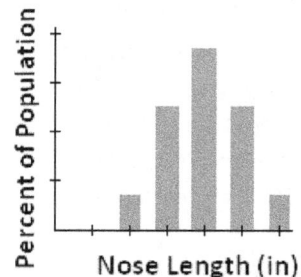

Suppose instead that the two loci contribute unevenly to the total nose length. For example, let's suppose that BB adds 5 inches, Bb adds 4 inches and bb adds nothing. Let's say that locus A contributes as it did before, 3 inches, 2 inches, or 1 inch, depending on what alleles were present. The phenotype distribution would then look like this:

AA/BB -> 3 + 5 = 8 inches, 6.25%
AA/Bb -> 3 + 4 = 7 inches, 12.50%
AA/bb -> 3 + 0 = 3 inches, 6.25%
Aa/BB -> 2 + 5 = 7 inches, 12.50%
Aa/Bb -> 2 + 4 = 6 inches, 25.00%
Aa/bb -> 2 + 0 = 2 inches, 12.50%
aa/BB -> 1 + 5 = 6 inches, 6.25%
aa/Bb -> 1 + 4 = 5 inches, 12.50%
aa/bb -> 1 + 0 = 1 inch, 6.25%

AA/BB   -> 8 inches, 6.25%
AA/Bb or Aa/BB   -> 7 inches, 25.00%
Aa/Bb or aa/BB   -> 6 inches, 31.25%
aa/Bb   -> 5 inches, 12.50%
AA/bb   -> 3 inches, 6.25%
Aa/bb   -> 2 inches, 12.50%
aa/bb   -> 1 inch, 6.25%

Genotype	Percent	Nose Length (in)
AA/BB	6.25%	8
AA/Bb or Aa/BB	25%	7
Aa/Bb or aa/BB	31.25%	6
aa/Bb	12.50%	5
AA/bb	6.25%	3
Aa/bb	12.50%	2
aa/bb	6.25%	1

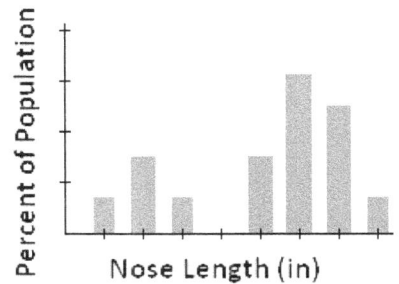

Let's do one more phenotype example from this. What happens if both loci behave in a dominant/recessive fashion? For instance, suppose that A is dominant over a: AA and Aa adds, say, 3 inches to the nose, whereas aa adds 1 inch. Let's say BB and Bb adds 3 inches and bb adds only 1 inch. The following phenotypes will then emerge (see graphic, following page).

AA/BB -> 3 + 3 = 6 inches, 6.25%   A-/B- -> 6 inches, 56.25%
AA/Bb -> 3 + 3 = 6 inches, 12.50%   A-/bb -> 4 inches, 18.75%
AA/bb -> 3 + 1 = 4 inches, 6.25%   aa/B- -> 4 inches, 18.57%
Aa/BB -> 3 + 3 = 6 inches, 12.50%   aa/bb -> 2 inches, 6.25%
Aa/Bb -> 3 + 3 = 6 inches, 25.00%
Aa/bb -> 3 + 1 = 4 inches, 12.50%
aa/BB -> 1 + 3 = 4 inches, 6.25%
aa/Bb -> 1 + 3 = 4 inches, 12.50%
aa/bb -> 1 + 1 = 2 inches, 6.25%

Genotype	Percent	Nose Length (in)
A-/B-	56.25%	6
A-/bb	18.75%	4
aa/B-	18.75%	4
aa/bb	6.25%	2

These look the same though!

56.25% : 18.75% : 18.75% : 6.25% ---> 9 : 3 : 3 : 1 ratio

Genotype	Percent	Nose Length (in)
A-/B-	56.25%	6
A-/bb or aa/B-	37.50%	4
aa/bb	6.25%	2

56.25% : 37.50% : 6.25% ---> 9 : 6 : 1 ratio

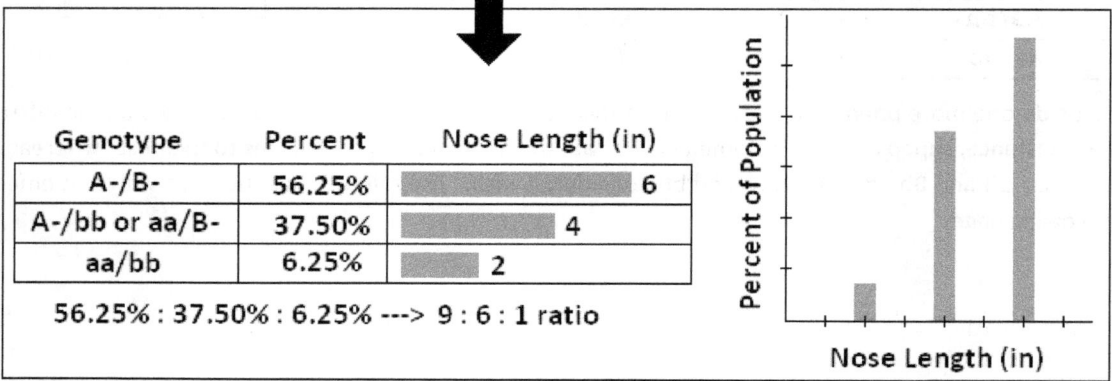

From the overall distribution, we see a 9 : 6 : 1 pattern emerge. This pattern is just the familiar 9 : 3 : 3 : 1 pattern in disguise – the same as the one we saw when examining the inheritance pattern of two dominant/recessive allele pairs.

The punch-line for this whole business is that when two loci work together to affect one trait, you end up with a funky spectrum of phenotypes. In the first case, the curve looks like a normal bell curve. When the two loci contribute unevenly to the trait, such as in the second case, you get a weird curve, and this particular example has two distinct regions: a large part of the populace will have big noses and a smaller part will have pathetic little noses. Finally, in the case of alleles that behave with dominant/recessive pattern, we saw the phenotypes collapse into a much smaller distribution, here 9 : 6 : 1.

In theory we could extend the argument to traits that are governed by three alleles, or four, or five, and so on. In those cases, we follow the same strategy. Figure out the number and probability of gametes (which can be complicated), fill in the Punnett square, and then group the resulting genotypes by how they get expressed. In the case of three alleles, for example, the Punnett square would be 8 × 8, because a genotype of, say, Aa/Bb/Cc can be segregated in eight different ways: A/B/C, A/B/c, A/b/C, A/b/c, a/B/C, a/B/c, a/b/C, a/b/c. So, a mating of two individuals having this genotype would require a Punnett square with eight possible gametes on the top and side. The probabilities also get a little more complicated, but the process is otherwise identical. In the end, you would end up with a smoother distribution curve. The more alleles involved in governing a trait, the more the distribution looks like a smooth curve rather than discrete columns.

Polygenics and phenotype distribution

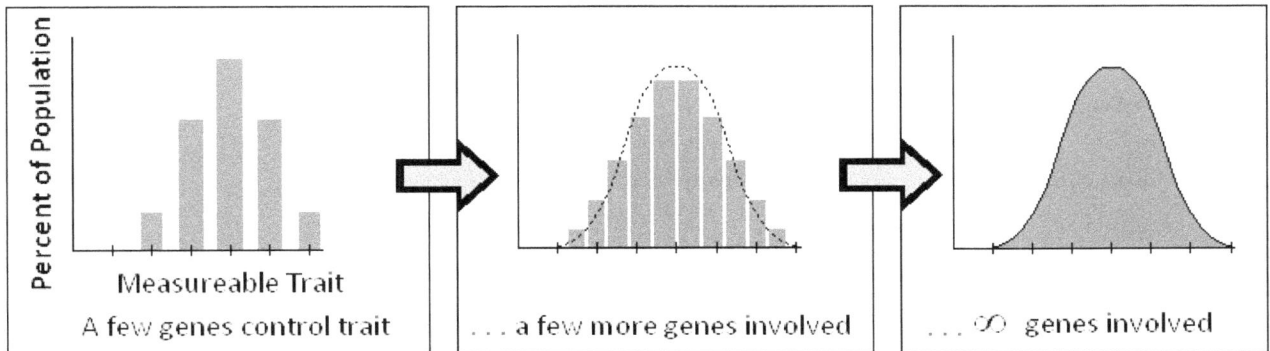

A few genes control trait — ...a few more genes involved — ...∞ genes involved

Finally, all these arguments can easily be expanded to cover the case of loci found on the same chromosome, as covered in the second half of the last unit. We use the same strategy: 1) find the gametes (here unevenly spread), 2) fill in the Punnett square and 3) group the genotypes by expression. If the two alleles are close together on the chromosome, then most of the gametes will have them segregated together, and a small fraction will have them swapped due to cross-over. The overall phenotype distribution will be similar to that of a single locus, with a small fraction of rare cross-over phenotypes. If the two loci are far apart on the chromosome, they will be inherited closer to how two loci on two different chromosomes get inherited. How closely they approach the single locus pattern versus the two loci pattern gives us an idea of how far apart the two loci are on the chromosome.

Complementary Genes

In biology, it is not uncommon to encounter polygenic traits that require at least one dominant allele at each locus. These kinds of systems are said to be **complementary**. In the simplest case of two loci, A and B, there must be at least one A allele and one B allele or the trait fails to appear. In other words, the appearance of either aa or bb will cause the trait to fail. If a trait depended on three loci, A, B and C, say, then there would have to be at least one A, one B, and one C present, or the trait fails. For example, if that system had Aa/bb/CC, it would fail, because it doesn't have at least one B. Important examples of these kinds of systems are most biochemical pathways, where each enzyme in the pathway is equally important. If even *one* enzyme in the pathway is coded by defective genes (say, aa instead of AA or Aa), then the *entire* pathway fails.

Let's see what would happen if we crossed an individual having a genotype of Aa/Bb with another having the same genotype. Once again, the gamete distribution and Punnett square is identical to that above, so we'll only concern ourselves with the phenotype distribution. After analysis, we have the following conclusions:

AA/BB -> Both enzymes function, 6.25%

AA/Bb -> Both enzymes function, 12.50%

AA/bb -> Pathway fails due to missing B enzyme 6.25%

Aa/BB -> Both enzymes function, 12.50%

Aa/Bb -> Both enzymes function, 25.00%

Aa/bb -> Pathway fails due to missing B enzyme, 12.50%

aa/BB -> Pathway fails due to missing A enzyme, 6.25%

aa/Bb -> Pathway fails due to missing A enzyme, 12.50%

aa/bb -> Pathway fails because both enzymes are missing, 6.25%

This collapses down into the following results:

A-/B-   -> Pathway works, 56.25%

A-/bb or aa/B- or aa/bb   -> Pathway fails, 43.75%

56.25% : 43.75% --> 9 : 7 ratio.

In other words, the system works/fails in a 9:7 ratio. This important ratio is used to confirm whether a system of two genes is complementary.

The 9 to 7 Ratio of Complementary Genes

A-/B-   aa/-- or --/bb

270

Summary

In this section, we discussed the inheritance of polygenic and complementary traits.

- When more than one gene contributes to a trait, the trait is said to be polygenic.

- A simple example of polygenic inheritance is when two alleles work in a dominant/recessive fashion to make a single trait. In that case, the distribution that results is 9 : 3 : 3 : 1, where the two 3's will look the same. This causes the 9 : 3 : 3 : 1 to collapse into a distribution of 9 : 6 : 1.

- If a polygenic trait relies on alleles that behave in a blending fashion, then the initial 9 : 6 : 1 pattern will spread out to a variety of curves. These can be bell-curved in shape, or can be funky with weird clumps and regions.

- The more genes that are involved in making a trait, the more the phenotype distribution will tend to spread out and get smooth, rather than appear as discreet columns.

- When a polygenic trait requires a dominant allele at each locus (i.e. at least one A, one B, one C, etc.), the alleles are said to be complementary. A system of two complementary genes will result in a phenotype distribution of 9 working : 7 failing. This is highly characteristic of complementary systems.

Unit 34

Other Genetics Terms

There are several more genetics terms that you will likely run into in a formal study of biology. Those terms are **epistasis**, **penetrance**, **expressivity**, **modifier gene** and **pleiotropism**. I'll list and explain each.

Epistasis

Epistasis is to multiple gene pairs as dominant/recessive is to individual pairs. In other words, one locus can override a second locus, regardless of the second locus' genotype.

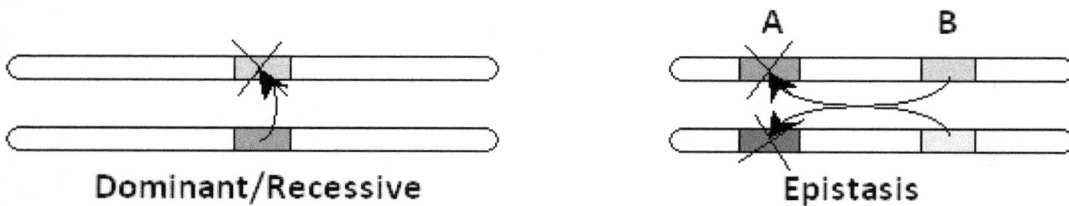

Dominant/Recessive   Epistasis

(The genes at locus B are "modifier genes"
for locus A, see term below)

Let's show this by example. Suppose we are studying a single pair of alleles, one for brown hair (B) and one for white hair (w). Further, suppose that brown is dominant and we're crossing a population of parents that are all homozygous for brown (BB). To be clear with our setup: *not one of the parents has the white allele – not a single* incidence. Given that situation, there is no reason to suspect that a child with white hair should *ever* be produced. Right?

Suppose every several generations or so, a white-haired child is produced, nonetheless. Technically this should not be possible since we started with only B alleles. What's going on?! One explanation is that new mutations are taking place (essentially turning a brown hair gene into a white gene). That is possible but unlikely. Mutations are rare, so you would not expect it to be happening on a regular basis. Another explanation – and one that is much more likely – is that *another* set of alleles (at a different locus) are having an effect on the brown-haired phenotype. Thinking about it logically, one might expect that the expression of brown hair would depend on two factors: 1) the production of pigment and 2) the pigment's transportation to the hair. If the genes required for the transportation of the pigment fail, the brown pigment will not be transported to the hair. So, even though an individual may be homozygous for the brown allele (BB), it will still end up with white hair. The second locus – the transportation locus – overrides the hair color locus. That is epistasis.

The main lesson is this: if you cross two homozygous individuals and get something you totally did not expect, then the phenotype in question depends on more than just that one allele pair. Something *else* is also involved, and – in the present example - its effects are standing out.

> **Epistasis** – The phenomenon where the phenotype of an allele pair is unexpectedly different because of the action of some other allele pair or pairs.

A real life example of epistasis is the appearance of albino mice in place of the usual gray or black color. The genes contributing to albinism are epistatic over the genes that create the color, so even though the appropriate color genes might be in the genome, the presence of the genes causing albinism overrides the color gene.

Penetrance

Penetrance is the percent of individuals who *should* have a certain phenotype (based on their genotype) and actually *do* have the phenotype. Let's continue the example we just used. Out of all the homozygous parents, one would not expect any children to have white hair – 100% should have brown hair. Suppose three out of every 100 individuals have white hair, for whatever reason. The penetrance for the brown hair phenotype is 97%. Penetrance of less than 100% indicates that there are more factors involved in the expression of a phenotype than just one allele pair. Epistasis is one reason, but environmental factors can play a part as well.

> **Penetrance** is the percent of the phenotype that actually "penetrate" the populace as a trait.

Expressivity

In contrast to penetrance, **expressivity** refers to the variation of the phenotype itself; it's like the quality of the gene expression. For example, some individuals with "brown" hair might actually have light brown hair, while others might have dark brown hair. If the penetrance is 100% (i.e. no unexpected white-haired offspring) then you can measure the expressivity by dividing up the phenotype into the various groups. For example, you might measure the expressivity in terms of percentage – 30% have light brown hair, 60% have mid-tones of brown, and 10% have dark brown hair. Like penetrance, variable expressivity indicates that other factors go into the expression of the phenotype, most likely polygenics or environmental.

> **Expressivity** is the variation of how the phenotype is expressed.

Modifier Genes

Modifier gene is a term that refers to the genes that alter the penetrance or expressivity of another gene. It's a generic term that can apply to most of the allele behavior we've talked about up to this point. For example, the "modifier genes" would be those genes that caused the 3% incidence of white hair in the population where you wouldn't expect white hair at all. "Modifier genes" might also be to blame for the variation in the color brown. You might expect an offspring to have dark-brown hair when it unexpectedly has light brown hair - something *modified* the color (assuming it wasn't something in the environment). When we talked about polygenic traits, any of the alleles in the group of genes that worked together to make a trait could be said to be a modifier gene of the other alleles. So, for example, if loci A and B both work together to contribute to the length of a snake, say, then we can say that the genes at loci B are modifier genes for the genes at A, and vice versa. If AA causes the snake to be one foot long, Aa causes it to be 3/4 foot long and aa makes it ½ foot long, then the genes at loci B "modify" that result by contributing their part to the overall length.

Note, however, that just because there is incomplete penetrance or variation in expressivity, it does not necessarily mean that other genes are to blame. There are lots of environmental factors that could go into altering the penetrance or expressivity of a gene, so just because some individuals have white hair or light brown hair, does not always mean there are modifier genes doing it. After all, if you encountered an individual on the street with neon green hair, you would not assume they had a strange set of genes; they've obviously dyed their hair. Hair dye is an environmental complication, not a modifier gene.

> A **modifier gene** is any gene that modifies the phenotype of another gene.

Pleiotropism

Pleiotropism is like the inverse of polygenics. Instead of a trait that is the outcome of multiple genes, it's a gene that causes multiple traits.

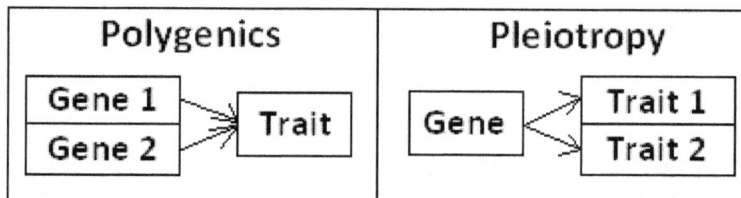

A classic example is from a series of experiments in the early twentieth century[69] involving the cross-mating of yellow and grey mice. Oddly enough, when a certain breed of heterozygous mice (Yg, "Y" for yellow fur and "g" for grey fur) were cross-mated, the result was a ratio of two yellow mice to one grey mouse. These results are unexpected. To see why, look at the following Punnett square:

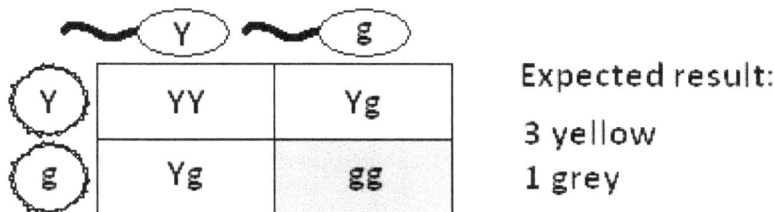

Expected result:
3 yellow
1 grey

From this Punnett square, you would expect a 3 to 1 ratio of yellow to grey offspring. The puzzle for the 2 : 1 ratio was solved by discovering that the gene in question, gene "Y", is a **lethal gene** to the mouse when homozygous: genotype YY is deadly. So, we would say that the allele Y is pleiotropic; it contributes two possible phenotypes: yellow hair *or* death before birth (when homozygous YY). Yes, in the broadest sense of the word, "dead" is a phenotype, and quite often its inclusion in the results is the only way to make sense of a medical study.

Another example of pleiotropism is the gene leading to the human disease phenylketonuria. In that disease, a mutation in the gene that makes the protein phenylalanine hydroxylase leads to the inability to break down a chemical called phenylalanine. This chemical begins to react in other ways and build up as various side products that are poisonous. These side products lead to multiple traits: brain damage, heart defects, low-birth weight, abnormally small heads, and behavioral problems – to name a few. The disease is autosomal[70] recessive, which means that both alleles must have the mutation (or any mutation that makes the protein ineffective) in order for the disease to happen. Either way, it's clear that the mutated gene is pleiotropic: it causes multiple traits.

> Pleiotropism is when one gene causes multiple traits. It's the inverse of polygenics.

---

[69] See the works of Lucien Cuénot.

[70] Recall that "autosomal" just means that the locus is on a non-sex chromosome.

Summary

In this unit, we discussed the various additional terms found in genetics.

- Epistasis is the process by which one pair of alleles overrides the phenotype of another, regardless of the other pair's genotype.

- Penetrance is a measure of the number of individuals in a population that have a genotype *and* its expected phenotype. If 100 individuals are expected to have brown hair, (based on their genotype), but only 90 have brown hair, then the penetrance of the trait is 90%. Penetrance less than 100% indicates some kind of complication, like another gene or environmental factor.

- Expressivity is the variation of a trait across a population. Variation in hair color from dark brown, to medium brown, to light brown is an example of expressivity. The presence of this variation is usually an indication that other factors are in play, such as other genes or environmental circumstances (up to and including the use of hair-dye).

- A modifier gene is any gene that modifies the phenotype of a second gene at a different locus. An example of a modifier gene would be one that alters the dark brown hair phenotype of another gene to light brown, or a gene that causes epistasis by shutting down the phenotype of another gene entirely. Remember, the modifier gene must be at a *different* locus than the gene it's modifying, otherwise it would just be considered a competing allele and be referred to in terms of dominant/recessive interaction.

- Pleiotropism is the phenomenon where one gene causes multiple traits. It is the inverse of polygenics.

In this section we cover:

1) Brief introduction to evolution, change and human philosophy
2) Fossil records
3) Radioisotope dating
4) Molecular and biochemical evidence
5) Morphological and embryological evidence
6) Biogeographical evidence
7) Microevolution and drug resistance

Unit 35

Introduction to Evolution

Up until now we've studied all the basic chemical and structural processes of life, and we've examined some of the basic patterns in which traits are inherited. Now, we embark on one of the most important topics of this book: **evolution**. In this unit, we're only going to give a qualitative definition of what evolution is and examine the evidence for it. In the next unit, we'll develop a more precise definition and then look deeper into the mechanisms that shape it.

What is Evolution?

In ancient times, mankind had a very static view of the Universe. We believed that all things never really changed. The planets, the stars, the continents, and even life . . . in the long run it stays constant. We believed that the Earth was at the center of all things and that everything else revolved around it. Further, we believed that humans occupied an elevated status among living beings: we were just beneath the deities and far above the lowly animals. People believed that all things were created at the same time. Our view of the Universe was very anthropomorphic, (i.e., human-centered).

We were made by the Great Tree and meant to inherit the Picnic Sandwich. All these things - including even the Mighty Creek and Meadow - revolve around us.

Hmmm... This sounds awfully ant-tropomorphic to me.

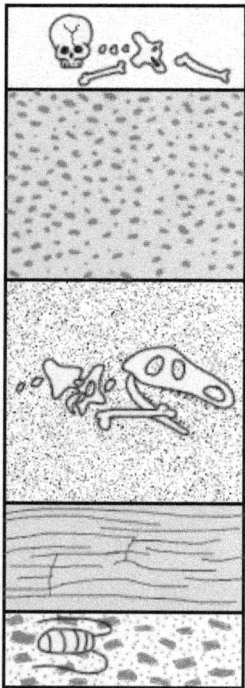

Fossils are always found in order of geological layers.

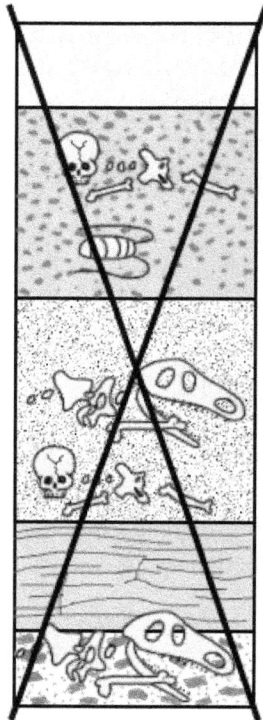

Fossils from different time periods are never found intermixed.

The path of science since those days has been a humbling one, as each of these beliefs has been shown to be highly unlikely, if not frankly incorrect. Not only have we discovered that the Earth is not at the center of the Universe,[71] but we've also found that we aren't above the animals at all. Everything about our biology works in nearly identical ways: we *are* animals. The only real significant difference, I might argue, is that we tend to think in more abstract ways. Further, millions upon millions of years of regularly layered fossils have suggested to us that not all species came into existence at the same time. On the contrary, the fossil record tells us a rather repetitive story; our version of life is only the most recent in a very long list of ever-changing sets of organisms. At the very start of the fossil record, we see life in very simple forms. The newer the fossil, the more complex it gets. After the dinosaurs, the slate seems to get mostly cleaned and the organisms get a little smaller, but are still complex. Overall, this gradual change in the structures of life is what we call evolution.

The basic idea behind evolution is rather simple: those organisms that have phenotypes that

---

[71] The Earth is incredibly far from it in fact. As the great philosopher Douglas Adams put it, we're in a very unfashionable end of our galaxy.

make them less able to survive and reproduce are less likely to leave offspring. Over time, the genotypes that control these less capable phenotypes become more and more rare, until eventually only the more "fit" genotypes remain. More often than not survival is for the "fittest" – which just means that those with the best genetics tend to have the best chance to survive and leave offspring. Then again, you don't necessarily have to be at the pinnacle of your species' genetics to have offspring, and in that sense evolution might be described as "survival of the fit-enough." Either way, change is introduced mainly in the form of 1) mutations in genes and 2) alterations in the environments, that both A) create new genotypes to be tested and B) alter the definition for what genotypes are best adapted. Essentially, mutations make new genes, and the ever changing environments are constantly changing their minds on what genes are best suited for survival. Over time, the species will gradually change. After millions and millions of years have passed, the species will no longer look or behave the same as it once did.

That's all well and good, but what other evidence is there that this process is really taking place? Well, in truth, the growing list of evidence has become a veritable mountain. First, I need to reiterate the fossil record. Ever since men have been digging up old dinosaur bones, they have always found the fossils to be consistent with the layer in which they find them. They *never* find human bones mixed in with dinosaur bones. The more ancient the creature is the deeper the fossil will be. Dinosaurs, in turn, are *never* found among the remains of those that came before. This leaves a strong chronological record – you line up the remains and follow the dots.

One may wonder whether this is just dumb luck, thinking that it's a matter of time before they find a fossil of a human and dinosaur together. Well, there is another way to date fossils to confirm the order of their appearance in geological layers. One such way is **radioisotope dating**. Not to be confused with some kind of atomic couple's tango, radioisotope dating is a method by which the amount of nuclear radiation emitted from the substance is measured and used to estimate the age of the sample. In the early sections of this book, we briefly discussed how some atomic nuclei are unstable. These particular nuclei often spit out pieces-parts in attempt to become more stable. One important example of this is carbon-14, an atom that has eight neutrons and six protons in its nucleus. It is formed on a regular basis in the upper atmosphere out of regular nitrogen (which has seven neutrons and seven protons) when the latter atom catches neutrons that constantly pelt the Earth from outer space. Once the nitrogen catches a neutron from the cosmic rays, it becomes unstable and spits out a proton to become carbon-14 (written ^{14}C).

Neutron from outer space + ^{14}N → ^{14}C + proton

Carbon-14 is also unstable, but it breaks down slowly and (like all nuclei) at a very predictable rate. For any given sample of the stuff, precisely half will have broken back down into nitrogen after roughly 5370 years.

At any given moment, a small fraction of the carbon in the air is in the carbon-14 form. It gets formed in the upper atmosphere and breaks down again somewhere else, cycling through each form on a constant basis. Well, once plants take in the carbon-14 and combine it into glucose[72] (see the photosynthesis sections of this book) it becomes trapped and no longer has access to the upper atmosphere. That means the amount of carbon-14 in a sample of biological tissue will slowly drop over time since it cannot get "recharged" in the upper atmosphere. So, the older the sample is, the less radiation will come from it as the carbon-14 slowly breaks down. Given that, scientists measure the

---

[72] From there it enters into the biomass when eaten by other organisms.

radioactivity remaining in a sample to get an estimate of how much carbon-14 is still left. Using this information, they can get a very reliable idea of how long ago the carbon got trapped. In other words, the radioactivity acts like a clock for the age of the fossil. The image below gives you an idea of how this works. The grey circles are the radioactive atoms. As time goes by, more and more break down into regular white circles. The number of grey circles that remains gives an idea of how old the sample is.

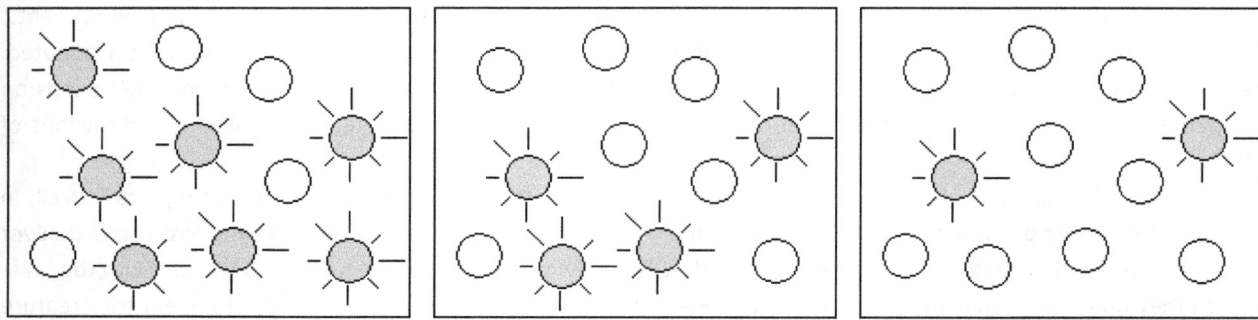

Young fossil   **Medium-aged fossil**   **Old fossil**

The point of the argument is simple: the fossil records always appear layered with the simplest forms of life at the most ancient levels. All other organisms are found at higher levels consistent with their fellow organisms. The fossils are never found to mix. Further, the radioactive dating of these fossils follows the same trend. The deepest fossils have the least radioactivity; they are very old. The dinosaur fossils have intermediate radioactivity; they are not as old. Human remains are always within, say, ten thousand years of age. Fossil remains of things that look *really close* to humans can be found before that. All of the radioisotope dating gives consistent results.

A consequence that immediately follows from radioisotope dating is that we are able to estimate the age of the Earth to be on the order of 4 to 5 billion years. This data is backed up by different radioisotope methods that use other kinds of radioactive nuclei, such as potassium-40, uranium, thorium, and so forth. This is in stark contrast to previous beliefs that the Earth is less than ten thousand years old and that all life was placed on it at roughly that time.

Some of you may not be impressed with all this physics, nuclear and geological mumbo-jumbo, so let's look at some more evidence. Consider the biochemical and molecular information. With only minor differences, life all works by the same reactions and patterns. The main biochemical processes in you and me have been consistently shown to be taking place in each new organism found. All organisms have cells. They all have ribosomes. They all have DNA or RNA, and it all works in the same way. The differences between one organism and another are often trivial in the greater scheme of things. Further, if you look at the similarities in genomes between organisms that look similar, you'll find the DNA to be strikingly similar as well. The more distant a species seems to be from a 2nd species, the more different its DNA will be as well. The fact that human DNA is *almost* identical to chimpanzee DNA is hard to overlook.

Next on the list of evidence are the similarities in skeletal structure. Even distantly related species show similar underlying bone patterns. Often cited examples are the bones of the forearm. If you examine the bones of the forelimb in a wide array of vertebrates,[73] you'll see that while the size of the bones differs, the overall anatomy is hauntingly similar. Each contains a single upper arm bone, two forearm bones, a number of small "hand" bones, followed by "finger" bones. Many have five fingers as well.

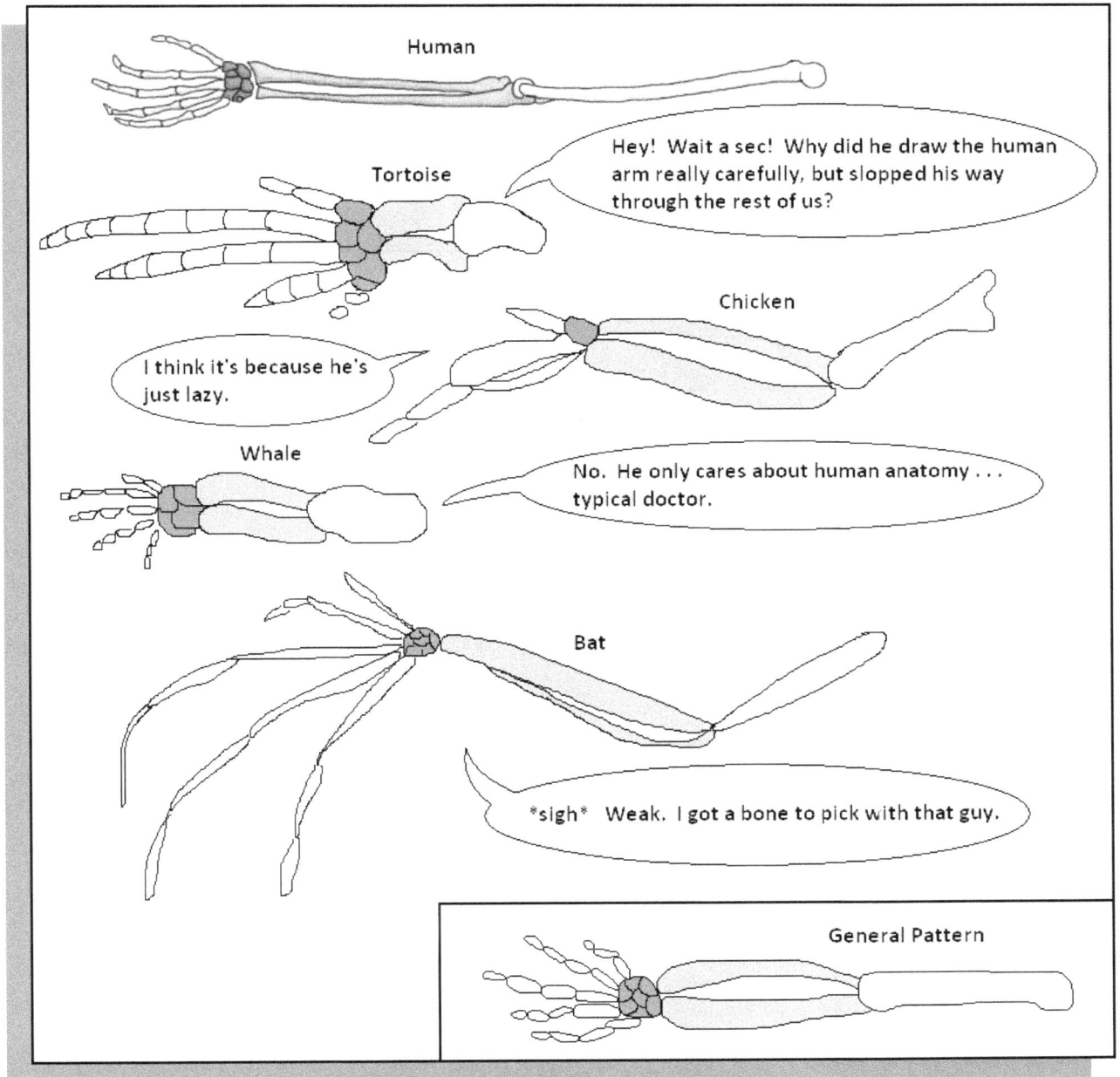

Human

Tortoise

Hey! Wait a sec! Why did he draw the human arm really carefully, but slopped his way through the rest of us?

Chicken

I think it's because he's just lazy.

Whale

No. He only cares about human anatomy . . . typical doctor.

Bat

sigh Weak. I got a bone to pick with that guy.

General Pattern

---

[73] Vertebrates are organisms that have a bony spine, (i.e. vertebral column down the middle back). These differ from organisms like my roommate, who appears to have no spine whatsoever.

The general pattern is shown in the bottom right part of the image. We see this pattern over and over in the fossil record for practically all land vertebrates having limbs, and it goes all the way back to a point about 350 million years ago with the appearance of a lobe-finned fish that could walk on land. From that point on, the pattern spread to all parts of the globe.

There is one more important thing to point out: there are organisms that have limb bones in places where they don't even have limbs! For example, some snakes and whales have tiny "vestigial" hind limb bones inside them, almost as if the limbs in question decided to stop growing early in the creature's life.

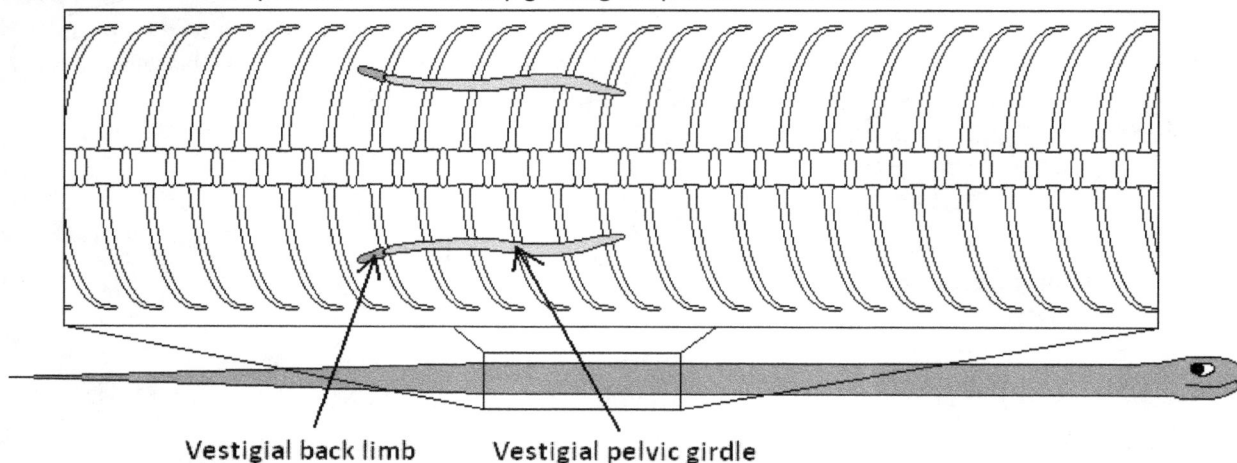

Vestigial back limb   Vestigial pelvic girdle

Why would a creature have hind-limb bones if it doesn't use them? The theory is that they evolved from primitive vertebrates that had limbs.[74] At some point, the ancestors of snakes and whales no longer needed legs at all, so now all that remains are tiny bones that are not used anymore. The fossil record is flush with all manner of intermediate body plans to support these ideas (i.e., snake-like creatures with smaller and smaller limbs, whale-like creatures with smaller and smaller hind-limbs, etc.).

Let's examine another area of evidence. Just after the egg and sperm combine during fertilization, the resulting fertilized cell undergoes a very dynamic period of growth and development called the **embryo** stage. While this is usually studied during the second semester of biology courses, (or even upper level courses), one observation from that stage gives us another clue about evolution. From what we can tell, all vertebrate animals go through the same initial stages; they look practically identical at first.

---

[74] The argument is something like this: The whale's ancestors lived along the coastal areas, and over millions and millions of years, their offspring that spent more and more time in the water were better suited for survival. Gradually, those that were suited to live in the water year-round became best suited for survival in that environment, up to the point that they stopped going on land. Now, one piece of evidence that they were once part of a land-walking species is the parts of the anatomy that still lingers but is no longer used.

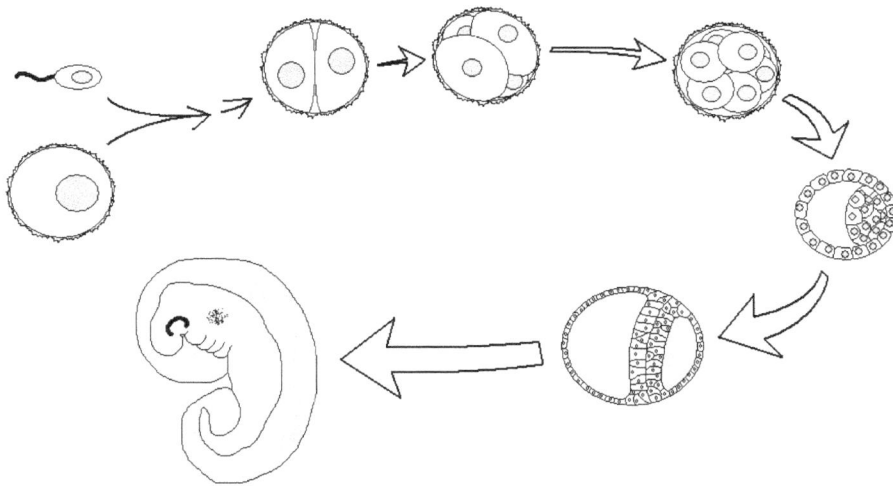

Fish? Bird? Shark? Whale? Bear? Cat?
Human? Weazel? Snake? Rat? Lawyer?

They all look like tiny aliens with tails. Even humans have tails at this stage, and the remnants of that tail stay with us as the tiny bones at the very bottom of your spinal columns . . . the ones just above your butt. Why would a human ever need to form a small tail? Why do all vertebrates start off looking practically the same? The prevailing explanation – based on countless fossil records, genetic evidence, geological studies, and other data – points to a common ancestor. As each new species developed from the original model, it carried with it the starting plan for development in the embryo stage. The more different two organisms are, the earlier they seem to depart from the original embryonic plan. For example, although fish start out looking the same as humans, they quickly diverge into their own growth patterns, whereas humans and chimpanzees look hauntingly similar for much longer in embryonic growth.

Let's go even further, this time examining **biogeographic** evidence. If species evolved from previous species, we would expect each geographic area of the world to have its own unique flavor of organisms, assuming humans hadn't tampered with the mix. Further, we would expect local areas to often have numerous variants of species that are closely related (e.g., hundreds of local kinds of butterflies, or beetles, or birds, etc.). In the cases where populations of creatures are separated from each other by long-term barriers, (such as an ocean), we would expect to see large variations between them.[75] The longer the separation, the more different the organisms might be.

Based on the geological evidence for continental drift, the major continents were all part of one big continent approximately 200-250 million years ago,[76] as discussed in Unit 43. Since then, the oceans have begun to slowly divide them, forming barriers for the movement of land animals. So do we find large differences in the animal populations from each of the continents? The answer, by and large, is yes. Each major land mass has a number of species that are specific to it, and most of these species have numerous local variants. If, however, you dig down far enough on more

---

[75] The exception to this would be animals capable of getting over the barriers, such as most ocean-going organisms, many birds, and anything transported by humans, such as horses.
[76] This uber-continent is often called Pangea.

283

than one continent, you'll likely find fossils of the same ancestor – from a time when the continents where still connected. Thus we could imagine that when the continents parted the ancestor on one land evolved one way, while the ancestor on the other land evolved a different way.

Still, there are animals that seem to be widespread across the globe. Consider frogs. They appear in all the major continents except Antarctica. Frogs come in a vast array of colors and sizes. Does this counteract our biogeographic prediction? Not at all! The frog's design is actually very old and primitive. Early frog fossils date back as far as 200+ million years – at around the time the continents were all still attached. Given their diversity, small size, and aquatic nature, it's not at all surprising that they are found just about everywhere. Further, it must be kept in mind that the same *kinds* of frogs are not found on each continent (again assuming humans haven't messed with things). For example, South American frogs are not found in China, or Russia, or England, or any other place other than South America, and vice versa, unless *we* import them there.

Bigger animals, such as koala bears, kangaroos giraffes, and buffalo have a much harder time getting from point A to point B. Once you put a barrier like an ocean in front of them, they're completely stuck. They stay on the same continent and diversify into a number of different breeds and species . . . until humans come along and muck everything up, that is. Of all our strengths and talents, our ability to completely screw things up is our paramount skill; try not to ignore the effects of humans on the world, especially when thinking about topics like evolution. After all, we didn't always have shopping malls, planes, boats, roads and cell phones surgically implanted in our ears.

Another key piece of evidence to present about the existence of evolution concerns the **microevolution** of viruses and bacteria. We see this every day in laboratories and fight against it in an endless battle in hospitals. Viruses and bacteria are constantly experiencing random mutations, and their rapid reproductive cycle acts to speed up the evolutionary process, testing each new generation against the antiviral and antibiotics we use. Every once in a while, a new mutation occurs that allows the virus or bacteria to be immune to the current anti-pathogen medicine. This new super bug gives rise to others that are like it, and before long the entire population is replaced by organisms that are no longer affected by the drug. This is evolution in action, made obvious because of the extremely rapid rate at which reproduction and mutations occur.

Most of the rest of the biological world does not reproduce quite so rapidly, but the mutations are still happening inside all other organisms just as they are in bacteria and viruses. Mutations are the driving force for trying out new genetic formula, and over the course of millions of years, these mutations slowly accumulate, causing gradual unperceivable change between one generation and the next. Arms slowly get longer or shorter. The shapes of eyes change, and ears get bigger or smaller. Tails get longer or vanish entirely. Flaps of skin attached to arms that help one generation fall gracefully from a tree get slowly larger until many generations later those who fall from the tree glide for long distances. Eventually, these flaps of skin become wings, allowing the descendants to fly from place to place. This is the way of things in biology: constant change. Like the changing Universe revolving around something *other* than us, the biological world is doing its own thing.

Summary

In this section, we discussed the basic ideas and evidence for evolution.

- Qualitatively speaking, evolution is the slow change in the number of surviving genotypes from one generation to the next, and is reflected by a change in the phenotypes. Mutations introduce new genotypes, which introduce new phentotypes that get tested over very long times, leading to slowly changing populations.

- The first evidence for evolution comes in the form of the fossil record. Fossil remains of animals are always found in the same order in the geological layers. The oldest organisms are very simple and are found in the deepest layers. The newer fossil remains are found in the layers that were laid down afterward. The most recent are highest in the geological layers. The fossil record is never found intermixed: human remains are never found with dinosaur remains, for example.

- Radioisotope dating involves measuring the radiation that is coming from a sample. The lower the radiation count is, the older the sample is. We can use radioisotope dating to confirm the age of fossil remains. The experiments confirm that organisms found in two different geological layers lived at different times – *in the order in which the layers were laid down*.

- Fossils and radioisotope dating has shown that not all organisms came into existence at the same time.

- Similarities in the fossil record also indicate a gradual change from one generation to the next, and by connecting the dots, we can get an excellent idea of how practically all species have evolved from primitive creatures.

- Cellular and biochemical methods have been used to measure the similarities in cellular function and DNA of today's organisms. These tests confirm yet again the degree of similarities suggested by the fossil record. They show that, for example, chimpanzees and humans are very similar compared to cows, or lizards, or fish. All of these organisms are more similar to each other genetically than they are to plants, and so forth.

- Morphological traits, such as bone structure, follow a very specific pattern from one species to another. The fossil record reiterates this pattern, and in the case of forearm bone structure, we see it all the way back to a primitive lobed-finned fish capable of walking for short distances on land.

- Embryological stages shows that all vertebrate animals undergo the same initial steps in development, until the point at which they all look like alien tadpoles. From that point, they diverge from each other, almost like a real-time reproduction of evolution right before our eyes.

- The unique organisms found from one continent to the next are another indicator of the evolutionary process. Primitive ancestors of two organisms were separated by a widening ocean, and from that point on the slow change that each experienced due to random mutation caused both to evolve in different directions. One might have become a black bear while the other might have become a polar bear, for example. There are few if any organisms that are truly trans-global . . . except humans of course. We've conquered the globe because our systematic inquiries into scientific processes have allowed us to create impressive transportation vehicles. This rather unexpected event is a first in evolution.

- Lastly, the process of evolution occurs in laboratories and in infections everyday as viruses and bacteria experience new mutations that get tested against the current antiviral and antibiotic medications being used. Out of the millions and millions of pathogens present in these situations, the odds are good that one will experience a mutation that will make it immune to the drug. It then gives rise to offspring that are immune. It's only a matter of time before the entire pathogen population evolves to be immune to our medication. This process of mutation is happening in all other organisms, only it happens at a much slower rate – slow enough that we can't perceive the change.

In this section we cover:

1) Natural Selection
2) Directional Selection
3) Stabilizing Selection
4) Destabilizing Selection
5) Sexual Selection
6) Speciation
7) Prerequisites for Natural Selection

Unit 36

Natural Selection

Now that we've looked at the evidence for evolution, let's begin to look at the details. Officially, ***evolution is the change in allele frequencies in a population from one generation to the next***. So, if allele p is present in 58% of individuals in one generation, but only 55% in the next, the population is said to have evolved. In this section, we will outline the major mechanisms that cause changes in allele frequencies, and most of these mechanisms fall under the category of **Natural Selection**.

Natural Selection

Natural Selection is the environmental bias in the survival of one phenotype over another. Suppose a new mutation occurs in a population of fish leading to a new larger tail. Presumably the fish with larger tails will be able to produce more propulsion through water than fish who did not inherit this new wonder mutation. What if the environment contains areas of rapid current that have bountiful food resources and/or spawning grounds? If so, "nature" will "select" for the fish that are better able to navigate the rapid current areas. Over time, the fish with larger tails will become more prevalent. This environmental factor leading to a change in allele frequency is called a **selection pressure** and is purely a result of random mutation and the physics of surviving in rapidly flowing water. This is Natural Selection at work, slowly shaping a species from one form to another, and over millions of years of this process, great change occurs.

The scientific community recognizes at least four types of Natural Selection:

1) **Directional Selection** – phenotypes evolve in one direction along a spectrum.
2) **Stabilizing Selection** – phenotypes evolve toward the middle of a spectrum.
3) **Disruptive Selection** – phenotypes evolve away from the middle, toward the extremes.
4) **Sexual Selection** – mate selection drives the evolution of gender specific phenotypes.

Directional Selection

The example I just provided is of directional selection, where a phenotype progresses along a spectrum to becoming more "extreme." In this instance, the fish tails got larger. One could imagine another mutation that makes the tails even larger still. The resulting fish would have an even bigger advantage over the previous fish. In this way, the fish evolve in a specific direction. If we dig up fossil bones and find the ancestors slowly evolving in a directional sense, we can guess that future generation might evolve further along the way as well. Bigger! Better! Faster! Right?

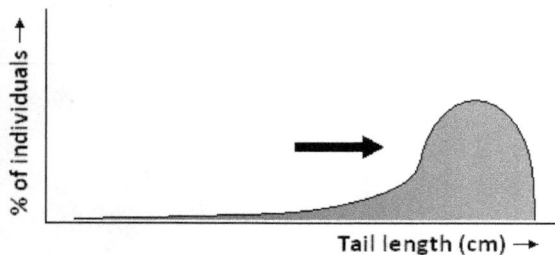

At this point, I would be doing you a disservice if I did not state a very important caveat that is often overlooked by society at large regarding the "directional" nature of evolution. Evolution can only result in organisms that are within the limits of the laws of physics. Life is no shortcut around Newtonian forces or thermodynamics. For example, while evolution might result in generations that slowly become stronger, the process *cannot* go past the physical parameters of flesh and bone. Beyond a certain threshold, calcium hydroxylapatite (the primary ionic material that makes bone) will snap and shatter. It does a species little good to evolve a muscle powerful enough to lift a building if the skeletal system is doomed to never be able to support it. At some point, the only way to get stronger is to evolve a more advanced skeletal system, such as a metal one. No instance of this has yet been discovered, and it is unlikely that it will. Always keep this in mind. *There is a limit to what evolution can do, and the physical laws determine that limit.* Also, directional selection is *not* a deliberate process, and bigger, faster, stronger and handsomer are not necessarily the goals of evolution. Evolution results in the organisms that are – simply put – able to survive and compete with each other. It does not create beings that transcend time and space or things that shoot lasers from their eyes. If you find such a creature, I'll eat my hat. That wouldn't be fun, trust me; my hat is rather old and gnarly,

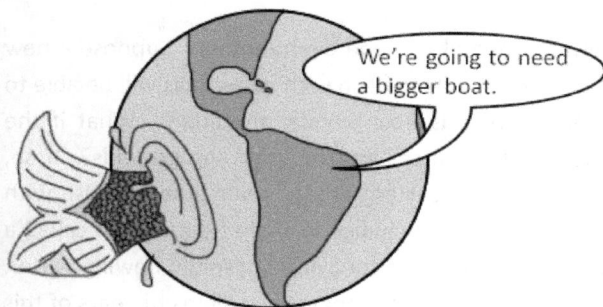

so I'm laying a lot on the line. No matter, if you were to come back two billion years from now (and the Earth is still here) you would probably see organisms just as ordinary (in terms of physics) as you have today. The only difference is that they would be *different*, but not super.

Lastly, the "pressure" to have larger tails does not cause the mutations to take place. The mutations are *random*. For every fish that has a mutation for a larger tail there are probably just as many inherited mutations for smaller tails, or mutations for goofier noses, or whatever else. Natural Selection does not "cause" any of these mutations to happen; Natural Selection simply steers the survival rates of organisms in such a way as to lead to a greater prevalence of organisms having mutations that help them survive and reproduce. It's like a giant filter, letting only those who can survive through.

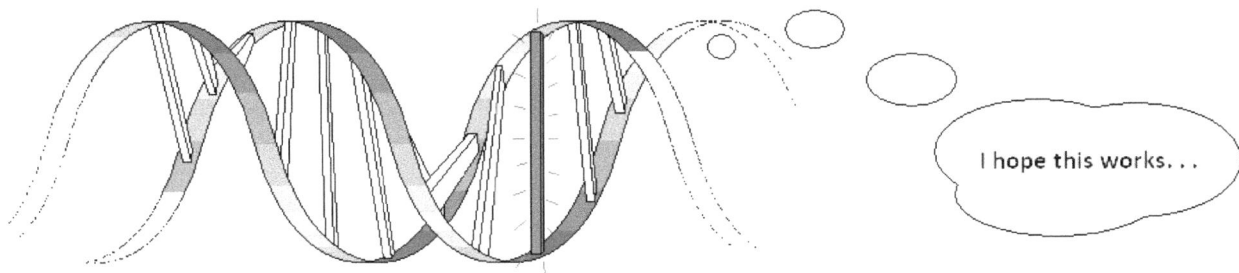

I hope this works. . .

Stabilizing Selection

The next kind of Natural Selection is stabilizing selection. Stabilizing selection is a process that causes a phenotype to stay within rigid parameters. Suppose we have a crab whose survival depends on body size in order to overpower its prey. This alone would apply a directional selection pressure that leads to larger and larger crabs. What if, however, crabs beyond a certain size are no longer able to fit under the typical kinds of cover provided by the environment? If a predator shows up, the really large crabs will be unable to hide; they'll get picked off. In this situation, the selection pressure would constantly lead to crabs that are of intermediate size.

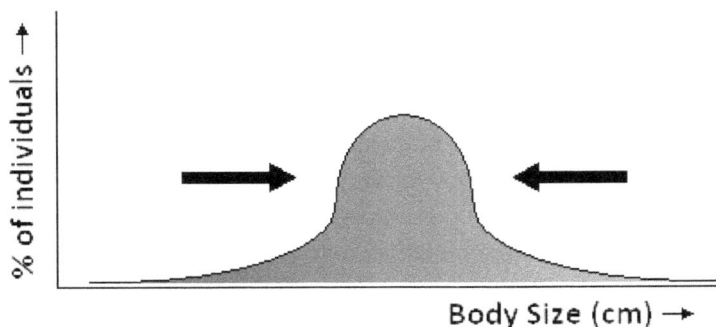

As another example, suppose you have a mouse that is living in a desert where the terrain is a deep red. The mice in that desert will blend in well if they are born with deep red fur. Mice that inherit "flawed" genes that cause their fur to be light red, or orange, or yellow, or razzmatazz even, will be less camouflaged and stand a greater chance of getting eaten. Given the situation, evolution will constantly result in a mouse having red fur; all others just can't make it. The red fur phenotype of the population is "stabilized" over evolutionary time.

Disruptive Selection

Disruptive selection happens when the middle range in a spectrum of phenotypes has less chance of survival and reproduction than the extremes. Suppose, for example, we have a population of fish that live in a coral reef. Further, suppose the predominant coral growing in the reef are either red or white. Given this case, the fish that are either red or white will have a good chance of survival because they will blend in well with their respectively colored coral. Fish that are intermediate colors (i.e. pink) will not be as well

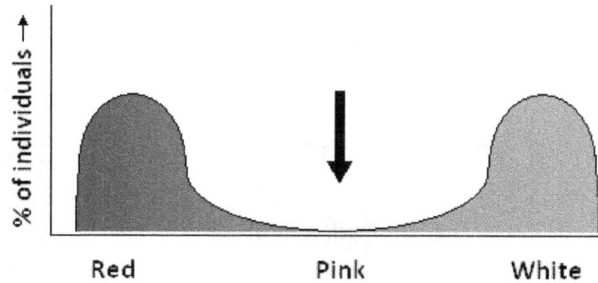

camouflaged and will get eaten by predators. They will also be wimpy; they're pink after all. Anyway, over time you will tend to see a split in the surviving phenotypes: some fish will be red and some will be white. Very few – if any – pink fish will survive. It's probably for the better, really; pink is such a silly color. Either way, this kind of no-man's land middle-ground effect is the result of disruptive selection.

Sexual Selection

Sexual selection occurs when picky mates look for particular traits in those competing for courtship. Traits that are selected tend to become more predominant in following generations. Common examples include healthy physique, unusual and extravagant plumage, book-writing skills and so forth. In most species, the female is the more selective gender. Why? The female is usually the one stuck with the task of gestating and watching over the young. In humans, as I'm sure you know, the female has to carry the fetus for approximately nine months. That is a *huge* investment in energy and resources. It also makes the female more prone to physical limitations during gestation, so in environments where predators roam around (like lions, sharks and politicians), pregnancy and young can lead to a greater chance of getting eaten. As a result, females have evolved a hyper-picky taste in suitors. They typically choose those that display the most potential in resources and physical prowess – things that assist her in providing for and defending the young.

This peculiar pickiness leads to some rather strange male phenotypes. For example, in many species of birds the males of the species will often have rather astonishing feather colors. Further, their plumage is often cumbersome as well. These feather characteristics give the male an expensive handicap. The bulky feathers tend to slow the bird (or otherwise impair its flying ability in some manner), and the bright colors attract predators. Yet the females consistently choose the males with the brightest, bulkiest feathers. Why? The theory is as follows: if a male can survive well with a handicap, it proves the rest of his genes are excellent. Being able to survive with such detrimental trappings must mean you are a hoss. What better way to show off how much of a high-roller you are than by giving the predator a head start and still coming out on top?

Other Mechanisms of Evolution

There are a few other mechanisms at work in evolution that are worth mentioning. In small populations, random fluctuations in the genes passed from one generation to the next can result in changes in allele frequencies. This is referred to as **genetic drift**. For example, suppose a population of six individuals (each having allele p and q) mate and produce offspring. For each breeding between partners, we draw a Punnett square. In each of those situations we would see a 25% chance that any one offspring would have pp genotype (versus pq and qq). If three offspring total were created, there would be a 1/64 (25% × 25% × 25%) chance that all the offspring would have only the p allele – the q allele will be lost spontaneously if this event occurs. This is an extreme example of genetic drift. Even without losing one allele due to probabilities, it is still likely that the two alleles will get inherited in different frequencies due to the really small numbers of individuals in a population. The punch line is that small populations tend to experience changes in allele frequencies from one generation to the next. It's not a deliberate process, but merely the effect of statistics.

The next mechanism that we sometimes see in evolution is the **bottleneck** or **founder effect**. This happens whenever there is a catastrophe that kills off almost all of a species or when only a handful of individuals from the species are used to colonize a new area. During either effect, several problems present themselves: 1) Some of the original population's alleles may have been lost if they are not present in the subset that survived or migrated; 2) Random genetic drift may cause more of the alleles to be lost since the new population is now very small; and 3) the remaining population is forced to **inbreed** in order to continue the species.

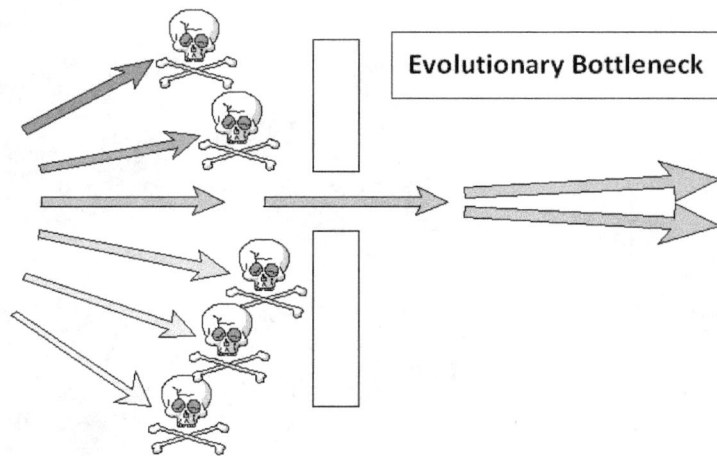

Evolutionary Bottleneck

Inbreeding – the act of mating with close relatives – is a socially unacceptable act for most human cultures and is not often seen in the animal world. The reason is that it leads to a genetic situation where there is a higher probability of congenital disease due to coupling of harmful recessive alleles. Most harmful alleles are relatively rare, but within any given individual in our society, (such as you or me for example . . . well, maybe not me) you'll likely find a few bad genes throughout the entire genome. If these alleles inside you are not partnered with a good allele on the other homologous chromosome, then you would have a genetic problem. You might have trouble with an enzyme in a chemical pathway, or be unable to produce a regulator protein and have a higher chance of cancer, for example. Either way, let us say that someone else (we'll stop picking on you) with a bad allele has several offspring. Each child has a 50% chance of inheriting the bad allele. So, if the original "bad allele" individual were to have six children, roughly three of the children will inherit the bad allele. Now, whereas the bad allele is rare within the population at large, it is *common* amongst these particular children. If these children then mated with each other, it is very possible that their respective children could inherit *two* copies of the bad allele, thereby leading to a congenital disease. In this situation, we only considered the possibility in which the original person had only *one* bad allele. In reality, most individuals carry several rare bad alleles at any given time. When you consider the odds of *any* of these alleles becoming homozygous because of inbreeding, we see that congenital diseases are not rare during inbreeding.

This kind of problem would become especially problematic if a species drops to just one male and one female. If the entire species is wiped out, except for one male and one female, then all the subsequence children must inbreed, leading to a genetic disaster and a good chance of instability in the species.

When all is said and done, anytime a species is brought to near extinction, its recovery may result in organisms that are not entirely the same as they once were, because no organism carries all the information that was once within the entire species. For example, suppose we sent a male and female astronaut to colonize Mars, and both happen to be Japanese. Further, suppose Earth blew up after they left so that the human race faced a bottleneck effect. All subsequent offspring of the two astronauts would be 1) Japanese, and 2) suffer high rates of congenital disease due to inescapable inbreeding.[77] The point is: if a species is limited to one male and one female, it will likely never be the same again. In fact, it may not even survive the first few rounds of reproduction.

Speciation

With all this talk of inbreeding and founder effect, a natural question might arise in the reader's mind: how does a new species even start if the initial stages require dangerous inbreeding? The answer to that question is that a new species does not start with only a *few* individuals. A new species forms gradually over millions of years as an *entire* population slowly changes. Each time a mutation occurs it spreads throughout the population, leading to small alterations. After millions of years, the gene pool of that species has so many mutations that it is fundamentally different from what it originally was. Suppose we have two populations of fish. Further, suppose the two populations were once a single population, but they got separated by a large mountain range. Over millions of years, each population will experience their *own* mutations. Now, if you compared the individuals of each population with those of the opposite population, you'd find a considerable amount of genetic difference. Further, if you tried to cross mate individuals with those of the other population, you'd likely find biochemical subtleties getting in the way: 1) chromosomal changes would lead to misalignment, 2) reproductive proteins wouldn't recognize and bind to each other very well, 3) mating rituals would differ leading to loss of sexual attraction and 4) physical differences in outer genitalia may complicate the matter even further, to name just a few problems.

At any rate, once you find that sexually healthy individuals from one population cannot mate and produce viable offspring (that can also reproduce normally), then they belong to *different* species. At some point in time, **speciation** occurred between the two isolated populations. In other words, a new species formed from the collective changes of the two populations, *not from the sudden appearance of a new and different individual*.

So if you ever hear someone say that evolution is wrong because they've never seen a monkey give birth to a human, you'll know they are grossly misinformed about how evolution works.

Prerequisites for Natural Selection

With all this talk about Natural Selection, there a few ground rules that we should probably cover in order to fully understand the prerequisites for Natural Selection. First and foremost, individuals have to produce at least as many children (if not more) than will survive to reproduce. If everyone in a population only produced one child, and every child died before it could reproduce, then there isn't much hope for that species, is there?

Second, Natural Selection works by applying a bias to the survival rates of different individuals based on their phenotypes, but in order to do that, the individuals *must be different*. Some of the individuals must be taller or shorter, faster or slower, have different color, etc. If everyone in the environment is identical, then Natural Selection won't lead to any differentiation between which of them is most likely to survive, and the species could not improve (become more fit) over time.

---

[77] This, of course, has absolutely nothing to do with the fact that the astronauts are Japanese. Any race of humans put in this position would face identical consequences of congenital disease and genetic bottlenecking.

Third, as we've just seen, Natural Selection works by selecting *for* or *against* certain physical traits. Individuals with poorly adaptive traits get wiped out; individuals with good traits tend to survive to produce more children. The only way the trait can increase in probability is if it is due to a genotype that gets inherited. In other words, traits that we acquire during our lives, like scars, illnesses, memories, behaviors, etc., do *not* get passed to the next generation because these traits are not recorded in the genes. The genes are the written record that gets passed along, and without it, the trait is forgotten. So, you can body-build all you want, but your sons will still be wimpy if they do not work out as well. You can pursue as many advanced degrees as you like, but without also studying and applying themselves, your children may still be dumber than a box of rocks. You can even become a politician, and your children might still turn out to be law-abiding and decent.

There is one final note to add to all this evolution business, and it's something that is very important to understand: individuals do *not* evolve (in the manner that is meant by science). You are what you are. You will not suddenly wake up with alleles that code for proteins to give you wings, or some other evolutionary change. Evolutionary change occurs between generations. You are stuck being whatever it is that you are, but your children may be slightly further along whatever evolutionary path that humans are following. The change will be imperceptibly small and the path unknown. Considering that most humans survive to reproductive age (at least in developed countries), it's hard to say what Natural Selection has in store for us.

Summary

In this section, we discussed the basic ideas of Natural Selection.

- Evolution is the change in allele frequencies in a population from one generation to the next.
- Natural Selection is the systematic bias in survival rates of individuals bearing different phenotypes. The genotypes that are responsible for successful phenotypes increase the odds of an individual's survival and reproduction, and so those genotypes tend to increase in frequency with each subsequent generation.
- Directional Selection is a type of Natural Selection in which phenotypes become progressively more extreme in character along a spectrum. For example, tails progressively get longer, or body height gets taller, etc.
- Stabilizing Selection is a type of Natural Selection in which phenotypes gradually tend toward the midpoint in a spectrum. For example, fur color might tend toward a middle color between two extremes, or body length might tend toward medium range values.
- Sexual Selection is a type of Natural Selection in which a mate will specifically choose certain phenotypes in the opposite gender. This often results in one gender having extravagant phenotypes. The most obvious example occurs in birds, where the male gender typically has unusual plumage and feather color. These characteristics tend to give male birds a handicap, so that only those whose overall genetics is excellent will survive. Females choose these birds to help guarantee that their offspring will inherit the excellent genetics.

- Genetic drift is a statistical phenomenon that occurs in small populations when the frequencies of alleles change randomly from one generation to the next, without any cause other than dumb-luck.
- The bottleneck or founder effect is when a small number of individuals from a larger population start a new population. The resulting individuals in the new population all share the subset of phenotypes that the founding individuals had. The rest of the phenotypes of the original population do not appear in the new population because the genotypes that led to them were not present in the founding individuals.
- Inbreeding can cause congenital diseases because rare harmful alleles are more likely to be shared by close genetic relatives. If these genetic relatives mate, the harmful alleles have a good chance of pairing up and being fully expressed. In contrast, the odds that two completely unrelated individuals will have the same harmful allele is very low, leading to a much lower chance of congenital disease.
- Speciation occurs when a population of individuals is divided by an impassable barrier. After many generations, the two resulting populations slowly acquire different mutations until the individuals from one population are no longer sexually compatible with the members of the other population. At that point, the two populations are considered different species.
- In order for Natural Selection to work, the following prerequisites must be met:
  - More children must be born than are killed off before reproducing.
  - Individuals in a population must be phenotypically different from each other.
  - The phenotypes that are selected for or against must be due to underlying genotypes. Otherwise there is no genetic record of that phenotype's level of success to pass on (or wipe out).
- Individuals do not evolve. Populations of individuals evolve from one generation to the next, and the change is often imperceptibly small.

Unit 37

The Hardy-Weinberg Equilibrium

The Definition of Evolution and the Hardy-Weinberg Equilibrium

As we saw in the last unit, evolution is the change in allele frequencies in a population. Any time the number of alleles for a given trait changes, evolution has occurred. Note that the definition of evolution does not imply that the change in frequency has to be permanent. Suppose allele p is present in 58% of individuals in the first generation, drops to 55% in the second, and then increases again to 58% in the third. In each generation, we would say the population evolved. You might think this is an indication that evolution would be a static process, tilting one way in one generation then back during the next, but keep in mind that thousands upon thousands of other loci may have changed allele frequencies at the same time. If you take the entire genome into account with each locus changing frequencies, then the odds are absurdly small that *each and every locus* will spontaneously shift *back* to a previous generation's frequency values. The more genes we add to the mix, the harder it is to keep the entire genetic sequence identical over many generations. If you add mutations into the mix, then it is practically impossible to keep the gene pool the same as time passes. Genetic ripples happen, and there is no life-guard.

Let's examine the argument from a different angle: Is it possible for two alleles to stay at the same frequency indefinitely? If we can devise a situation in which allele frequencies can be locked into place, we can refute the entire notion of evolution. This topic was tackled by the mathematicians Godfrey Hardy and Wilhelm Weinberg, and their conclusion runs like this:

If you examine a population having one of two possible alleles, then the alleles in question, p and q, will obey the following relationships:

1) $p + q = 1$
2) $p^2 + 2pq + q^2 = 1$

Furthermore, the frequencies p and q will stay the same provided **all** of the following conditions are met:

A) No mutations are occurring
B) Natural Selection is not occurring

296

C) The population is infinitely large

D) All members in the population participate in breeding

E) All mating is completely random

F) Everyone produces the same number of offspring

G) There is no migration in or out of the population in question.

First, let's consider the equations. Equation (1) just says that the sum of the frequencies is 1 (which is the same as 100%). That just means if p is at 57%, then q will be 43%. This equation is true if we are examining a locus that has only two possible alleles in the population, here p and q. In other words, it implies that there is no other allele "r," say in addition to p and q. The argument works if there are more than two alleles, but it is needlessly complex for our consideration here. We will assume a locus with only two alleles to keep the point clear and understandable. So, if 57% of the loci have allele p, then the other 43% must have allele q.

The second equation is slightly more nebulous, but it's just the result of crossing the frequencies during mating. In that event, equation (1) is multiplied by itself to become

$$(p + q)(p + q) = 1 \times 1.$$

If we expand this equation by multiplying it out, we get:

$$p^2 + 2pq + q^2 = 1$$

Hidden in this is our Punnett square: the p^2 (which is the same as pp) is the probability of the top-left box. The q^2 is the probability of the bottom-right box, and the pq's are the two other boxes in the Punnett square. So basically equation 2 just says that the probability of the individual boxes of the Punnett square adds up to 100%. This, of course, is something we've encountered numerous times before in the heredity units. To summarize the equations:

$p + q = 1$	tells us about the current population.
$p^2 + 2pq + q^2 = 1$	tells us what happens when the population reproduces.

Next, let's step through each condition of the Hardy-Weinberg Principle. Here is the big test: **If all of these conditions apply to a population, then that population will not evolve**. If even one condition fails, evolution – the change in allele frequencies – will occur.

A) No mutations. This one is probably obvious. If mutations are occurring then it means a new allele will appear in the population, say allele r. If allele r is present then the combined frequencies of p and q will drop. For instance, frequencies of p and q of 70% and 30% might drop to, say, 69.99% and 29.99%, with the frequency of the new allele, r, being 0.02%. Since the frequencies changed, evolution has occurred. Therefore, *no mutation can ever take place **or** evolution must occur*. Unfortunately, mutations are impossible to stop.[78] As we saw in previous

----

[78] The impossibility of stopping mutations has a lot to do with the Second Law of Thermodynamics (the frequently quoted "order to disorder" rule of science). Basically, any sequence of DNA will undergo change because of this law. This is ironic, since mutations drive the wondrous variety of

chapters, mutations occur at a rate of about one in one million base pairs. Some species have a somewhat lower rate. Some have a higher rate. Unless the mutation rate drops to zero (again impossible) evolution must occur.

Given this impossibility, the rest of the conditions in the Hardy-Weinberg equilibrium are just overkill, but for the sake of thoroughness, let's examine the rest.

B) No Natural Selection. This statement means that Nature has to be fair, killing everything with equal probability. We know from common experience that Nature is not fair. For example, the "runts" of each litter are kicked out of the nest in many species. The slowest individuals are the first to get run down by the lion. Freak accidents kill perfectly healthy specimens. That's life. At any rate, since Nature is not fair, it means that some organisms are enjoying Nature's favor. That influences which genes make it to the next generation, which of course changes the allele frequencies and brings about evolution.

C) Infinitely large population. This one is not very obvious, but what it basically says is that small populations suffer from random genetic fluctuations (remember genetic drift?). Suppose that we have a small population of, say, four individuals. Further, say these individuals have p and q alleles at 50% each. Dumb luck alone might cause all the children of the four individuals to inherit all p alleles. The result is that the q alleles basically vanish from the population due to freak randomness. If the population is larger, like say 100 individuals, then the statistical probability of such extreme random fluctuations are much smaller. The larger the population is, the smaller the fluctuations. The only way to remove the random fluctuations entirely is by having an infinitely large population. The only way to have an infinitely large population is to have an infinitely large planet, which of course is impossible. Therefore, genetic drift alone will lead to evolution.

D) Everyone must breed. This condition says that everyone needs to participate in the breeding process and that no one can have reproductive problems. If only a fraction of the individuals breed, for whatever reason, it reflects either 1) unfairness in Nature or 2) a possible difference in behavior due to the genes involved. Genes that cause an individual not to breed will get weeded out, changing the frequencies of the alleles. In addition, subsets of a population may not necessarily have the same frequency of alleles as the larger population. For example, if the population has allele frequencies of 50% p, 50% q, but the part of the population that is actually mating has frequencies of 40% and 60%, then this will cause the next generation to tend to have allele frequencies closer to 40% and 60%, leading to a change in the overall allele frequency of the population. Any of these effects results in allele frequency change, which is evolution.

E) Mating must be random. There can be no "hot" guy or "hot" girl. If mating is *not* completely random[79] it reflects unfairness in Nature *or* sexual bias of those selecting mates. Unfortunately, not everyone is as attractive as biology book authors; some people just don't attract members of the opposite sex as well as others. This means their genes will be less likely to be passed to the next generation. This brings a change in allele frequencies and causes evolution.

F) Equal numbers of offspring. If there is a difference in the number of offspring from one couple to the next, then those producing more offspring will cause the population's alleles to tend to shift toward their particular frequencies. For example, suppose those having the q allele tend to produce less offspring. Over time, this will cause

---

life. In a sense, disorder (i.e., chaos) is pushing evolution. Disorder has shaped our race into what it is, and we are kindly returning the favor to Nature.

[79] . . . such as if there exist differences in the sexual desirability of mates within the population.

the number of q alleles in the population to drop, which, of course, drops q's frequency. The change in allele frequency causes evolution to occur.

G) No migration. This is another no-brainer. If organisms are entering or leaving the population, then they take their p's and q's with them, thus changing the allele frequency of the population and bringing about evolution.

All of these conditions must be met before a population of organisms stays the same from one generation to the next. If even *one* of those conditions is violated, evolution *must* occur. What would life be like if all of these conditions were met? Well, we would all be identical people stuck in an infinitely large world swarming with infinitely many other individuals from every possible species. There would be no hot guys or hot girls, and we would all be forced to breed with whatever mate came along. Further, we would all have to give birth to the same number of offspring – NO MORE NO LESS! We would never be able to leave the planet, and no one would ever be allowed in – NO EXCEPTIONS! Last, and *certainly* not least, the laws of physics would have to ambiguously ignore the physical material we are made out of – from the tiniest of particles to the most impressive proteins, cells and organs – so that the genetic information that holds our design *never changes through mutation or accident*. Thermodynamics could do whatever it wanted everywhere else, provided it didn't burden us with its annoying tendency to change our DNA – after all, that wouldn't be fair if one of us got a mutation.

This would all have to be true for life *not* to evolve.

Now imagine the alien who shows up at our door after travelling millions of light years and observing the *same* physical laws *everywhere*, only to be faced with an inexplicable magical barrier around Earth that kept him out and kept in a weird pocket of physical laws that behaved queerly to ensure that everyone was treated identically. Imagine a world where nothing ever changed in a Universe *constantly* changing.

The truth is, we live in the Universe and have to play by its rules. Nature and its processes of life, though beautiful and fascinating, are *not* fair. Some people die in plane crashes. Some people get diseases. Some people never marry. Some people never have children, and some get stuck with lousy roommates. These are the realities we experience as creatures of physical matter. When you stop and consider it, randomness and change are the only unchanging facts of life. All you can do is play the hand you were given and hope the next generation is better off as a result.

Summary

In this section, we discussed the Hardy-Weinberg Equilibrium.

- Populations that have two possible alleles for a given locus can be described by the following equations:
  - $p + q = 1$
  - $p^2 + 2pq + q^2 = 1$

  where p and q are the frequencies of the alleles in question. Populations that have more than two alleles for any given locus have other equations that describe them, but the end argument turns out identical.
- If the values of p and q change from one generation to the next, evolution has occurred. For these values to remain the same, *all* the following conditions must be met:
  - There can be no mutations. (This alone violates the laws of thermodynamics)
  - Natural Selection cannot occur. (The laws of physics must behave so that we all live identically)
  - The population must be infinitely large. (We would need an infinitely sized planet)
  - Everyone must breed. (No one can have reproductive problems)
  - Mates must be chosen randomly. (No hot guy or girl)
  - Everyone must produce the same number of children. (If your neighbor has fifteen kids, then you have catching up to do)
  - No one can enter or leave the population. (Requires an iron-strict quarantine)
- A *single* violation of the above conditions dooms a species to change (i.e. evolve) from one generation to the next. The first condition provides new genetic material that causes new phenotypes to gradually creep into the population. The first four conditions are impossible to defy, and therefore evolution – as defined as the change in gene frequencies – *must* occur.

In this section we cover:

1) The theoretical beginnings of life
2) The Second Law of Thermodynamics and its role in life

Unit 38

How Life Started

Let's summarize everything we've found in this book. Particles come together to make atoms, and they do so because of quantum mechanics. The Earth is swarming with these atoms; it is covered in carbon, oxygen, nitrogen and hydrogen, all jockeying with each other for the perfect number of electrons. Sometimes they steal electrons and sometimes they share them, forming small molecules. Our planet is also pelted on a regular basis by an unfathomable amount of energy from the sun. That energy keeps these small molecules in a constant state of activity, and through this blitz of activity, the small molecules come together in very typical reactions to make mind-bogglingly-complex larger molecules. Some of the simplest of these complex molecules include amino acids, nucleotides and fatty acids.[80]

Panning back in our view, we see that DNA codes for proteins which guide these small chemicals to react to form larger biomolecules: carbohydrates, lipids, more protein and more DNA. The complete set of a DNA's proteins help to hold all these resulting molecules together into a coherently functioning cell – all of which acts to shield the DNA from the outside.

All the while, DNA is constantly experiencing errors, and we call these errors mutations. Most mutations are silent. The majority of the rest can cripple a protein, making the cell unstable and leading to the death of the entire organism. Still, a small percentage of the mutations lead to better protein versions. Better protein means a better functioning cell and better protected DNA. These new DNA/protein versions do better than the others and flourish. Over millions and millions of years, this process molds the population of DNA/protein systems into very interesting shapes and sizes, even leading to multiple cells working together for the same goal of survival.

To this array of chemistry and chemical accidents, we add the restrictions applied by the environment, which "selects" the DNA/protein systems that will survive and removes those that will not. These environmental restrictions we call Natural Selection.

That's all well and good, but we are unfortunately down to one final problem. DNA cannot exist without protein, and protein cannot exist without DNA. We have a chicken and egg scenario, and unless we find a solution, the entire thing will remain metaphysically unsettling.

As we move to attempt a possible solution to this dilemma, I must impress upon you a reminder of the thing that has given us all the technology we enjoy: we started this investigation with the **Principle of Natural Causality**. For

---

[80] Shockingly, these chemicals have been shown to spontaneously form in primordial conditions – that is, they were capable of forming on Earth during the volcanic, storm-ravished early times.

as unlikely as you would believe that all this could come into existence naturally, that principle forces us to keep looking for a natural explanation . . . if nothing else but for the sake of thoroughness. For as long as humans strive to better themselves, you can be certain that the drive to find a natural explanation will never stop. So, we may as well look into it.

As it turns out, the path we're about to take is rather interesting, but to fully grasp the argument, we'll need to step out of our biology comfort zone and look at the problem in a manner that you are probably not used to seeing. We're going to look at life in a completely new way, with our feet firmly planted in the arena of basic physics. Don't panic. We'll leave most of the math out of it. More specifically, we will focus on one particular law that can guide us to unravel this mystery of which molecule probably came first. That law is the Second Law of Thermodynamics. Without it, you are scientifically blind.

The Second Law of Thermodynamics

Thermodynamics is the study of how energy flows through physical systems. As you're about to see, however, it is more than just a look at energy. It is also a study of physical arrangements – a concept typically wedded to words like **chaos** (or more properly **entropy**) and **order**. The Second Law in particular deals with these things. It says that a system will tend toward a physical arrangement that has the greatest number of possible **microstates**. This is a somewhat complex area, so perhaps the best way to teach this is through example. Suppose we have one of those little plastic games that have the little balls in them. You know the ones; the goal is to try to arrange them in some particular order or get them into specific holes. Well, suppose we have such a plastic toy that contains a ten by ten grid and three balls, as shown here:

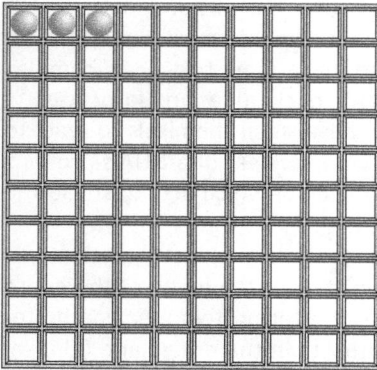

If you were to flip this toy over, the ball bearings would hop around and settle into new positions. In how many different ways could they possibly settle? Well, assuming the three bearings were identical, then the first one could settle into 100 different possible positions. The second ball could then settle into any of the remaining 99, and the last ball could settle into any of the remaining 98. That gives us 100 x 99 x 98 = 970200 ways that the three bearings could settle into the grid. Now, let us suppose we have 100,000 of these toys, and they ALL start with the ball bearings in the first three positions (just like the image to the left). Taken as a whole, the entire system is said to have only *one* arrangement – that of the first three positions. In other words, each toy has the same combination. The entire set is extremely ordered (there in only one specific state present in all individual games). Now suppose we set these 100,000 plastic toys on a vibrating table. If we came back in, say, one hour, we would find that very few (if any) remain in the original arrangement. In fact, the odds are good that each individual toy will have a completely different arrangement.

302

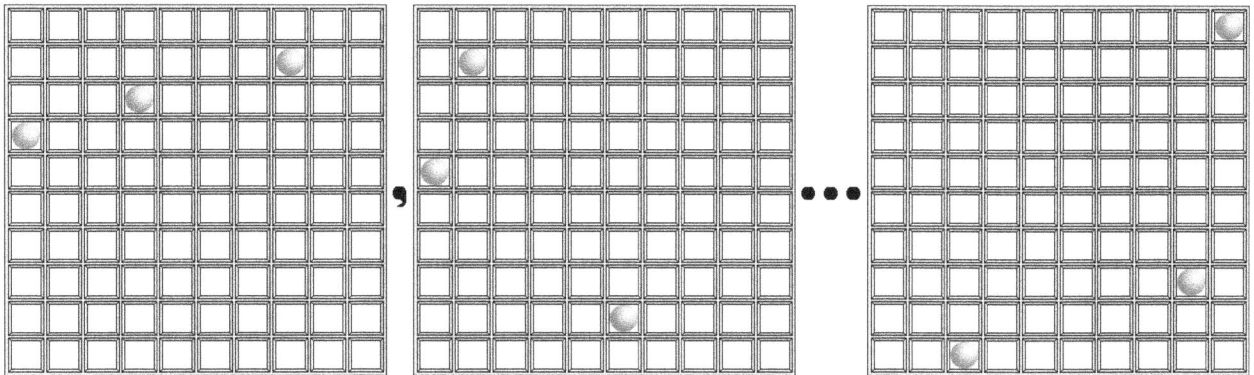

To put this another way, the system went from one possible combination (all in the top left three positions) to somewhere near 100,000 different possible combinations. So, in the jargon of thermodynamics: the system has tended toward an increase in the number of microstates. In regular-person talk: it experienced an increase in the number of combinations.

So what, right? Well, the reverse of this would be completely absurd. You would never expect to come back in another hour and find that all the toys spontaneously landed back in their original arrangements where the bearings occupied the first three slots. If it did, you would suspect that something was awry. In fact, if you were to watch such a weird process take place, you could only conclude that you were seeing it go backwards in time. This is the Second Law of Thermodynamics. All things tend toward a greater number of combinations and it is this process that helps us, as humans, to distinguish the directionality of time. We know by experience when we are looking at a tape being played in reverse; things just look like they are going in the wrong direction, and that is because we all have an inherent (albeit undeveloped) understanding of the Second Law.[81]

It is important to understand that the validity of the Second Law rests on *statistical numbers*. If you only have *one* plastic toy, then it would not be terribly surprising to come back and find it in its original position. After all, the original position is just as likely as any other position if you set it on a vibrating table. With more than one plastic toy, however, the odds of finding them *all* in the same position are astronomically small. Since each toy is doing its own thing, the odds are much better of finding them all in their own respective positions. This means that over time, each physical object changes in its own way, leading to the observed increase in combinations.

Let's take a few real life examples. Suppose we have a thousand drinking glasses sitting on a shelf in California. All these drinking glasses are identical. Now suppose there is an earthquake that knocks all the glasses onto the floor. Each will shatter into its own unique pattern of glass. To find two that shattered in the *exact same way* would be preposterous, if not downright weird. So, we say the system as a whole (the entire set of 1000 glasses) went from one possible microstate (the glasses all being identical) to 1000 microstates (each glass having its own arrangement of fragments). Further, we would not expect to come back some time later to find that the glasses have spontaneously hopped back together and leapt back onto the shelf. If we did see this we could only conclude that someone cleaned up the mess and replaced the glasses with new ones.

The identical argument could be applied to *you*. Suppose we have an infinite number of copies of you. Over time, each copy of you will change in its own way. We go from one combination to an infinite number of other

---

[81] A more robust way to state the Second Law is that it forces *closed* systems to become more disordered. In other words, nothing from the outside can interfere with the system. The Second Law is still acting on systems that are open, however. In the latter case, though, when order "spontaneously" happens, we seem to know instinctively to look for the thing causing it. *In the absence of any cause for order*, all systems – open or closed – tend toward disorder (increases in combination).

combinations. You can also think of this in a different way: since you can't know how your particular copy will change, you are faced with nearly an infinite number of possible rearrangements as you get older. To live forever, your DNA would need to code for a nearly infinite number of repair proteins. DNA is finite though, so no matter how much you might dream, you will eventually fall apart with age. It is the nature of reality. *You are rearranging, even as we speak.*

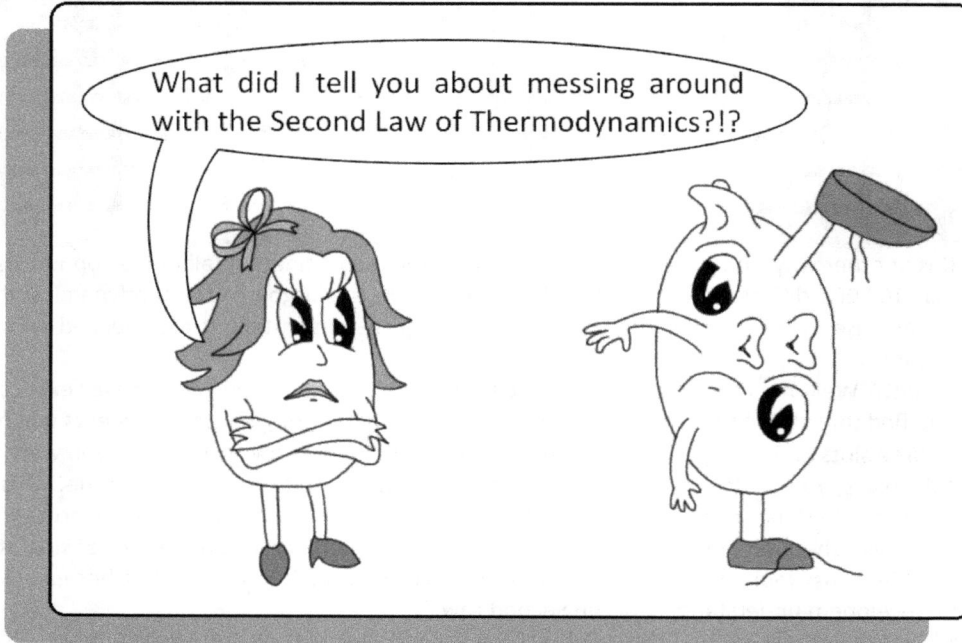

> What did I tell you about messing around with the Second Law of Thermodynamics?!?

Anyway, let's apply this to DNA, since that is the direction of our ultimate goal. Suppose we have 1,000,000 identical strands of DNA, each one 100 adenine's long as shown here:[82]

Hopefully, you can see where I'm headed. These million strands will each experience their own mutations so that over time we'll end up with one million *different* strands. As always, one combination shifts to become many combinations. DNA is not excluded from the Second Law. *That* is the driving force behind evolution, a *physical law.*

---

[82] The other strand would of course be just 100 thymidine bases. Since the structure of the partner strand can be determined by the first, we can ignore the second strand in our analysis to keep things simple.

	#1	#2	#3	#4	#5	#6	#7	#8	#9	#10	#11		#98	#99	#100
1)	G	A	C	G	A	A	T	T	A	T	G	...	T	C	C
2)	T	G	C	A	T	C	A	C	T	G	C	...	A	T	A
3)	A	C	T	A	C	G	T	A	T	C	C	...	A	A	G
⋮															
999,999)	G	T	A	C	T	T	C	T	A	C	T	...	A	T	T
1,000,000)	A	G	C	A	T	C	G	G	T	A	A	...	C	G	C

To further illustrate this, let's show the original combination as just a single dot, and all the millions of different combinations arranged in a circle around it, as if they all started in the center and spread out in their own directions. A few sample combinations are marked on the circle below.

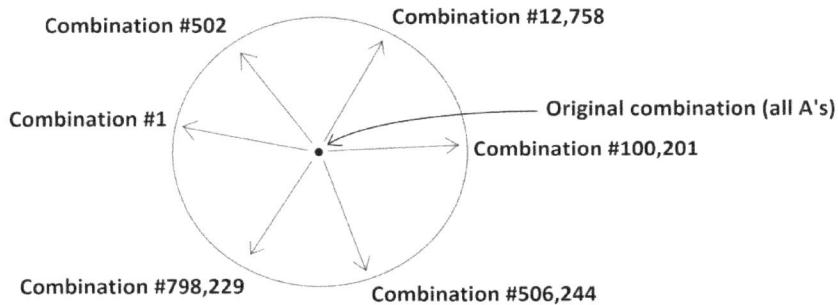

The same argument can be made starting with one DNA strand. The DNA will begin replicating itself, and over the course of millions and millions of years, the total collection of all errors will result in a massive array of different possible combinations. We can draw this in a circle, just as we did before. Here is where nature enters the picture and complicates things considerably. From our original DNA, we would expect millions of different pieces of DNA to emerge, but Nature unknowingly interferes by applying environmental forces that destroy the arrangements that are not viable for keeping an organism alive to reproduce and spread the genetic material. The result is a circle that looks more like a scraggy plant, with each branch representing a species. (In the following image, each branch ends in a major group of species).

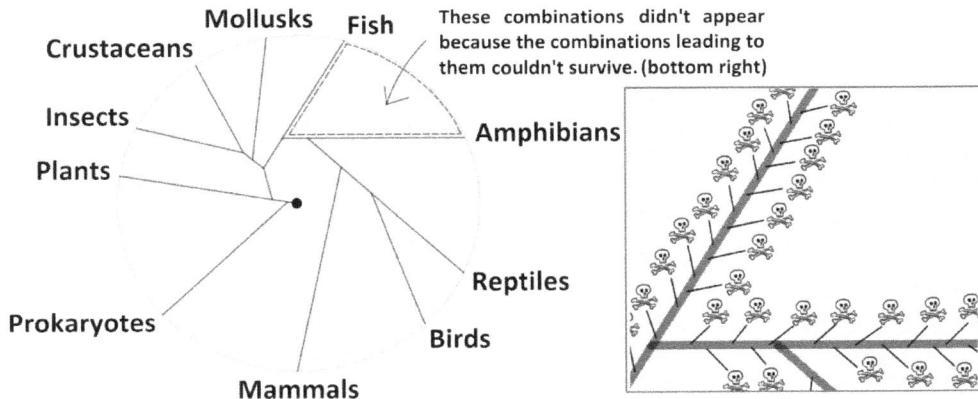

The punch-line is that this process gives us the illusion of design. The reality is just the opposite: the species were not designed, but simply exist because they are left standing after nature destroyed the rest. The entire process is just an aspect of a disordering physical law, where only a limited number of survivors make the final cut.

Weird as this may seem, it still doesn't answer our chicken and egg scenario, so let's dig a little deeper. . .

Let's gradually translate what we know about DNA into a general outline. First statement: *DNA interacts with atoms in proteins and solitary nucleotides, and uses this interaction to replicate.* This can be generalized by a second statement: *DNA reacts with atoms in molecules OUTSIDE ITSELF, and uses this interaction to replicate.* This can be collapsed into a third statement: **A self-replicating molecule interacts with external atoms to create more copies of itself.** It may be a strange way of saying it, but DNA can be described by the last statement.

The last statement suggests that we can achieve the same effects of evolution with *any self-replicating molecule.* So, let's try that. Self-replicating "molecule A" causes atoms outside itself to rearrange so that they also become molecule A. Provided it happens on a relatively small scale, this does not violate any physical laws,[83] but the Second Law will still act upon this set of reactions to cause mistakes to happen. A modified version of molecule A will form. Let's call it "molecule B." If molecule B is better at making copies of itself, then molecule B will out-compete molecule A. Before long, molecule B will win over the resources and become the dominant reaction, but continue making mistakes at the same time. Sooner or later, molecule C will show up on the scene, which will then take over because of greater efficiency, and so forth.

What might greater efficiency mean? Well, suppose millions of years down the line, a descendent of this reaction inherits the ability to use some of the nearby molecules as a "shield" to reduce the chance of accidental destruction. This improvement makes it more adaptable than the molecules that came before. The replicating molecule now has a *team* of molecules that are involved in the process of survival. Millions of more years will pass, and before you know it, replicating molecules now employ a more complex array of defensive molecules. The process becomes more and more elaborate, driven purely because the Second Law forces imperfect replication. The vast majority of these screw-ups fail to become better (and usually fade away) but once in a while an improvement accidentally comes along.

This scenario is all that is required to lead to life as we know it, and it says nothing about DNA or protein. All that is needed is a self-replicating molecule and the Second Law of Thermodynamics. The self-replicating molecule could have gradually become the modern day DNA, with the other molecules gradually becoming the proteins.

In this vein, it is important to understand that each replicating molecule inherits the defensive chemicals that its parent molecule used. Even today, no DNA molecule is booted from the cell and told to make its own fortunes. *When life replicates, it gives the new DNA a set of old proteins to start with.* The same could be said of multicellular organisms, such as ourselves. When we are born, our DNA is given a body to start with, and our parents (mostly mom) helped us put that body together. We could continue the analogy beyond birth. After we are born, our parents helped us to survive. After all, essentially none of us would make it as infants in the woods, and if the entire species decided to take that strategy, we would vanish from the Earth. To go *even* further, very few of us would be able to survive if our civilization's infrastructure collapsed. Not many know how to plow the land and grow food. Yet our infrastructure cannot exist without us, and we are almost at the point where we cannot exist without it. Imagine an alien visiting our hyper-advanced civilization a thousand years from now, when survival without our infrastructure had

---

[83] For example, if we look at the little plastic toys we talked about earlier. It is possible for small numbers of them to accidently land into the same position as each other. The point is that you would not expect *all* of them to do so. Physical systems *tend* toward increases in combinations, and at very large scales, local decreases in combination can happen due to "dumb luck" (a statistically unlikely but *possible* event).

become practically impossible.[84] The alien might conclude that 'what came first: man or machine?' was a chicken and egg scenario: people can't live without machines, but machines can't be built without people!

At any rate, the punch-line for this theoretical approach is that neither DNA *nor* protein necessarily had to come first. A "self-replicating" molecule could have been first on the scene. This may then have evolved into the DNA / protein interaction that continues to this day. The protein, in turn, helps maintain the other chemical types, such as the phospholipids that form the cell membrane "shield" around the entire thing. All these things are a continuation of a long legacy of gradually increasing complexity that results because of the Second Law of Thermodynamics.

But is all of this even possible? I mean, what are the odds? We are only one planet. What are the odds that this nebulous "self-replicating molecule" business would happen here? Well, to take a proper stab at this question, we have to step even further from our comfort zone and think about the big, BIG picture. . .

A Galaxy of Galaxies

It has been estimated that there are a total of two billion galaxies in the *known* Universe.[85] For effect, that is *2,000,000,000*, and *that's only in the part of the Universe that we know about*. The odds are very good that there are many more galaxies beyond our sight. If each galaxy is like our own, then each galaxy has a *minimum* of 200 billion stars. That gives us 2,000,000,000 x 200,000,000,000 = 400,000,000,000,000,000,000 stars, minimum. Suppose that, say, 0.1 % of all those stars have planets with similar orbits and composition as Earth (i.e. with oxygen, carbon, nitrogen, hydrogen, sulfur, etc.).[86] That gives us 400,000,000,000,000,000 Earth-like planets. If we say that the average molecular mass of molecules on an Earth-like planet is, say, 330,000 g/mol,[87] then each planet represents roughly a total of 1×10^{47} molecules.[88] How many molecules would be on all those planets combined? Well, despite annoying my publisher, I will list this out in purely numerical form to drive the point home. The total number of such molecules in the Universe would be:

40,000,000,000,000,000,000,000,000,000,000,000,000,000,000,000,000,000,000,000,000,000 possible molecules

Only *one* of those needs to display the ability to accidently self-replicate. By the way, all the numbers I gave are very *conservative*. The odds are likely that this number is much, MUCH higher. We only need *one* to match this requirement, and all the consequences of evolution will follow. As we're about to see, amino acids and many of the other basic biomolecules actually do *spontaneously* form in primordial conditions, so – by the Principle of Natural Causality alone – we can find all the parts of living reactions just *sitting around* waiting for something to happen.

"Wait!" You might say. "We only have one planet to consider: Earth! It makes no difference how big the Universe is!" This, at first glance, is a reasonable argument, but if we had developed on some other planet, we would

---

[84] This could happen if, say, the Earth continues to be transformed to the point where we can no longer farm for food in the old-fashioned way, or if far too few knew how to farm (and defend themselves) if chaos broke out. Alternately, it could happen if we colonize another planet that is only habitable because of technology (Mars?). In such a case, a catastrophic destruction of infrastructure would kill everyone.

[85] The numbers in this section vary depending on the source, but regardless of these differences, the outcome of the argument does not change significantly.

[86] This, actually, is quite reasonable. Oxygen, nitrogen, carbon, hydrogen, and sulfur are very stable and common, based on nuclear physics and the observations of the known stars.

[87] The rough equivalent of a molecule with the same mass as 100,000 DNA nucleotides – a *very* generous number. We could increase or drop this number and the end argument is still a monstrously big number.

[88] For those of you who have not had much chemistry, this calculation will make much more sense after an initial course on that material. Basically, each "mole" (abbreviated "mol") represents 6.22×10^{23} molecules. That is the same as 622,000,000,000,000,000,000,000 molecules. If each "mole" weighs 330,000 grams, then for each 330,000 grams, there are 6.22×10^{23} molecules. If the Earth is 6×10^{27} grams, then there are (6×10^{27} / 330,000) x 6.22×10^{23} of such molecules on the planet. The total is given above.

be sitting on that planet having this discussion instead. The point here is that a planet like Earth is not unlikely; equivalently, life developing by accidental chemical reactions is not unlikely either. If we take Earth alone, however, then the number of molecules playing in the game of life-starting odds are:

100,000,000,000,000,000,000,000,000,000,000,000,000,000 possible molecules on Earth.

Still, this nebulous "self-replicating" molecule is rather undefined and – despite the numbers leaning in its favor – an undefined molecule is unsettling. So, let's zoom in on a *very* good candidate molecule to show how this can be a very real explanation for how we got here.

RNA and the Primordial Earth

Our solar system is somewhere between 10-20 billion years old. The general pattern of its formation is as follows: The major celestial bodies - the sun and all its planets – formed from huge collections of space dust that collapsed inward due to gravity. This created roughly solid spheres, each with variable amounts of atmospheric gases remaining at their surfaces. These massive celestial bodies collapsed further inward, and the immense pressures and radioactivity caused their inner cores to heat up. The details depend on many factors, and the planets of our solar system are each somewhat different as a result. In the case of the sun, for example, the amount of material involved caused the heat to become so great that nuclear reactions began to take place, spilling light and other forms of electromagnetic radiation across the other forming planets.

Let's consider the Earth. As the outside of the Earth cooled it hardened, but the intense pressures from the core caused volcanos to sprout from cracks, spewing massive amounts of hot gases outward. This created the Earth's primitive atmosphere, which contained mostly H_2O, CO, CO_2, HCN, and H_2S. The huge amounts of water caused regular downpours, which etched away at the surface and dragged crystalline salt into the growing oceans.

It may help to understand that the natural tendency of gas is to settle downward because of gravity, but constant electromagnetic radiation from the sun keeps it very energetic, bouncing its molecules back up into the sky. As these molecules move upward, they tend to pull charges with them. This causes a build-up of a charge difference between the ground and sky. When that build-up becomes too great, lightning results – which is just electricity racing between the sky and Earth in attempt to neutralize the difference in charge. At any rate, because of a variety of factors and the vast amount of different gases on primitive Earth, our planet experienced a very violent electrical childhood.

This gives us a system of gases choked-full of hydrogen, carbon, oxygen, nitrogen, and sulfur, each of which is regularly exposed to massive amounts of solar, geothermal, and electrical energy. Under these conditions, it has been repeatedly shown that these simple atoms and molecules are capable of forming into amino acids and nucleic acids,[89] as well as a variety of other organic compounds. These complex compounds would have been too heavy to float around as a gas, so they would have gotten dragged into the oceans, lakes and streams of primordial Earth. To take the argument further, the Earth is estimated to be 4.5-5 billion years old, while the earliest forms of life appeared somewhere in the ball park of 3.5-4 billion years. That means the Earth would have had 0.5-1.5 *billion* years to simmer, slowly building up the concentration of complex organic molecules. That's a *lot* of organic molecules forming naturally over a very *long* time.

From here, there are a number of possibilities. Consider what occurs along the ocean shores, where regular formation of temporary ponds would have caused populations of these molecules to become trapped. The sun's energy would have warmed up such ponds causing them to begin evaporating. As the ponds get smaller, the organic molecules in them become more complex and get pushed together. Under the conditions of heated ponds such as

---

[89] In particular, check out the research of Stanley L. Miller, circa 1953.

these, basic amino acids spontaneously polymerize to form proteins.[90] Yep. Long before life was even a dream, its proteins were already here on Earth. Further, experiments have verified that many of the other biological macromolecules can form under primordial conditions as well. Once water floods into the drying ponds again, all these complex heavy molecules would get swept back out into sea. So, not only are the basic building blocks of life being formed, so are the bigger ones as well.

In addition to shoreline ponds, it can also be shown the clay has a catalytic effect, causing small populations of organic molecules to polymerize. Clay is not rare, and if anything it would have made the ponds of the previous paragraph more adept at creating biopolymers. Lastly, the formation of hydrophobic organic molecules[91] leads to the spontaneous formation of little cell-like bubbles, where the hydrophobic molecules line up in a very regular way (just like phospholipids). Along the surface of these molecules, inside and out, there tends to be yet another catalytic effect, leading to the increased formation of polymers. So, the thought is that small molecules would manage to get inside these bubbles and get attached to growing macromolecules. These big macromolecules would be too large to leave the "cell", resulting in a localized build-up of biomolecules, like little primitive communities. Further, as these little bubbles get bigger and bigger, they have been shown to split apart naturally, forming "daughter" bubbles over and over again. So, not only can they cause localized concentrations of biological macromolecules to form, they also act like primitive replicating cells. We see all this happen with just regular chemistry and physics . . . and remember we only need one of these molecules to have accidental self-replication abilities. That brings us to the punch-line.

We have already synthesized RNA molecules that are capable of self-replication. In Unit 11 of this book, I mentioned that RNA molecules have been shown to play catalytic roles, and the spontaneous formation of such biomolecules during the *vast* stretches of time on primitive Earth forms an excellent base upon which life could have started. As such, one of the major theories on how life started runs thusly: Life began with the accidental formation of a self-replicating RNA molecule, which gradually led to the development of DNA as the central genetic molecule during the course of billions of years. This slowly increased in complexity due to random errors forced upon the reaction by the Second Law of Thermodynamics.

Closing Remarks for this Section

Throughout this book, we have examined how life works, from its particles, atoms and molecules, to its energy use, to its complex structures. This investigation led us to a chicken and egg dilemma about which molecule came first: DNA or protein. The physical laws of thermodynamics allowed us to formulate an excellent potential answer to that problem in the form of a progenitor to the entire thing: a self-replicating molecule. We looked up at the stars and counted the astounding number of potential Earth-like planets. From that, we found that the total number of possible molecules in the Universe that we had to start with were so many that it *probably* doesn't even have a name for the number.[92] Out of that astral mix, we only need one molecule physically capable of accidental self-replication. If it happened on *any* of those planets, then *that* is the planet where we would have been reading this book. So, Earth is not unlikely. To that end, we have managed to put together self-replicating RNA molecules in labs after a few hundred years of tinkering with chemistry. Nature, on the other hand, has been putting RNA together for several billion years. So, all of biology boils down to this question: Throughout the Universe, (Earth included), what are the odds that this self-replicating RNA molecule was one of the more than 40,000,000,000,000,000,000,000,000,000,000,000,000,000,000,000,000,000 possible molecules?

If it was, then that molecule may be our oldest ancestor.

---

[90] For example, have a look at the work of scientist Sidney W. Fox.

[91] . . . which can also be seen in many asteroid samples, by the way.

[92] I'm sure mathematicians have a name for numbers this large, but to be honest, naming it would do no justice to its staggering effect on the odds of life's spontaneous birth. I'm sure I will be approached at some point and told the name of the number, but I'm deliberately leaving it out. Getting lost in semantics is one our race's impressively sad skills.

Summary

In this section, we discussed theories on how life started, while relying on the Principle of Natural Causality

- The Second Law of Thermodynamics tells us that all physical systems will tend to change in a way that increases the total number of combinations (whenever possible). This can be seen if you compare more than one identical system over time: each one tends to change in a way that is unique to it. If they remain the same, or converge to become identical, you'll need to look for a reason for the exception.
- The Second Law of Thermodynamics acts upon DNA to cause mutations, so that one copy of DNA will lead to a huge number of different DNA combinations over the course of millions of years.
- The species we see today are the lucky survivors in a process where nature is constantly removing genetic combinations that cannot survive. The result is the illusion of design in what appears to be part of a disordering phenomenon.
- For the process of evolution to begin, all that is required is a self-replicating molecule. This molecule will replicate imperfectly (due to the influence of the Second Law of Thermodynamics encouraging the formation of new combinations), and these imperfections occasionally produce a molecule that is better than the ones that came before. From here, the theory is that the improvements continue until external molecules are incorporated in the process as "shields" for the self-replicating molecule. Eventually, the process leads to highly complex systems, such as cells and multicellular bodies.
- Out of the astronomical number of planets with Earth-like orbits and composition in the Universe, only one self-replicating molecule is needed to start this process of life. The thought is that our planet happened to be one of potentially many planets on which this kind of molecule was accidently formed.
- In primordial Earth conditions, simple inorganic molecules have been shown to spontaneously form many of the complex biomolecules necessary for life. RNA polymers are among those that can be spontaneously formed.
- An RNA polymer that can self-replicate has been made in labs, suggesting that such a molecule could have been formed accidently and spontaneously at some point during the billions of years in which Earth was accumulating organic compounds.

In this section we cover:

1) Basic ecological definitions
2) Population growth models
3) Factors affecting population growth

Unit 39

Ecology I: Population Size

Populations, Communities, Ecosystems and the Biosphere

Until now, we've been up to our knickers in the specifics of chemicals, cells, inheritance and finally evolution. Now we make a quick transition to how groups of organisms tend to interact with each other. This area of study is broadly known as **ecology**, and it includes many factors found in physical reality and how these factors influence the survival of a species.

Before we get all bothered over the details, we need to lay down a few definitions. First is the definition of **population**. We saw this word before in the early parts of this book, but that feels like years ago, so let's rehash it: a population is defined as a localized number of individuals belonging to the same species. So the individuals of the species of fungus left in my fridge by my roommate represent one population of that species of fungus. The same species found in *your* fridge (left there presumably by your roommate, of course) is a different population of that fungus.

Next on the list of definitions is the **community**. The community is the group of interacting species within a given area. So, the fungus, the bacteria and whatever else he left in my fridge all comprise a community of different species, all interacting with each other. Your local forest is a great example of a community of living organisms – a large variety of different species interacting with each other.

Getting one step deeper into the definitions, we now encounter the **ecosystem**, which is the community plus the physical nonliving structures in and around the community. So, the insides of my fridge – the shelves, the lining, the light he broke, and so forth – are sadly a part of the ecosystem growing in my fridge. I would love to add cleansing materials to that ecosystem, but I honestly fear from my life; I'd rather the fungus remain out of *my* immediate ecosystem, after all.

The example of my fridge is rather small-scale. Usually all these terms are used to define much larger examples. A forest contains many populations, since there are many different species. The forest is a community,

311

since there are many populations within it. The forest also represents an ecosystem, because it typically contains nonliving elements, such as rocks, water, air and dirt, among other things.

To wrap up all these definitions, all the ecosystems on the planet are contained within the **biosphere**. The biosphere encompasses the relatively thin shell of the Earth's surface. Most life on earth is contained there. The deeper you go in the Earth, the less life you encounter. Likewise, above the clouds the incidence of life drops off as well. Since there is only one Earth,[93] it means there is only one biosphere. The biosphere is the highest level of organization in biology, and it is virtually a closed system in terms of materials. That means that atoms rarely escape from it into outer space. The only thing that enters and leaves the biosphere is energy: the Sun's energy comes in and heat radiates out. All other materials cycle around over and over again.

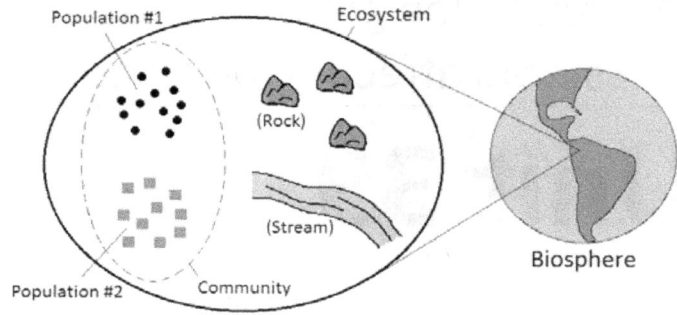

Population Shape and Size

Now that we know that a population is a localized collection of individuals that are all of the same species, let's dig a little deeper into the subject. Specifically, let's look at how populations appear and how they grow or shrink in size. We'll have to be somewhat general and do a bit of hand-waving . . . after all, animals and plants come in a huge variety of shapes, sizes and living styles.

One particular aspect of populations that scientists are often concerned about is how spread out or how "distributed" a population is. Are the individuals living closely together? Are they spread out and solitary? Is the distribution uniform or random? The answers vary, but how individuals are distributed often falls into one of three categories, shown in the image that follows.

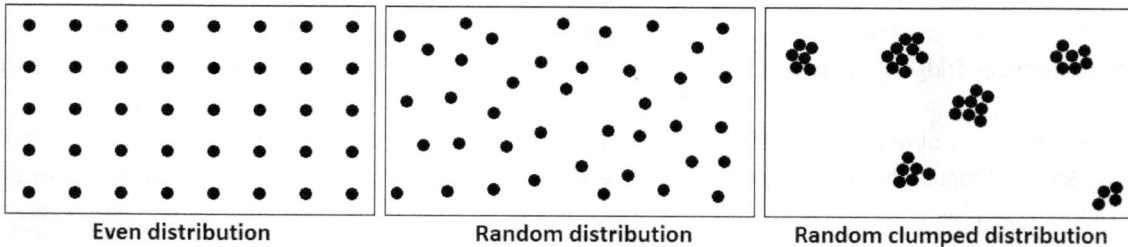

| Even distribution | Random distribution | Random clumped distribution |

Of the distribution types shown, the random distribution and randomly clumped distributions seem to be more common than evenly spread-out populations. Evenly spread-out populations are unlikely, because the physical environment is rarely uniform enough to allow animals to spread out evenly. As such, nonsocial species spread out

---

[93] I'm sure this is a very reassuring thought to all the aliens out there in the cosmos, each holding us at arm's length. They probably complain that their roommate left us here to ruin a perfectly good solar system.

wherever their environment allows, creating a random distribution. Social organisms usually clump together,[94] and the clumps are randomly distributed based on where the environment allows. In addition, clumping tends to be where major resources can be found, or where there is a minimum of environmental hazards, and so forth. Consider human cities, for example. Our cities tend to be associated with oceans or major water ways, primarily because water constitutes a major resource and means of mass transit (especially in the old days).

On the other hand, clumping isn't always an indication of social behavior. Take the example of water again: if there's a water hole in the desert or other arid place, you'll often see a bunch of organisms clumped around it, regardless of how social they would be otherwise.

Another strong motivator for clumping is reproduction. Asexual reproduction results in offspring that are nearby, such as might occur in many fungi. Similarly, sexual reproduction results in clumping because of the need to find a mate. In the latter case, some nonsocial populations are clumped only during mating seasons, after which they return to being spread out.

Here, a writer in his natural habitat prepares to defend his watering hole against a deer.

Other basic aspects of populations that scientists are often concerned about are population size and total biomass of the species[95]. Let's focus on the population size, or simply: the number of individuals in the population. How do we measure this? Consider plants first. They are usually not particularly hard to count, and to accomplish the task we often use a technique called the **quadrat** method. This entails dividing up the region into squares and then counting the number of organisms within a random sampling of squares. Once you have counted a sufficient number of squares, you can assume the other squares are similar. So, if 20 out of 100 squares each have an average of 15 plants in them, then we can assume that the total number of plants in the region is roughly 100*15 = 1500.

Now, in theory this method works for animals as well, especially those that are relatively slow or those that do not migrate, like snails or spiders and several of my relatives . . . but what about the fast moving animals? How in the world do we count *them*? They are constantly moving this way and that, migrating and scattering, especially the moment *we* show up.[96]

---

[94] ... with exception to grumpy bio book writers, but I suppose that falls under nonsocial.

[95] Usually directly related to the number of organisms in a population, the total biomass just represents how much mass the population has. This piece of information can sometimes tell you about the nutritional state of the population. If there are a lot of individuals in the population, but the total mass is surprisingly low, it suggests the population is starving. At other times, it becomes too difficult to distinguish 'individuals' of the species, so the total biomass serves as a more efficient measurement to estimate population size.

[96] ... which proves they are smarter than we give them credit for.

Well, one method, called **marked capture** (or mark and recapture[97]) involves the initial capture and tagging of a bunch of animals in the wild. At a later time, animals in the region are captured again, and the percentage of those wearing the tag gives an indication of how many animals are in the region. For example, suppose we captured and marked 100 fish within a lake. Two days later, we captured 100 more and only 10% of those were tagged. From that, we can guess that one out of ten fish in the lake are tagged. Since there are 100 tagged fish in the lake, there must be around 1000 fish total. The disadvantage of this method is that it does not take into account animals born, animals that die, or animals that migrate in and out of the area.[98] You can use this method on all kinds of critters, from frogs, to deer, to rabbits, to the children in your neighborhood.[99] Using this technique, I've managed to determine that there are roughly 2000 children in my neighborhood, and with more advanced techniques I've determined that most of them are clustered in the next apartment. Will the one with the blond hair *please* give the one with the brown hair his toy back? All the humanity!

Now we have a basic idea of how we measure the size of a population, but why would we want to know this, anyway? Well, by knowing the size of the population and observing how the size changes over time, we can get an idea of how populations are affected by their environment – by other species, by physical forces, etc. If species A increases in number (for whatever reason) leading to a decrease in species B and C, it suggests that there may be an association between species A and species B and C. This can spur more research that leads to more discovery. In effect, measuring population size is a way for us to trace *possible* causes and effects in an ecosystem, with suitable hand-waving, of course. Think of measuring population size as being as fundamental to an ecologist as a ruler is to a builder . . . at least that's what the tv cartoons next door suggest. It's a lesson so vital that they play it around the clock.

Before we knock ourselves out by the little bits of math that are about to follow (ack!), let's jump to what we know about common population growths. They usually start out increasing in size and then at some point they stop growing as fast. Eventually, they stop increasing in size altogether. This kind of model is called logistic growth. It over-simplifies reality, but does well for an approximation. It looks roughly like the following image.

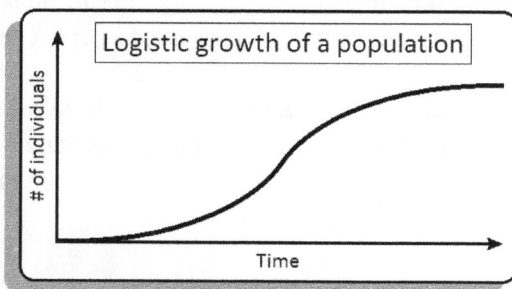

Logistic growth of a population

Let's see if we can come up with a math equation to match this. Why? Well, having an equation with lots of fancy letters in it often helps us to target factors in reality that are influencing the trend we are observing. For example, suppose we know that the trend follows an equation that has two variables in it. This suggests that there could be two primary factors in reality we should look for. Sometimes the act of trying to derive an equation based on our assumptions teaches us about the errors of our assumptions. Either way, let's give it a shot and see what happens.

First, let's imagine a situation in which everything is ideal for growth. For example, what do you suppose would happen if, say, each organism gave rise to two offspring, and all the offspring survived to reproduce as well?

---

[97] There are a bunch of variants to this name. Tom-AY-to, Tom-AH-toh, I say.

[98] In this particular case, animal death is probably the most relevant, assuming no fish were born in the short time between capture, and also assuming none packed their little fish bags and migrated to another lake. I try to keep an open mind.

[99] Relax. You can tag them harmlessly by giving them goofy hats. What were you thinking?

Further, what if none ever died or moved away? The first individual would give rise to two more. These two would give rise to four. Four begets eight, which begets sixteen, and so forth. Before long, the growth would turn into a ridiculous exponential explosion covering the earth in whatever organism we are talking about, and knowing my luck they would all move next door to me.

Now, if a population really did grow this way, it would follow an equation something like this:

$$N_{new} = 2N_{current}$$

This means: the number of new individuals, N_{new}, is twice as much as the current number of individuals, $N_{current}$. This form of the equation is not incredibly useful though, because it only churns out numbers one generation at a time. To do this properly, we'll need to describe the equation using some calculus and differential equations . . . which is WAY outside the scope of this book, thankfully. So, for the moment, you'll need to trust me that the above equation yields the next one after dropping it into a mathemagic box and transforming it: *POOF*

$$\Delta N/\Delta t = rN$$

Don't panic. The little triangles just mean "change in," so the above equation says the "change in the number of individuals over a period of time equals rN," where r is a number. In this case, r is just 2. The graph for such a growth is to the right.

The graph accurately represents how such an ideal population would grow over time. Unfortunately, this does not match our original graph, so we'll need to dig a little deeper. Most of the time, the environment is usually full of predators, parasites, harsh weather, illnesses, diseases, mother-in-laws, final exams, taxmen and a whole slew of other miserable things. All of these things can bring death,[100] so instead of considering only the births, let's also consider the deaths as well. We replace "r" with "b-d," or "births minus deaths":

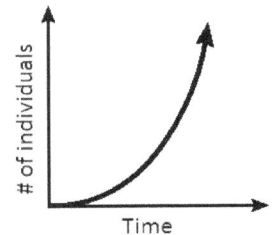

$$\Delta N/\Delta t = (b - d)N$$

So, if there are 2 births per parent and each parent died, then b-d would 2-1 = 1, causing $\Delta N/\Delta t$ to be equal to N. That means the population would increase by a size equal to whatever generation it was, designated as N. Like the first situation, this also leads to exponential growth, but much less severely. What about if the rate of births equals the rate of deaths? In this case b-d = 0, and $\Delta N/\Delta t = 0$, meaning the population does not change size. It has reached equilibrium. Finally, what if the deaths outnumber the births? If this is so, then b-d < 0, and the number of individuals decreases per generation.

---

[100] Ok, maybe final exams don't bring death, but they certainly feel like it. Surviving the Board exams was nothing short of a Darwinian marvel for me.

To summarize:

Births exceed deaths	-> population increases
Births equal deaths	-> population stays same size
Deaths exceed births	-> population decreases in size

Curves for $\Delta N/\Delta t = (b-d)N$

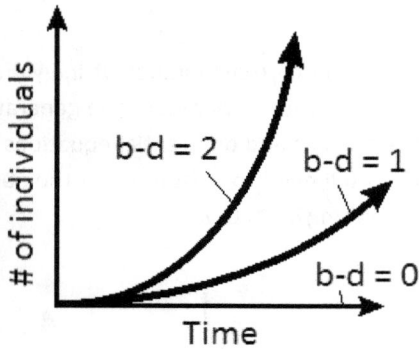

b-d = 2

b-d = 1

b-d = 0

# of individuals

Time

I've included the nifty little graph to the left to show how the population size changes over time for several different values of b-d. Notice how the rate of growth gets smaller and smaller as b-d gets closer to zero.

Still, none of these reflect our initial graph! Even after taking into account deaths, we can't seem to achieve our goal. So, instead of horsing around looking for the equation, let's do what scientists often do and borrow an equation for mathematicians and see how far off we were.[101] The equation looks like this:

$$\Delta N/\Delta t = r_{max}(K-N)/K \ * N$$

If you graph it, you get our original graph. I've posted it again below with a few extra details.

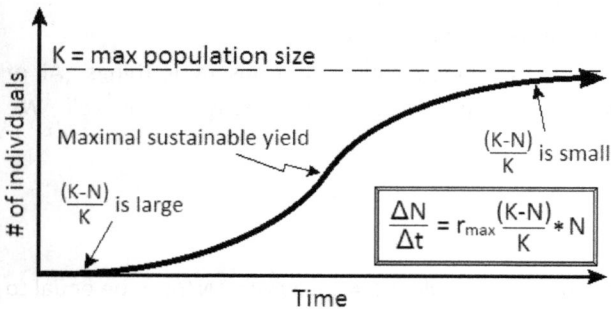

K = max population size

# of individuals

Maximal sustainable yield

$\frac{(K-N)}{K}$ is large

$\frac{(K-N)}{K}$ is small

$$\frac{\Delta N}{\Delta t} = r_{max}\frac{(K-N)}{K} * N$$

Time

I know what you are thinking: what is the "$r_{max}(K-N)/K$" gobble-dee-goop? Since a hazy film may be forming over your eyes, I'll try to simplify this mess. All of these equations follow the pattern $\Delta N/\Delta t = xN$. In our first attempt, we tried using just r where x is. That led to an exponential growth, which doesn't match the target graph. We then tried b-d in place of x, to reflect the births *and* deaths. That *still* led to exponent growth (positive and negative) or to zero growth. Now we are trying $r_{max}(K-N)/K$ in place of x. Here, the r_{max} is the maximum rate of growth. It is called the **intrinsic rate of increase** and it represents the rate of growth under ideal conditions. Next, the K refers to the maximum number of individuals that the population appears to be able to hold, designated by the horizontal dotted line. N is the number of individuals in the population at any given period of time. This equation works. Why? It succeeds because *the value of (K-N)/N changes with time.* Specifically, as the number of individuals, N, approaches the maximum, K, the top part of the (K-N)/K approaches zero. This forces $\Delta N/\Delta t$ to approach zero as well, which means that the population growth slows down as it approaches the top of the graph.

---

[101] I snuck into the math department last night and found the equation under an old coffee cup. I doubt they'll miss it. This particular equation is often called the Verhulst equation, after Pierre-François Verhulst.

Since the equation works because (K-N)/N changes with time, this suggests that our (b-d) attempt would have worked if b and d changed with time. That is the first lesson we get from our math challenge: the rate of births and deaths both change in time. The second lesson is that they seem to be related to the maximum number of individuals the environment can hold. The closer the population gets toward this maximum size, the greater the number of deaths and/or smaller the rate of births. When the population is small, births clearly outnumber the deaths. When the population is near its maximum, the number of deaths becomes roughly equal to the number of births. Smack dab in the middle of the graph is a very brief point in which the rate of increase is at its highest (a point called the **maximal sustainable yield**).

When all is said and done, this analysis tells us that the *rate of births and deaths are related to the population size*. That leads us to look for factors that relate population size to birth and death rates.

Consequences of High Density Populations

How might population size affect birth and death rates? Well, as the number of individuals within a population increases, it slowly becomes harder and harder for each individual to survive. The first – and probably most obvious – factor is competition for resources. Think of it this way: if your school cafeteria was your sole source of nutrients and it began running out of food, you'd have to fight your fellow students for the last scraps.[102] You would starve if you couldn't compete. Nature works in a similar way. The limit on natural resources acts as a built-in mechanism to keep the number of organisms in check. It usually works by killing off the weaker members of the group because these individuals are the least able to compete.

The second reason that deaths increase with population size is that parasitism and disease increase in incidence. The more organisms that exist in a population, the more targets there will be for bacteria and viruses. After all, the population of organisms *represents the primary resource* for bacteria and viruses! If the population increases, then the bacteria and viruses have an over-abundance of resources at their disposal, which leads to an increase in growth of these pathogens. Once again, all this parasitism and disease usually leads to the death of weaker members of the population.

Next is predation. As the population gets larger, it experiences more death due to predators (assuming they are around). This last factor can be quite deceiving, however, because the predators and prey in a stable ecosystem usually exist in proportion to each other.[103] For example, for every 100 deer, you might have, say, 1 wolf. For every 1000 citizens, you'll have one lawyer. The more the predator consumes, the larger the size of the population needs to be. That's probably why there are only a handful of presidential candidates per 100 million people . . . but that's just a guess. I threw out my TV years ago. I get most of my news via Pony Express.

Anyway, if the population got ten times as big, the predator population will also get ten times as big, assuming no other complications. This in and of itself will not impose a limit on the population's growth. In fact, strength is often found in numbers (assuming the population is social). The more members of a group, the more eyes

---

[102] Well, that's perhaps a bad example. From the school cafeterias I've been to, the less food they had the happier everyone seemed to be.

[103] Or more correctly, they tend to fluctuate around a proportion in a cyclic fashion. If the prey increases in number, the predators will slowly increase in number, after an initial delay. If the prey drops in number, the excess predators will linger for a time then drop off as well.

there will be to watch for danger. It's a lot easier to sit and eat when you have people guarding your borders. When predation does occur, it usually takes the weakest members of the group. It's like the old adage: You don't need to run the fastest; you only need to run faster than the slowest person in the group. That's evolution for you - survival of the fit-enough, and most of us have family members that are living proof.

While disease, parasitism and predation certainly play their part, the absolute limit on the size of the population ultimately falls to limits placed by the physical environment, usually from the availability of resources and space. In fact, when it comes to these things, organisms often have to compete against more than just other members of its own species. Quite often, one species will have to compete with another, creating a situation referred to as **interspecies competition**.

The degree to which two species will compete with each other for survival generally comes down to how similar their **niches** are. What is a niche? Well, put simply, the niche is a catch-all term that basically describes all the aspects of their living strategy. It encompasses things like: what they eat, where they nest, where they mate, the time of day they are active, their resistance to environmental change, the pub they frequent, and so forth. If two species use the same nesting grounds at the same time, say, then they will tend to butt heads a lot. If they eat the same things, then they will tend to fight a lot, especially after frequenting their aforementioned pub. Anyway, if two species are too similar in their niche, one of four possible outcomes usually results:

1) One species will eventually go extinct.
2) The two competing species will gradually become segregated to two separate geographic locations. In effect, they establish a boundary. On one side, species A will have defeated species B. On the other side, the opposite has occurred. This kind of phenomenon is often referred to as **allopatriation**[104]. They start as **sympatric**[105] and become **allopatric**.
3) One species will give the other a sound defeat, but will experience periodic mass death by regular environmental forces. This temporarily allows the weaker species to rebuild its numbers, setting the stage for the next round. This kind of battle tends to be a temporary and unstable condition. Eventually, one species will usually win, resulting in (1) or (2).
4) The two competing species will gradually acquire changes in lifestyle that minimizes the amount of niche overlap. This may occur through change in behavior or through evolution. For example, one might switch the kind of animal it hunts, or it may evolve a change in beak size so that it can eat a different kind of seeds. If they nest in the same area, the two species might evolve a change in nesting sites. Sometimes, they can even evolve **mutualism**, where the two species begin working together to help each other survive. Insert happy fuzzy feeling here.

All of these outcomes alleviate the strain of competition between two species, and these patterns are strong enough that it has led to a hard-and-fast rule in biology: *no two species can occupy the same niche*. Of these outcomes, mutualism is perhaps the most fascinating and deserves a closer look.

---

[104] This means, roughly, "different nations."
[105] "Occupying the same nation"

Mutualism

Mutualism is a phenomenon in which two species have evolved to coordinate their niches so that both benefit. A classic example of this includes mitochondria and chloroplasts living inside eukaryotic cells. Believed to once have been free-living, the mitochondria and chloroplasts were probably parasites on early eukaryotic cells, invading one cell and then moving on to the next. At some point, both parasite and prey evolved a genetic relationship in which the mitochondria and chloroplasts began assisting the eukaryotic cell, which in turn gave the mitochondria and chloroplast a home. Other examples (usually covered in the second semester of biology) can be found in the study of botany, such as a number of mutualistic relationships involving plant roots. These include the nitrogen-fixing bacteria and fungal mycorrhizae. These latter two organisms live in and/or around plant roots, benefiting the plant while receiving benefits from the plant in return. Presumably the start of this relationship was not lovey-dovey though, no doubt taking millions of years of cellular warfare before a truce was drawn.

Either way, let's get back to the point. For the reasons given above, high density populations are often disadvantageous to the species in question. Organisms in overcrowded areas sometimes leave the population.[106] This emigration behavior is very risky and very energetically costly though. As such, migration is usually only done when it's *more costly to stay than to leave*. Overcrowding can also spark internal regulation of reproduction as well, where mating pairs will sometimes forgo reproducing because resources have become too sparse. For example, long term restriction of calories (i.e. starvation) is well known to cause a lot of species to abstain from reproducing. The reason for this is quite logical: if there aren't enough calories around to sustain the current generation, then newborns will not survive and the entire process will only spread out the calories even more. You *and* your children could die, so the best bet is not to mate at all until the conditions get better.

Advantages of High Density Populations

Despite all this business of death and destruction in high density populations, there are definite advantages to having large numbers of cohorts. In plants, for example, an increase in density often leads to a reduction in wind. In addition, dense forests are usually better able to minimize temperature and moisture fluctuations, to the benefit of all within the forest. For social animals, as we mentioned above, an increase in the members of the society can lead to more complex interactions and greater cooperation. Not only are more members of the society looking out for common enemies, but more are able to take on specialized roles within the population. Humans are an excellent example of this concept, and despite the frequency of doomsdayers who point to the logistic graph and predict our extinction by overcrowding, I quickly point to our technological advantages. While we have a keen knack for mucking up environments, we have just as keen a possibility of getting off this rock to muck up other planets as well. The future is brighter than ever!

---

[106] . . . excluding bullheaded individuals, such as yours truly.

Well, that about covers our initial look at ecology. Now that we've discussed the basics of populations, population size, and the effects of overcrowding, let's move on to examine the effects of changing environments. See you up ahead.

Summary

In this chapter, we covered the basics of ecology:

- Broadly, ecology is the study of how organisms interact with their environment and with each other.
- A population is a group of individuals within the same species living in a geographic region.
- A community is a group of populations (of different species) living in a geographic region.
- An ecosystem is a community plus the physical objects in it, such as rocks, streams, wind, etc.
- The biosphere is the shell around the Earth that contains all forms of life. It is materially a closed system, meaning that practically nothing escapes it into outer space. Energy enters in the form of the Sun's light and leaves in the form of heat that radiates out into space.
- Populations can have even distribution, random distribution, or randomly clumped distribution. Of these, the random distribution and randomly clumped distributions are the most common. Even distribution is rare because the physical environment is usually not uniform enough to allow this form of distribution. Random clumping is generally caused by the location of physical resources. It is often an indication of social behavior, but not always.
- Measuring population size is important to ecologists because it helps them trace possible cause and effect through communities of organisms. It is an essential measurement in their field. There are two major ways that populations are measured:
  - Quadrat method – the region is divided into squares and organisms within a random sampling of squares are counted. The rest of the squares are assumed to hold a similar number of individuals. The total number of organisms is assumed to be the average number of individuals in the sampled squares times the total number of squares.
  - Marked capture – a subset of individuals within a population are captured, marked and released. Shortly later, another sampling of individuals is captured, and the number of individuals in the second set that are marked gives an indication of how many individuals are in the entire population.
- Population growth often follows a logistic model, described by the equation:
  - $\Delta N/\Delta t = r_{max}(K-N)/K * N$
    Where $\Delta N/\Delta t$ is the change in the population's size, r_{max} is the rate under ideal conditions, K is the maximum number of individuals the environment can support, and N is the population size at any given point in time. This equation tells us that the growth of the population depends on how close it gets to the maximum size it can have. It also tells us that the rate of births and rate of deaths

change over time. Specifically, the rate of deaths tends to increase as the population nears overcrowding.

- Overcrowding leads to increased death due to several primary factors:
  - Limit of natural resource
  - Parasitism and disease
  - Predators

  Of these factors, the first is usually the main determinant of the maximum size a population can be.
- The niche is a catch-all term that describes the lifestyle of the species in question. It includes what the species eats, where it nests, the method of mating, the times of year it is active and so forth. The total number of variables encompassed by the niche is enormous and generally unknowable.
- Overcrowding usually results in the death of the weakest members, but it also can spark competition between species as well. This is called interspecies competition. Two species having highly overlapping niches will result in fierce competition. The result of this competition usually falls into 1 of 4 categories:
  - One species will go extinct.
  - The two species will become allopatric.
  - One species will nearly win, but will be periodically weakened by external forces, during which time the losing species will regain ground. The process will then repeat.
  - The niches of the two species will change to minimize competition. This occurs by elective changes in behavior or through evolution. Sometimes the two species will evolve a mutualistic behavior that benefits both species.
- Mutualism is where two species evolve to work closely together, to the benefit of both species.
- Overcrowding can sometimes lead to emigration, where it becomes less costly to move away than it is to stay. Overcrowding can also lead to internal regulation of breeding. Caloric restriction, presumably through competition for food, is one example in which many species reduce their breeding to compensate for the lack of resources.
- The primary benefit of high density populations is through the potential to work together through social interaction. In plants, high density can foster favorable microenvironments that reduce wind, temperature and moisture fluctuations.

Unit 40

Ecology II: Ecological Change

The Changing Environment vs. Reproductive Strategy

In the previous discussions, we attempted to derive an equation to show how population size changes with time. Our initial attempts floundered because we didn't include birth and death rates that *change* in time. Then, we corrected the problem by using an equation that included a complicated term that had K in it. Now, we are quickly confronted with the same conundrum. K *too* changes in time, or to put it in a less sciency way: the maximum amount of individuals that can live in an environment tends to go up and down. One year, the local woods might be able to support a max of 500 deer, but the next year some kind of blight might go through and kill most of the vegetation,

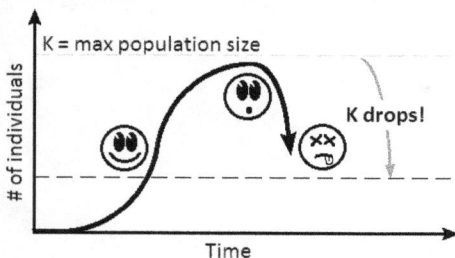

dropping the maximum number to, say, 100 or so. This is reflected in the graph by a drop in the dotted horizontal line. The curve that represents the population then takes a dive in order to catch up with this change.

Common causes of such population crashes are severe unexpected weather, fires, sudden floods, volcanic eruptions and large scale human alterations of the environment, but I suppose they could also include meteors, alien invasions, the advent of reality TV, and the end of the Mayan calendar.[107] Deaths resulting from these kinds of

---

[107] I live under a rock, but as I understand it: the Mayans used up all their paper while making their really long calendar. The shortage of paper was supposed to cause the world to end somehow. According to experts, this end-of-the-world publishing oversight was supposed to occur before *these* bio books were to be published. In that regard, it's unfortunate that the world didn't end; one publishing disaster would have helped to avoid another.

things are often referred to as **density-independent**, because they are not caused by the density of the population.

Density-independent processes don't necessarily have to be extreme, despite the pseudo-ludicrous list I just gave you. Often they are an inherent characteristic of an environment, so much so that it's probably safer to think of the population limits (the K we talked about) as being fundamentally variable. After all, nothing stays the same forever. A slight change in water availability from one year to the next could spark a cascade of events that drops the maximum deer population from 500 to 450, or increase it from 500 to 550, say. It's subtle but present nonetheless.

This fundamental variability in the environment has led to a number of different evolutionary strategies in organisms. I'll point out a few of the more distinctive patterns, though I will hazard a warning that these come with a fair amount of hand-waving. Exceptions occur, after all.

Tiny animals tend to reproduce faster

Being small has its advantages, but being able to resist changes in the environment *isn't* one of them. Since environmental change is ironically one of the few constants in reality, many small organisms reproduce rapidly – if not explosively – as a countermeasure to unexpected mass death. This is usually most obvious in insects, where a single brood might lead to millions of newborn bugs. The vast majority of these die before making it to reproductive age. For those that survive, life is a ticking time bomb, making quick reproduction all the more imperative. In terms of the mathematics we talked about previously, small animals tend to have a very high r_{max} value. Organisms that follow this pattern might get wiped out easily, but since they reproduce so quickly, they can make a rapid comeback.

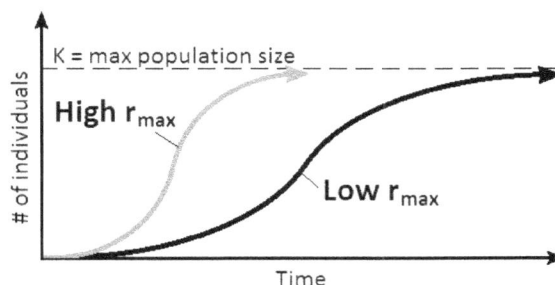

Large animals tend to reproduce slower

Larger animals take much longer time to grow than small animals, but they are much more resistant to environmental changes. Whereas a quick frost might kill legions of mosquitos, it will be little more than an uncomfortable chill for deer. As such, larger animals do not need to compensate for unexpected mass death by reproducing crazy-fast. They can take their time. In fact, since it sometimes takes a very long time to grow, parents often have to focus large amounts of time and energy rearing offspring. As such, it's usually more beneficial for them to have few children and focus their resources on them. The r_{max} for such organisms is very low.

Although organisms that follow the low r_{max} lifestyle might be more stable against adversity, there are disadvantages. Their low reproductive rate means that their populations do not recover very easily from disaster. Sometimes this fact becomes painfully clear after we use pesticides to wipe out pests, leading to a drastic decrease in the bigger predators.[108] The pests can make a rapid recovery, whereas the larger predators may take a long time to bolster their numbers.

---

[108] … since the pests are the food source for the predators. Kill the pests and the predators have nothing to eat!

These two reproductive patterns are sometimes described in terms of survivorship curves, which represent the probability of surviving vs. age, as shown here:

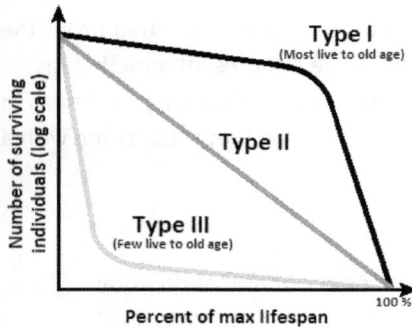

Type I
(Most live to old age)

Type II

Type III
(Few live to old age)

Number of surviving individuals (log scale)

Percent of max lifespan

100%

You'll notice in the image that I've drawn three curves, labeled Type I, II and III. The Type I curves represent organisms that generally reproduce slowly and live to old age. Many human populations – especially those of Western societies – are represented by this curve. Their lifespans are long, and most children survive into late adulthood as a result of improved living conditions, better nutrition, the relative lack of predators, and advances in medical technology. When the conditions are right, we linger to a ripe and grumpy old age – much to the dismay of everything else.

The Type III curve represents organisms that produce a lot of young, few surviving to old age. The example of the insects I mentioned above is an extreme version of Type III, but organisms that follow this pattern are not necessarily all tiny. In fact, most animals follow type III curves. It's also very common among plants as well, which produce countless saplings that usually do not survive to adulthood. Type III curves teach us that being young is not always a good thing.

Type II represents those in the middle, which reproduce at an intermediate rate and whose death is not necessarily predictable based on age. Their lifespans are more favorable than Type III, but not quite as good as Type I. Examples include a number of birds, coral, hydra and biology book writers.

Changing Ecosystems

Just now, we talked about how environments can change, leading to alterations in the maximum number of individuals living there and to differences in reproductive strategy. Now, we look at the inverse situation in which the species living in an area change their environment. Specifically, we'll look at how plants and animals conquer newly exposed ground and how their actions transform the area. The result is a relatively predictable pattern of changes, called **succession**. There are two broad types of ecological succession: primary succession and secondary succession. Primary succession involves the deposition of new soil. Secondary succession involves conquering a region that already has soil laid down. It's probably easier to learn by example, so we'll start with the temperate land next to oceans. Let's head to the beach!

Primary Succession

Succession from Beaches

No body of water stays put forever. Oceans and lakes move over great periods of time, such as decades, centuries or longer, rivaling some of the slowest processes known: plate tectonics, mountain formation, tv commercials, etc. In the process of water recession, beach is exposed. If nature did nothing with the extra beaches that were left behind, the Earth would be littered with waterless-beaches. In reality, all this leftover ground becomes new real estate for plant growth. In fact, as soon as the water stops pelting that space, a number of species stand ready to pounce. On the other hand, the sand on beaches is not suitable for most land plants, so the first to move in tend to be small and sparse. In many cases, the plants that root there only live for part of the year, with each year bringing a new crop of annuals. With them come a few miscellaneous species of bugs, crustaceans and small rodents, (mice, rats, seasonal tourists, and so forth). It may not seem like much, but the importance of this austere stage in succession is that generations upon generations of little plants and animals slowly lay down a basic matting of **soil**. Soil is just a collection of plant and animal body parts that are gradually breaking down. It contains all kinds of debris and dead things from previous generations, like leaves, branches, bug limbs, bones and feces. All of these things are choke-full of nutrients that make it ideal for future plant growth. So, as the upper beaches acquire more and more soil, more robust plants are eventually able to take hold.

Further up on the beach, we encounter plants from an older stage in this process. This is where the beach *used to be* many years before, and here these plants usually live year-round. Their persistent nature serves to bind the sand and soil together, and they become the home for even more bugs, rodents and similar creatures. Behind this area of plant growth (and further up the beach) you'll find the first shrubberies and bushes. These build more bulk than the grasses and annuals closer to the water, and as a result they rapidly deposit more soil than any of the previous plant types.

Just beyond the bushes, you'll sometimes find a small patch of pine woods. Many pine trees and other related plants grow rather quickly, but are not as energetically efficient as the more advanced broadleaf plants. So, shortly after the pines begin rapidly growing, the hardwood broadleaf trees will sometimes move in. As these latter plants grow, they begin to shade and kill the pines. As such, just beyond the pine trees you'll see hardwood trees and a young forest. This last community of plants is the final stage of succession.

That is the characteristic pattern of succession: sparse annuals lead to sparse perennials, which give in to bushes and shrubs, which are taken over by small pines tree, which in turn are conquered by broadleaf forests. Walking from the water's edge toward the forest, you get a peek into the recent past.

Succession from Ponds

A similar pattern can be observed when a pond slowly recedes over decades.[109] Let's start with the center of the pond and move closer to shore. In the middle of the pond, it is too deep for any plants, except for algae. Cyanobacteria can grow here as well. These two organism groups grow and deposit sediment at the bottom as they die. This slowly makes the pond shallower.[110] As you get closer to the shore of the pond, you see a slow buildup of rugged vascular plants that are able to grow in shallow water. Next, nearest the shore, appear the broadleaf floating plants, such as lilies. These tend to shade and kill the deeper water vascular plants, restricting them to the inner parts of the pond. As generations of broadleaf floaters die, they contribute in turn to the further buildup of biological material at the bottom. Soon, this sediment makes the water shallow enough to allow the rigid cattails and other stalked plants to grow. This rapid build-up of soil around the pond gives terrestrial plants a chance to take hold. When this happens, the water content typically drops low enough that the stalked plants are no longer able to grow.

---

[109] Assuming more water isn't constantly added, which is sometimes the case.

[110] Those of you who are familiar with Archimedes Principle may point out that the water level should just rise if stuff is deposited at the bottom. You are quite correct, but increasing the water level will also spread the pond out, which acts to make it shallower. Also, and just as importantly, all

This gives way to grasses and other small land plants, which continue transforming the soil and decreasing water content. Further away from the pond, you see bushes and other shrubbery. Further still you see a small strip of pines and then hardwood trees farthest out, just like before.

Throughout this process, all the plant life is drawing water from the pond, and as we saw in earlier sections, this water is used to make glucose. The glucose is then used – in part – to make cellulose. It is this cellulose that sinks to the bottom when plants die. So, in a sense, the plants turn the pond into dirt. Not all the cellulose sinks to the bottom of course. Some is eaten by animals, or carried away to make nests, or whatever. Either way, the water in the pond slowly recedes until the middle becomes nothing more than a marsh and then eventually just a meadow. When all of this is done, the plants that need lots of water die off and give way to plants that acquire water by pulling it from the soil. If you come back decades later, all that excess water will be gone and nothing will remain but a hardwood forest. If you come back even later, it will have become a parking deck, although I'm not entirely sure how that happens. In all my years of wisdom, I have begun to suspect humans for this last transformation.

Succession on Volcanic rocks and other newly exposed rock surfaces

Whenever natural forces leave rocky surfaces lying about, plants move in as quickly as possible. The first to take advantage of these new perches are lychens,[111] which can often take hold in tiny cracks. As they die, they deposit a buildup of dead material in the rocky depressions. Once the buildup is significant enough, mosses can start growing. As before, these also die and contribute to further buildup over many generations. Eventually, when the spot has been transformed enough, ferns and small vascular plants move in. Throughout this process, each generation chisels away at the rock's surface. This widens the cracks and breaks the rock down. The process can be rather slow depending on how hard the rock is, but you will eventually see larger and larger plants move in until the new rock surface is covered by soil and hardwood trees. At other times, some of the cracks of the new rock surface can be rather large, and within these cracks you see a rapid progression of succession, until a hardwood tree or two is growing from it.

Limits of Primary Succession

In all the above examples of primary succession, the main theme is the buildup of soil. This is the bottleneck in the process, because soil buildup can sometimes take decades to occur, depending on what other physical forces are at work. For instance, if wind and water are constantly beating down on a surface, that surface will experience slower soil buildup and greater runoff. As such, that area may take much longer to progress through the stages of succession. In the worst case scenario, it may not even progress at all.

Another factor that contributes to this process is the availability of water. In regions were water is lacking, such as in deserts, succession will not occur in this manner; rather, it usually stops with sparse and rugged vegetation, the kinds that are the most efficient at acquiring and holding water. Also relevant in the discussion is the climate of the ecosystem we are considering. In fact, the above discussions are only generalized arch-types for succession. No

---

that water is also *being used up* by the plants at the same time. This latter part is probably the more important process, leading to a gradual decrease in water depth and a decrease in the pond's diameter.

[111] Well, actually the first to move in are bacteria. Bacteria are literally everywhere and *almost* as prevalent as coffee shops. To spot bacteria on newly cooled lava rock is not surprising.

two locations on the Earth are identical; the kinds of species that partake in the succession of one area may not be the same as those of another. Arctic areas, for example, rarely progress beyond the pine stage, as these plants happen to be particularly resistant to cold. In tropical areas, the number of plants and animals that participate may be so diverse and grow so explosively that it's difficult to argue that discrete stages exist at all.

Secondary Succession

In secondary succession, an area of earth is cleared of pre-existing and well-established vegetation, such as might happen if humans cut down a forest or if a fire rips through some woods. After the vegetation is cleared, you have a region that *already* has soil in it. In this case, what happens is usually a free-for-all for whichever plants are capable of moving in and growing the fastest. At first, you'll see grasses and bushes take over, as well as some pine trees. As the broadleaf trees begin to take hold, they will begin to shade the plants under them, which will then die off. After ten, twenty, or thirty years, the area will be covered in a hardwood forest again . . . ripe for the appearance of a parking lot or golf course.

As before, the above limiting factors of water and climate also apply, making the individual instances of secondary succession different from each other.

Climax Ecosystems

Regardless of how a hardwood forest gets to where it is, it does not usually progress through any other stage of succession. It is a **climax community,**[112] which means – broadly speaking – that the community does not change much from one generation to another. This should *not* be taken to mean that they don't change at all though! On the contrary, hardwood trees grow and die all the time. Fifty years from now, you will have a hard time recognizing the forest as it is today. Not only will the *placement* of trees change, but so will the *kinds* of hardwood trees, which do not necessarily follow any predictive pattern. A patch of hickory trees today might give way to a big elm in twenty years, which in turn might yield its space to an ash tree fifty years from now. Change is constant, but regardless of the smaller changes, a hardwood forest usually stays a hardwood forest. This is equally true of other kinds of climax communities. Cacti growing in a desert one year may be different from those of another year, but if you come back in fifty years, the region will probably still be a desert with cacti. Unless major environmental forces move in to alter the basic conditions, the climax ecosystem will likely stay there indefinitely.

Other Patterns of Succession

As we have seen, succession is marked by a gradual buildup of soil and slow progression to larger kinds of plants, where the final plant type is usually the hardwood[113]. Other patterns of succession can be seen as well. As succession gets underway, we often see a gradual increase in shade, a decrease in wind, and an increase in soil

---

[112] The idea of a climax community is a very old one and has fallen into disfavor by many ecologists, mostly because of differences in details that exist between one community and another. Here, we are only interested in the prototype – the idealistic model, if you will – of succession. If nothing else, this discussion has historical relevance.

[113] . . . or whichever kind of vegetation is the dominant form for the climate in question.

nutrient content. In addition, we typically see a rapid increase in the number species that take part in the transformation process. This process increases then slows as succession continues. The *kinds* of species changes as well. Then, once the climax ecosystem has arrived, the total number of plant and animal species usually stabilizes and doesn't change much.

As succession progresses, the amount of energy that is harvested by photosynthesis slowly increases as each stage's plants become more numerous and more efficient. When hardwoods take over, the efficiency in photosynthesis rapidly increases. Since photosynthesis results in the buildup of cellulose, we also see a drastic increase in the total biomass once hardwoods take over.

During succession, a gradual development of *vertical* ecosystems also occurs, especially as the climax ecosystem arrives. Much like the progression of human settlements, the first stages are near the ground, and then each successive stage gets taller and more complex. In the climax ecosystem such as a jungle or temperate forest, the vertical height of the ecosystem can be several hundred feet high. All of this altitude provides great opportunity for other plant and animal species to make their homes. This increases the complexity of microsystems within the forest. Complexity, in turn, tends to lead to greater stability of the species within the ecosystem. Let's have a look at that before closing up this unit.

Ecosystem Stability

We've just seen examples of how ecosystems tend to change. In the presence of sufficient moisture, bare areas slowly accrue more and more soil from generations of progressively larger plants. This leads almost certainly to a climax ecosystem, with many complex microsystems within it. We also joked around about how these turn into parking lots and golf courses, but transformations such as these can give us a segue into the factors that actually change ecosystems. What factors make ecosystems resistant to change? What does it take to alter an ecosystem? What trends can we find?

Well, first and foremost, each ecosystem is unique with a unique collection of physical matter and a unique collection of species. Whatever may destabilize one ecosystem may not destabilize another. As with many areas of biology and ecology, there is an adequate amount of hand-waving. The topic can be very complex, so let's just skim the surface of ecosystem stability.

Early theories held that ecosystems that had lots of species in them were more resilient to change. The real trend seems to be the opposite. Communities with a *small* number of interacting species seem to be better able to withstand change, arriving at a new equilibrium shortly after the change is introduced. For the more species-numerous communities, repercussions resulting from an initial change can bounce around the ecosystem for generations before arriving at a new status-quo. On the other hand, ecosystems with increased physical complexity seem to lend to more species stability. For example, let's suppose you are attempting to avoid your professor because he has an assignment for you. In which environment would you be more successful at avoiding him: in a flat plain or a complex canyon with many hiding spots? Obviously, it is much easier to avoid a professor if you are in a canyon with oodles of cracks and crannies. If you are standing on a flat plain, the odds are good he will spot you, and then you're hosed. Likewise, animals that make their homes in complex physical environments seem more resilient to change than those who live in less complex areas. As always, the rule is a very soft and general one, so don't complain to me if the professor still finds you even after you hide in the canyon. If the strategy works, however, feel free to credit me for the idea.

Regardless of the complexity of the ecosystem, sometimes the loss of a single species can have profound effects on the remaining ecosystem. Such a key species is called a **keystone species**, and we can't always tell which species that is in advance. Think of it as the ecosystem's Achilles Heel. For example, let's suppose an insect species lives in an ecosystem – on a college campus maybe. Further, let's say that the insects bug us, so we develop a chemical to kill them. Unbeknownst to us, four other species are predators of the insect species: small mice, squirrels, frogs and bats. Further, these four species are preyed upon by other species, such as cats, dogs, freshmen and biology writers.[114] When we spray to kill the insects, all four small rodent species are nearly wiped out, which in turn drops the populations of cats, dogs, freshmen and biology writers. In this example, we would call the insect species a keystone species, because their removal caused disastrous change to the rest of the ecosystem.

Apart from killing off a keystone species or turning the place into a parking deck, how else might a stable ecosystem experience change? Well, generally speaking, anything that alters the average level of moisture, temperature, wind and other weather conditions has the potential to change an ecosystem, provided the change lasts long enough. In such an event, the species that take over are those who are best adapted to the new conditions. For example, suppose the Earth experiences global nuclear war. Many claim that the dust kicked up into the atmosphere by the nuclear ruckus would linger for years. If that's true, much less sunlight would reach the ground. This would drop the planet's total plant biomass. The drop in the biomass would result in a drop in herbivores, which would in turn lead to a drop in predators. When all is done, only small plants and animals would likely survive the event. After all, small plants have less structural overhead requiring energy, and small animals require less plants to survive. On the other hand, there is also the possible danger that the oxygen content might fall dangerously low due to the rapid loss of plant life. If so, this could lead to the extinction of just about all forms of life except for the most devious and crafty organisms. . . there would be nothing left but bacteria and lawyers. It's a terrifying possibility, but it's also hard to say what would actually happen. Let's hope we never have a chance to test the theory out.

Summary

In this unit, we looked at many of the cause and effect factors of changing environments, as well as the factors that stabilize them.

- Environmental change is common, and when it occurs it often changes the maximum number of individuals within a species that it can sustain. This is denoted by a K that gets smaller. When this happens, any population whose size is higher than their K value will experience mass-death, until the population size drops below K. If K gets larger, the population will begin to increase in size until it approaches the new K value.

---

[114] The latter two naturally capture small rodents in order to supplement their Raman noodle diet. It's like I always say: "A rodent a day keeps the doctor away, until you catch the plague."

- Small animals are more sensitive to environmental change than large animals. To counteract the tendency of mass-death by environmental instability, small animals often adopt a high r_{max} reproductive strategy, which means they reproduce very quickly. They have lots of young and/or have very short childhoods. Some large animals (and most plants) also have a high r_{max}, so the "small species = fast reproduction" is not a perfect rule.

- Large animals have more resistance to environmental fluctuations. As such they can take their time to reproduce. They often have a low r_{max}, which just means they reproduce slowly. They have low numbers of offspring and/or long maturation periods. As with the "small species = fast reproduction rule," the "large species = slow reproduction" is also not written in stone. It is just a general tendency.

- The topic of reproductive strategies is often complemented by the presentation of survivorship curves. Type I organisms typically have low r_{max} values (reproduce slowly) and usually live to old age. Type II animals live to intermediate ages, and Type III tend to die very young. Type III species are often associated with having high r_{max} values.

- Whenever a physical region is laid bare, a series of ecological changes called succession will slowly transform the barren spot to an ecosystem that contains a dominant set of plant species that are appropriate for that climate. This final stage is often called the climax community. For deserts, it will usually be a community of cacti. For temperature regions, it will usually be deciduous hardwoods. For artic regions, it will often be a pine forest, and so forth.

- There are two types of succession: primary and secondary. Primary succession is slow and marked by the gradual buildup of soil. Secondary succession is rapid and characterized by a pre-existing layer of soil.
  - When lakes and ocean recede, the beach is first claimed by small sparse vegetation. These help build soil. Next to take over are more robust grasses that live year round (along with whatever insects and rodents make the area their home). Next to arrive are bushes and shrubberies. These give way to pine trees and finally to deciduous hardwood. These events imply an ample supply of moisture and appropriate temperature.
  - Ponds can often recede over the course of decades or centuries. From inside to outside, the species encountered are: bacteria and algae, underwater vascular plants, broadleaf floating plants, rigid cattails and other stalked plants, grasses, bushes, pines and the hardwoods. Each group lays down sediment, making the pond shallower and eventually making the conditions appropriate for the next group to follow. With enough time, the pond will disappear into a meadow and finally be taken over by the deciduous forest.
  - When new rocks are exposed, lychens and bacteria will grow in the cracks. They deposit the initial soil. Soon mosses will begin to grow, followed by plants such as ferns. Eventually, vascular plants and hardwoods will take root. By this time, the entire surface of the rock will typically be covered in soil and plant growth.
  - When a forest is cleared and laid barren, the vegetation will rapidly move back in, starting with small grasses and shrubs, followed by bushes and pines and then by hardwood trees. The progression is much faster than with primary succession.

- During the course of succession: shade increases, temperature fluctuation decreases, soil water moisture increases, wind drops and the number of species in the region gradually increases. Also, the net amount of energy captured by photosynthesis drastically increases, especially at the stage of the hardwood forest. Finally, the complexity of microenvironments slowly increases as well.
- Ecosystems that have a large number of interacting species seem less resilient to change. The initial disruption is sometimes slow to move through the ecosystem, via each species-to-species interaction. Sometimes the change will affect many generations before a new equilibrium is established.
- Ecosystems that have small numbers of interacting species seem to be a little more resilient to change. Sometimes the initial change can spark an initial and dramatic set of events, followed by a rapid return to equilibrium.
- The more physically complex an environment is the more stable any given species within the environment will tend to be. This is especially true for prey species because they have more places from which to escape a predator.
- Sometimes, drastic change can occur in an environment if an important species is wiped out. Such a species is often called the keystone species, and its disappearance is often followed by the disappearance of many other species in the ecosystem.

In this section we cover:

1) Theoretical basis for social behavior
2) Social disinclination for: murder, cheating (theft), etc.
3) Altruism and self-sacrifice
4) Conditions causing social misbehaviors

Unit 41

Ecology III: Animal Behavior

Social Organization and Altruism

As we saw a few units earlier, living in crowded places has many disadvantages. Disease, parasitism, competition, roommates leaving moldy things in the sink and neighbors with lousy music can make your day rough. On the other hand, there are also many advantages to it as well, especially for those species that work together. Cooperation can turn a rag-tag band of organisms into an efficient machine. For instance, a cooperating group can form a strong defense against other groups, and they can better resist fluctuation in physical forces. Further, through the systematic use of scouts, more food can be found for all. In order for such cooperation to occur, however, the group must display some degree of **social behavior**. Consequently, we now turn our attention to this phenomenon. Studying this is *very* tricky though! It's fraught with hand-waving and mud-slinging, primarily because it's not always easy to nail down causes. Are social behaviors learned? Are they ingrained in our genetics? Evidence can be found to support either claim, but here our specific focus will be on genetic theories, because these fall under the purview of biology. Learned aspects of behavior are better covered in a course of psychology or sociology.

As we progress through this unit, please keep in mind that we are treading on theoretical grounds developed through *many* observations of animal societies, *especially* humans. What follows should not be read as scientifically proven or as an indication that cultural learning is unimportant in the final product of social behavior.

A genetic basis?

There are many social species out there in the biosphere, and it's very easy to argue that each one lives in an environment that is fundamentally different from the others. Still, there are clear behavioral patterns that seem to emerge from each population, regardless of their 'culture' and upbringing. The mere existence of similar social patterns across such culturally diverse groups lends a strong indication for the genetic basis of social behavior.

Perhaps a better way to state this is as follows: in order for a species to be social, it seems that it must first have a *genetic tendency* to do so. Super, but what does *that* mean? Well, under standard conditions the social organism will tend to behave in a manner that will foster group stability, and it doesn't always need to be taught how to do it.

'*Under standard conditions*' is the critical part to understand. No matter how strong the patterns, *everyone* has their limit before they snap. I point this out explicitly, because one of the first counter-arguments to these theories is usually something like: "Yes, but how about when so-and-so killed all those people?" *Genetic tendency* just establishes favorable odds. In a society of millions of people, with camera crews ready to pounce and an internet that shoots out information like a machine gun, even .01% social deviation would make it seem as though our society is falling apart. On the other hand, if only .01% deviated, then it would mean that 99.99% conformed. In any other area of science, numbers like that would be highly indicative that a theory was not far from the mark. At any rate, we'll get more into social misbehavior a little later, when "genetic tendency" will hopefully make more sense.

At the heart of the evolution of behavior is the notion of the **cost/benefit analysis**. If an action costs more than the benefit, then organisms that have a genetic tendency to choose that action will be at a statistical *disadvantage* in survival and reproduction. If the action in question costs very little but has great benefit, then those with the genetic tendency to take such an action will be at a statistical *advantage*.

For example, let's suppose there exists a species of big purple balloon-like creatures. Throughout the course of millions of years, let's say that anytime a different species encountered these balloons, two outcomes were possible. If the animal species was nice to the balloons, the balloons would give them food. If the species was mean to the balloons, the balloons would pop and release a deadly toxin that killed the rude animal. After millions and millions of years, the animals that were genetically prone to automatically be mean to the balloons would tend to get killed off, while those with the genetic tendency to be nice to the balloons would increase in number. Eventually, the animal species would simply 'know' at birth to be nice to the big purple balloons, because very few of those who 'knew' to be mean survived. The nature of the balloons would be recorded in their DNA, much like the nature of gravity, water and so forth. Like our basic ability to smile, niceness to big purple balloons would not need to be taught to offspring.

In the above example, the evolved species would have a genetic tendency to be social toward the big purple balloons. In that example, being nice carried little cost and yielded great benefit. Being mean had little benefit and great cost. Now, let's take on a more realistic example. Instead of balloons, let's use other members of the same species. In other words, let's say you have a number of options when encountering another of your kind out in the wild. What does evolution have to say about millions of years of such interactions? Would there be a genetic tendency to choose one action over another?

I suppose the answer depends on the species, so let's pick on a few. Let's say you are a Great White Shark. You are the biggest, baddest dude in the ocean. You don't particularly need help to survive. You take as you please and go where you like. When you encounter another Great White Shark, you have very little to gain by befriending the other shark. In fact, the other shark is probably just going to get in the way. On the other hand, attacking the other shark isn't necessarily a good idea either, because you could be seriously hurt or killed in the process. The best course of action is to swim away and ignore the other shark. Cost/benefit analysis would lead to a relatively reclusive species, since those who chose this non-confrontational path tended to survive with the least cost and highest benefit.

What about the lanky, goofy and slow human species? You certainly are *not* the baddest dude in the jungle. You can't outrun the lions or the deer, and you're often stuck eating twigs and berries. So, what is the cost/benefit analysis of the outcomes when confronting another member of your species? Suppose you both had the genetic tendency to attack each other on sight. In that case, one or both of you could get hurt or killed, and neither of you would be any closer to catching a deer or defending against a lion. You would face an uncertain future with possible protein malnutrition. What if the genetic tendency was to ignore each other? In that case, you could at least avoid getting hurt, but you still wouldn't have much to gain. The last option seems to offer the greatest benefit. Those who are genetically inclined to cooperate stand to gain a far more bountiful future.

Ok, great, so 'cooperation' genes win the gamble and after millions of years you tend to naturally clump into groups. What other genetic tendencies would need to be in place to strengthen the social interactions? Let's look at a few.

1. *Do not murder your tribesmen*

While it *is* true that murdering another member of your social group (tribe, pack, whatever) *will* reduce your competition for food, mates and resources, it *also* leads to inherent instability in the group. After all, if you're genetically inclined to kill other people, then others might be just as inclined to kill you. You might have a better pick of mates, say, but you would have to spend a great deal of energy watching your back. The constant expenditure of energy watching each other closely would reduce the effectiveness of the group, leading to lowered mutual defense, decreased regulation of physical forces, less efficient scouts and so forth. In effect, murder is just a step backwards to where we were before: no help in catching deer or defending against lions, and nothing to eat but twigs and berries.

So, while the benefit of murder would be *less competition*, the cost would be great *social instability* for the entire group. Individuals whose genes made them inclined to commit murder were less likely to work effectively in groups and were much more prone to die alone. As such, under standard conditions, millions of years of cost/benefit analysis would lead to humans that rarely commit murder.

> Genetic Social Rule #1:
>
> Do not kill other tribesmen unless you absolutely have to, because it 1) leads to increased risk of reciprocal injury or death and 2) it undermines the effectiveness of the group working together. Ineffective groups threaten everyone's survival.

2. *Cheating (i.e. theft)*

In order for the group to succeed, each individual's work must yield a positive gain to the group. Consider a small factory of 25 people. Suppose each day of work nets five dollars per person. Altogether, the group makes 25 x $5 = $125 per day. Suppose you are cheating and consuming more than you make;[115] let's say each day you consume three dollars. The group's total earnings would drop to $117 dollars per day.[116] Your cheating makes the group less effective. On the other hand, the group is *still* making a profit, and as long as they don't catch you, your actions will benefit *you*. This is a slippery slope though, because if everyone did it, then the group would be wildly inefficient.

---

[115] Tisk Tisk!

[116] 5$ -> -3$ is a net loss of 8$.

Clearly, there must be a genetic tendency not to cheat. . . one that is less absolute than the tendency not to commit murder.

Ok. Let's say your neighbor cheated and got caught. What should be done about it?

It's easy to argue that the group should just cast your neighbor out,[117] but the confrontation with him will cost you energy, and he might fight back! That means you risk injury and death in confronting him. Not only that, but just because he cheated this time, it doesn't mean he will not be helpful later. So, whether or not he is confronted (and punished) depends on the cost/benefit of the confrontation. If he stole a dime, it's not worth doing anything about it. The hassle of confrontation would cost you more than the benefit. If he stole a week's worth of food from your child, that's different. When all is said and done, the genetic tendency of social groups is to have occasional small-impact theft. What do I mean by "small-impact theft?" Well, if you stole a million dollars from a village, you would practically cripple the community. If you stole a million dollars from New York City, the city will barely be affected. Getting away with theft is pointless if your actions destroy the group; you'd be back to eating twigs and berries and catching your own deer!

Anyway, our second rule is:

Genetic Social Rule #2:

Cheating and stealing from your tribesmen should be discouraged as much as possible. Once in a while, the cost of theft will be low and the benefits great – leading to occasional theft. In catching cheaters, react only if the confrontation has low cost and great benefit.

The punch-line from this rule is that theft cannot be stomped out entirely. It will always be there, waiting for the right opportunity. Nature is always looking for opportunity, and your best defense is not to be idealistic. It's like the old saying: *everyone* has their price.

3. *Mating Order and Social Structure*

Mating is vital to the continued propagation of genes. On the other hand, the mating rituals of countless species show very high costs. Once again the cost/benefit analysis comes into play, leading to a very tricky balancing act. For example, it does you very little good to spend all your energy shooting for the 'hotty' in the group if she's clearly out of your league. After all, she's *also* looking for the best genes in her potential mate. So, no matter how mad your video game skills may be, she's probably not going to think that your top score in *Space Boogers from Mars* is a reflection of good genes. Nor will she believe that your level 78 dark-elf time-wizard/knight-captain from the mystical land of Mugpooyoo is an indication of sperm prowess. So, while the benefit of acquiring the perfect 'hotty' may be great, the cost of a completely futile attempt will be greater. You may have to settle for the next-best mate... or risk having no mate at all. The same happens in other animal populations.[118] The net result is a tendency for animal species to practice very structured, tiered mating groups. The highest-desired individuals mate with each

---

[117] . . . oh what a wonderful day that would be!

[118] . . . except that *their* dark-elf time-wizards totally suck.

other, the next-highest will pair with each other, and so forth all the way down to the genetically weakest, with several members of my distant family below that.

Many species establish mating pairs or groups that rarely change from year to year. Why? Simple, it saves a lot of time and energy. Seriously: flowers, dinner and a bottle of wine can get expensive. It is often even more costly to attempt to steal a mate from an established mating pair or group. In the worst case scenario, you may end up in a fatal or disabling confrontation with the mate-defending rival. The best case scenario, in contrast, is usually the least likely; your rival will probably not back away without physical rebuff ... and yes, he already knows how awesome your time-wizard is. So, to put it bluntly, trying to take your neighbor's wife is rarely genetically efficient... under *standard* conditions, that is.

Genetic Social Rule #3:

    In the vast majority of cases, leave your neighbor's mate alone. Unless the circumstances yield low cost and great benefit, you will likely not gain from the attempt. Those in the past who tried usually spent too much with little or no gain, resulting in less offspring.

---

4. *Altruism*

The first three rules give us the genetic tendency not to do certain bad things to each other. Society is more than the sum of don't-do-bad-things though. We clearly do good things for each other as well. How can we explain this? We begin, in part, by a concept known as altruism.

Altruism is a term that refers to the social perception of 'selfless giving.' It often appears in the form of helping strangers, such as holding doors open, picking up someone's coat, and so forth. Many people consider altruism as strong evidence *against* evolutionary theory because of the belief that 'do-gooders' have nothing to gain from the action. Surprisingly, the reality appears to be the opposite. It takes a bit of mental contortion to understand the arguments though, so bear with me. Again, these are not proven theories, so don't get your knickers in a bind.

As before, we start from the genetic cost/benefit analysis. In the case of altruism what we often find is that the most costly acts of altruism are reserved for those we *perceive* to be closest to us. If we perceive that someone is close to us, we are more likely to loan them our car, our laptop, our money or whatever. The less familiar someone is, the more reserved we are. For instance, opening the door for a complete stranger carries practically no cost, as does picking up their coat. The benefit of these actions is that you may look more socially acceptable to a potential mate. Also, you may benefit from reciprocal altruism in the future. In effect, opening the door for someone is an advertisement of your willingness to be a "team player." On the other hand, you aren't quite as likely to commit high-cost altruism on a stranger. You wouldn't, for example, give them a key to your house in case they needed a place to rest. The benefit of this action is vague at best, and the costs could be grave.

Why are we more helpful to those closest to us? Well, in our distant past, we lived in small tribes and the genes that led to social altruism could easily have been shared by numerous members of each tribe. When an altruism gene caused one person to help another, the odds were good that the recipient of the 'do-good' action also carried a copy of the *same* altruism gene. Since that person had the same gene, he or she would also be more likely to help in return. This interaction increased the collective odds of survival for those with altruism genes. Over the course of millions of years, this amplification could cause altruism genes to become extremely common.

336

Parents with "Help-Tribesmen" Gene   **Parents with "Help-Tribesmen" Gene**

Helps →

← Helps

Cost: Time, Energy and Materials
Benefit: Cooperation!
    Saved Time, Energy and
    Materials

Cost: Time, Energy and Materials
Benefit: Cooperation!
    Saved Time, Energy and
    Materials

Offspring

Number of "Help-Tribesmen" genes tends to increase

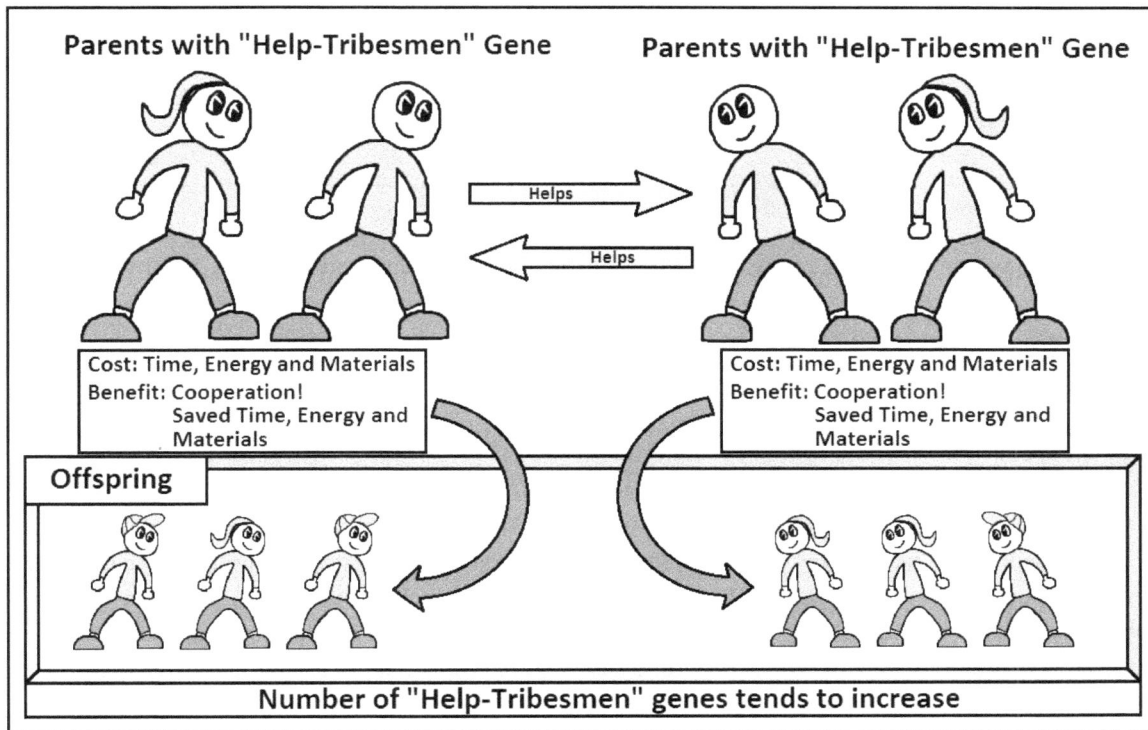

That said, this little trick of genes helping identical copies *can* be a gamble, because there is technically no absolute guarantee that another member of the population *definitely* has an identical altruism gene. So, those genes that focused altruism preferentially on close relatives tended to win the bet, because those who are genetically related are more likely to share the same copies of genes.

"Oh sure, Dr. Ickes, but why do we help our friends when we *clearly* know they are not genetically related? Answer that smarty pants!" Ok, simmer down. There is no need to attack the putative intellectual attributes of my trousers. We're getting a little deeper into theory. To properly understand this phenomenon, it's important to put yourself in the position of an altruism gene in a small low-technology tribe. You'd *like* to focus on close genetic relatives, but you have no eyes or ears; you have no lab tests. You have no choice but to rely on the perceptions of the individual you are in. Think of it this way: if you feel "close to" somebody, then it usually means 1) there is a history of successful social interaction with that person, and 2) since you live in a small tribe,[119] there is a *very* good chance the person has genes in common with you. After millions of years, the feeling of "being close to someone" becomes an excellent first-line approximation for genetic relatedness and/or social reliability. If nothing else, it labels that person as having good "social credit." Helping them is low cost/high benefit. In contrast, before the advent of science, *knowing* genetic relatedness was not always reliable or even possible[120] - unless the other person was *really*

---

[119] Our genes don't know we live in communities of millions of people! Shhh! I won't say anything if you won't.

[120] Except for mother-child, few family relationships carry guarantees of actual genetic relatedness. Even siblings can have very little genes in common. The mother-child bond is often cited as one of the strongest familial bonds in human nature, and this can be understood in light of certainty in genetic relatedness. It is one of the few instances in which one person *knows* without a doubt that another is related: the mother carried and gave birth to the child. For the child, in contrast, the relationship is not technically certain. For example, they could have been stolen at

337

different, in which case you could definitively say they were not related. Feeling close to someone *and* having a strong *hunch* they are genetically related results in the lowest cost/benefit analysis.[121] The odds are very good that helping these individuals will pay off very well, because for millions of years your social genes have benefited from such gambles. So, here is rule #4:

Genetic Social Rule #4:

     Help those you feel close to, *especially* if you believe they are close family. The greater the degree of relatedness, the greater the odds they carry the same 'helping genes' as you. Even if they are not related, they've proven themselves to be reliable and will be likely to reciprocate help if you need it in the future.

---

5.   *Self-Sacrifice*

The existence of self-sacrifice – or giving up one's life to save others – is another aspect of social behavior that many point to as a strong argument against evolutionary theory. This behavior, however, can be lumped in as an extreme example of altruism. In many cases of self-sacrifice in the animal world, what we sometimes see is that the oldest members of a family unit will typically be the ones who commit the act. For example, in prairie dogs, one member of the family (usually the eldest female) will stand watch for predators. If a predator is spotted that appears to be moving toward the group, the female will issue a loud call. This naturally will get the attention of the predator, sometimes leading to the death of the female. Here we can understand this otherwise perplexing predator warning by realizing that the genes that lead to this behavior are probably shared by the others in the family. One gene saves ten other copies of itself. Further, since the eldest female is beyond reproductive age, it maximizes the ability of the other altruism genes to reproduce by preserving the reproductive-age members.

Another very striking example of this can be found in any insect species in which all reproductive potential is focused on the queen of a hive. In these species, the brood creates thousands of sterile workers, all of which will readily die to protect the queen. All these offspring are undeniably the children of the queen. The genetic relatedness is beyond question. Since the drones cannot reproduce – and because the queen is carrying the same genes – they contribute to the survival of their genes by dying in defense of the queen.

Genetic Social Rule: #5

     If increasing the numbers of the altruistic gene requires sacrifice of one copy, then the action – no matter how regrettable it *may* seem – is warranted. Copies that exist in bodies that are beyond reproductive age or in bodies that cannot reproduce are the best suited for this role.

---

birth. This is rare though, so usually the child reciprocates an equivalent bond. The father-child bond, on the other hand, is sometimes less strong, because there is always the possibility that the kid could be from some other male. All aside, family bonds still carry a much higher possibility of true genetic relatedness. The closer someone is to someone else on a family tree, the more likely the odds of altruism.

[121] There is, of course, the possibility of *not* being close to someone and yet *knowing* they are closely related. You might help such a person a little more than you would the guy on the street, but you still wouldn't extend your fullest altruism to them. This is one way in which the genetic gamble that "feeling close to someone" = "they are probably related" breaks down.

6. *Us vs. Them*

When we began talking about altruism, we noted the idea of the perception of relatedness. In theory, altruism and social genes work by amplifying their numbers through repeated reciprocal assistance to each other. To do so, they must rely on their perception of closeness to other people. The greater the feeling of closeness, the greater the odds of committing acts of altruism. What about in the opposite extreme? What about a stranger that shares no cultural, religious, political, linguistic or physical features with you? In a small tribe, these things are strong indications of non-relatedness. Without genes in common, altruism genes would be spending a great deal of energy and materials helping a stranger with no guarantee of benefit. This expenditure of resources results in a disadvantage to reproduce, lowering the survival probability of such a generous gene. In other words, there is no inherent tendency for these genes to amplify, leading to the genetic tendency for xenophobia in biological societies.

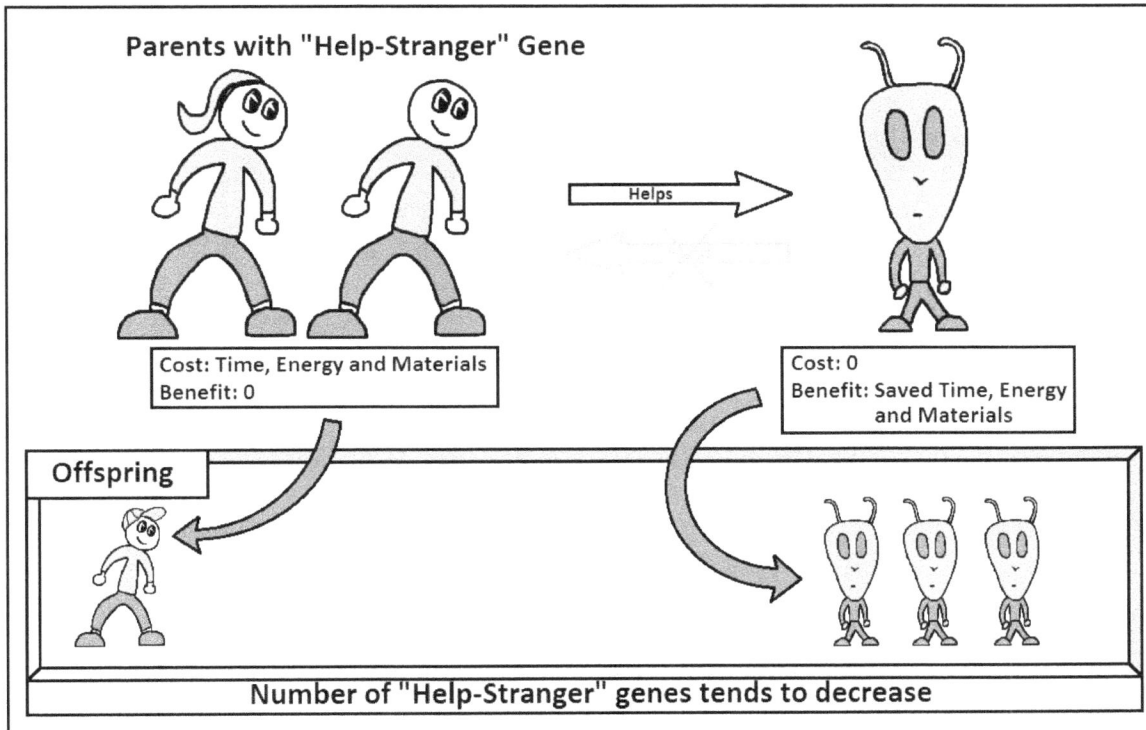

Parents with "Help-Stranger" Gene

Helps

Cost: Time, Energy and Materials
Benefit: 0

Cost: 0
Benefit: Saved Time, Energy and Materials

Offspring

Number of "Help-Stranger" genes tends to decrease

The first five social rules apply to actions toward fellow tribesmen: Do not commit murder *to your tribesmen*. Do not steal *from your tribesmen*, etc. When it comes to a complete stranger, these rules become less certain. The stranger may appear threatening to you. Alternately, they may show very little sign of interest in cooperating, and/or may represent clear competition for survival. Violence, especially in light of linguistic differences, is often regrettably unavoidable. In the best-case, and thankfully the most common scenario, you and the stranger will regard each other with suspicion and will avoid breaking the above rules in order prevent escalation of violence. In this event, small low cost acts of altruism usually occur as a token of willingness to negotiate. In effect, it's a way to test the waters to see if the other individual has compatible altruism and social genes.

Distribution of Expression

All physical traits have distributions. Among brown-haired people, for instance, a range of colors can be found, from the nearly-blond to the nearly-black. If behaviors are rooted in genetic tendency, then they too will show a distribution of expression. There will be those who follow the genetic tendencies without fail and those who skim the edge. There will even be those who blatantly ignore the rules altogether. Most will adhere.

What might lead to such a distribution in behavior? Well, for starters, a mutation in any gene responsible for the basis of social behavior could result in abnormal behavior. Such individuals might be 'a little off' or might appear mentally ill.[122] Alternately, people living through horrific experiences may no longer be able to distinguish standard from nonstandard conditions. Or, they may no longer be able to identify with any 'tribe.' Thoughts such as "no one understands what I experienced," or "I no longer belong here," may alter their perception of closeness to others and could put them at risk for straying from the genetic rules that would otherwise influence them to help other people. At other times, humans may break their genetic tendencies because they feel they have no other choice. In fact, I'd wager that 'receiving orders' from someone who miscalculated rule #6 is probably one of the biggest causes for social misbehavior. There are probably lots of reasons in which social behavior can break down. Human cultures are very complex, leaving lots of room for conjecture.

Regardless of the reasons, any behavior that leads to instability of a social group is often – and rather broadly – referred to a **sociopathic** behavior. We will not go much into depth, because this is covered more thoroughly in a course like psychology, except to say that the symptoms can range in severity from being perpetually rude, to locking yourself in a grungy apartment and writing biology books, to something much more severe like deliberately injuring or killing others. Some symptoms are obvious while others are not. In fact, not everyone will always agree that a certain behavior is sociopathic, which makes this area of study particularly murky.

Closing Remarks for Altruism and Social Behavior

When I first studied the genetic theories behind social behavior, I was immediately struck by how well they parallel those propagated by major human cultures. In this latter respect, the historical view is that such rules (rules of morality, if you will) were given to humanity by external forces. Behavioral ecology tells us the opposite: our perception of morality is merely a reflection of the internal drives that direct our social behaviors. These internal drives exist because without them we wouldn't have survived the tribal days. We put to writing what we feel is fair and then go to great lengths to teach such rules to our children, without realizing that most of these teachings may be unnecessary. If the theories are correct, then telling little Timmy not to kill other humans is probably as pointless as

---

[122] Conversely, just because you might think someone is a 'little off' or mentally ill, that is not an indication that they have some kind of mutation. It is best to leave the diagnosis of another person's behavior to someone specialized in such a field.

instructing him not to leap into a crocodile pit. In that light, I've never come upon someone having a discussion where the line "Sorry, madam, but my son is still in his murderous stage" was ever uttered. In all likeliness, I never will.

On the other hand, terrible things admittedly *do* happen, and people sometimes do *really* bad things. When these things happen, behavioral ecology offers us several logical places with which to analyze what went wrong. Does the offender feel a lack of (or damaged) kinship with the person wronged? Does the offender have a mental illness? Has the offender experienced a traumatic event? Were there extenuating circumstances? All these considerations help to make sense of our human behaviors. In contrast, simply labeling someone as 'bad' does not help to solve the situation nor prevent it from reoccurring. Such branding only serves to make things worse.

If we are to move forward as a species, understanding who we are – and how *similar* we are – is critically important. Millions of years of evolution have made it clear that teamwork is the way to succeed, and recent genetic evidence suggests that any two humans picked at random are about 99.9% genetically identical. That means we are much more related than we think. Sadly, our genes 'think' we live in small simple tribes, where less is to be gained by helping those who look different. The reality is much more upbeat: billions of people live on the planet, and we have achieved technological wonders beyond imagination. If we can work together we accomplish even more. We can go much further than catching deer and avoiding lions; we can cure diseases, establish bases on the Moon and on Mars, find cleaner fuel, and maybe even eradicate telemarketers ... The future is at our feet, but to get to it we'll need to reach passed our genes.

Summary

In this unit, we explored the basic genetic theories regarding the evolution of social behavior.

- The benefit of social behavior is that cooperation may result in greater benefit than if individuals worked alone. This is relevant for any species in which more is to be gained by cooperating.
- Social behavior shows several general patterns across almost all social species. Since these tendencies develop despite differences in culture, it suggests a genetic basis for at least some social behaviors.
- In order for a species to be social, it seems that it must first have a *genetic tendency* to do so under standard conditions.
- Practically all social behaviors can be understood in the context of a cost/benefit analysis. If something has low cost/high benefit, then it pays to carry through with the action. Conversely, if something has high cost and low benefit, it is best to avoid the action. In theory, cost/benefit analysis shapes the frequency of genes that give organisms the tendency to choose one way or the other. If one action consistently has high cost and low benefit, then genes influencing organisms to take this action will slowly drop in frequency over generations, due to the inherent disadvantage of taking that high cost/low benefit action. If another action is consistently low cost and high benefit, then genes causing organisms to take this action will increase in frequency.
- Cooperation requires stable social interaction, which fosters the following rules:

- o Do not murder your fellow tribesmen.
- o Do not steal from them, unless it is a minor theft and you can get away with it.
- o Do not challenge your tribesmen for his/her mate, unless the cost/benefit analysis of the confrontation is very favorable.
- Altruism is the perceived act of selflessness toward another individual.
  - o High cost altruism is rare between two individuals that have no prior acquaintance.
  - o High cost altruism is common between two individuals who have long-standing and positive history of social interaction, especially if both individuals believe they are genetically related.
  - o In theory, altruism is due to genes that cause individuals to tend to help other members in their tribe, because the other members tend to have the same altruism genes. The other members then reciprocate. This reciprocal cooperation results in low cost high benefit interactions that tend to amplify the number of altruism genes from one generation to the next. Altruism genes "try" to target close relatives as much as possible, because close relatives have the highest chance of having the same genes (and thus ensuring the amplification phenomenon).
  - o Altruism genes must rely on the perception of the organism carrying them. True genetic relatedness is often difficult to tell for certain, so a feeling of close kinship is a close approximation. Throughout the bulk of human evolution, we lived in small tribes. As such, a feeling of close kinship was usually an excellent indication of "social credit," with high potential for reciprocal altruism and/or altruism gene amplification.
  - o Self-sacrifice is generally considered an extreme example of altruism, where one gene copy sacrifices itself to preserve many others. This often occurs within the eldest individuals or those incapable of reproduction.
- Xenophobic social behavior may have developed because major differences between individuals were a sure sign of genetic non-relatedness and thus represented a very poor chance of gene altruism amplification. Strangers are often offered low cost altruism as a test to see if they are capable of reciprocating. If so, the cost/benefit analysis of helping such individuals will become more favorable, but in theory never as favorable as someone you believe is genetically related.
- When social rules break down, they often do so between strangers (i.e. the xenophobic social behavior just mentioned). Other possible causes may be mutations in social genes (difficult to prove though), traumatic events that alter the perception of kinship, nonstandard conditions in which someone feels threatened, situations in which someone has no choice (receiving orders during war) and so forth.

Unit 42

Biogeography I

So far, our discussion of ecology has focused on populations. We talked about how populations grow, how they are affected by their environments, how their environments are affected by them, and how they interact socially. Now we are going to transition to talking about more of the non-living components of the biosphere. Specifically, we're going to have a cursory look into a field called **biogeography**, which basically studies how biology and geography interact with each other. In this unit, we'll have a look at the driving factors behind rain and wind. We'll also trace the Sun's energy through biological populations and examine how elements such as carbon, oxygen, nitrogen and phosphorus are cycled through nature. Finally, near the end of this unit, we'll take a closer look at soil. In the next unit, we'll start to examine the major kinds of ecosystems on Earth.

The Planet's Electric Bill

Our planet is the third one from the Sun and gets practically all of its energy from it. When light hits Earth, the planet heats up and some of this heat gets radiated back into space.[123] The energy that doesn't get radiated is stored as potential energy of many forms, like chemical bonds, evaporated water and so forth. The Sun powers nearly everything around you. If it's something you like, thank the Sun. If it's something you don't like, then blame it. The Sun is paramount to almost all life on Earth, and if it went out, our society would rapidly fade into a forgotten history, silently cursing our failure to pay the galactic electric bill. For so long as that great big bulb in the sky is burning, we'll keep getting energy, and that energy will continue to push strange natural phenomena.

---

[123] Apparently, we do a lousy job insulating the place.

Major Patterns of Wind and Rain

The Sun does not shine equally on all parts of the Earth. See, despite common belief, the world is round.[124] That means part of it faces the Sun while other parts face other directions. This leads to light that beats down strongest on the equatorial region of the planet and weakest at the poles. As we mentioned before, light energy makes heat; heat makes water evaporate. So, at the equator you tend to have a great deal of water rising up into the air. As air rises it cools. Cool air cannot hold water as well as warm air, causing it to lose its water, which turns into rain. This is why the equator receives a lot of rain. Further, the now-dry air drifts north and south because of the air still rising up below it. As it moves north and south, the dry air finally begins to fall back to Earth at roughly the latitude of the Tropics of Cancer and Capricorn, causing these regions to be unusually dry. If we look at a globe, we consequently see many of the world's driest areas at the tropic lines.

The Earth   The Sun

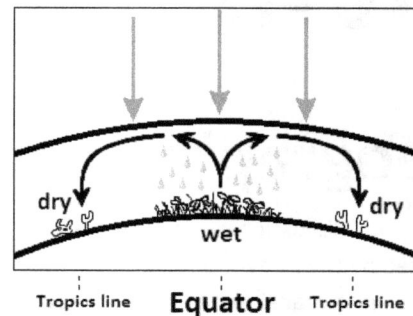

dry   wet   dry

Tropics line   **Equator**   Tropics line

"Hogwash, Dr. Ickes! Not all of the tropic lines are dry! You're not tricking me with your Earth-is-round science mumbo-jumbo!" Very well, you are partially correct: not all of the regions of Tropics of Cancer and Capricorn are dry. The reason for this is due to wind patterns, which are complicated by numerous factors. Let's ease ourselves into this scuffle and get our feet wet with a discussion on ocean breezes as a first example. Land heats and cools much quicker than water.[125] During the day, the land tends to be much warmer than the ocean. This makes air above land warmer than air over water. It rises up and drags sea air in to fill the space left over. So, during the day, the wind comes in from the sea. At night, the land cools much quicker than water, so that the inverse tends to happen; breezes tend to go out to sea.

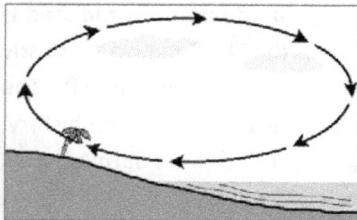

Sea breeze comes inland during the day.

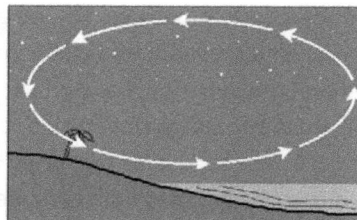

Breezes move out to sea at night.

Sea breezes tend to be a local effect and larger wind systems can mask them quite easily, if the conditions are right. So, let's zoom out to a more global view. The Sun's light causes heat to accumulate. Heated air doesn't just rise; it also experiences an increase in pressure. High air pressure will cause net movement of air toward low pressure areas. This, however, is complicated by the rotation of the Earth. Wind at the equator is spinning at the same speed as the

---

[124] This is an audacious claim, I know, but bear with me.
[125] We saw this way back in our early days in biology. Water heats and cools slowly due to hydrogen bonds. This makes it a relatively good insulator against temperature fluctuations.

344

equator, but if it moves north or south it will encounter parts of the Earth that have slower rotational speeds. The difference in speeds causes wind to deflect in large circles hundreds of miles in diameter, a phenomenon known as the **Coriolis Effect**. So, if a high pressure area near the equator attempts to move toward a low pressure area elsewhere, it will begin to deflect and spin around the low pressure area, instead of reaching it. The further the pressure system moves toward the poles, the greater the spinning. This is the premise behind hurricanes.

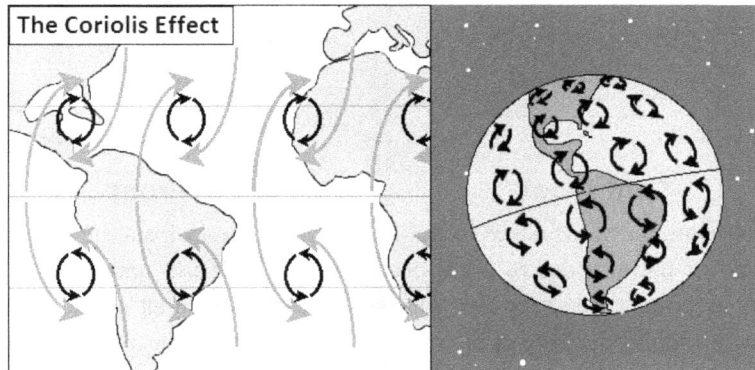

The Coriolis Effect

Another surprising culprit in the production of wind is the Moon. The Moon exerts a gravitational force. It may seem imperceptible to us, but to liquids and gasses the effect is significant enough to cause motion. Air and water move toward the part of the Earth that is nearest the Moon, leading to water currents,[126] tides and wind.

Finally, we must consider the placement of mountains. These rocky, physical barriers tend to influence the course of air currents by creating wind channels along the surface of the Earth. In addition, any time wind travels *over* a mountain range, it must rise and cool. The cool air dumps its moisture on the near side of the mountain. When it comes down over the top of the mountain, it is dry. As a result, the near side of the mountain is often wet and covered in vegetation, while the far side of the mountain is typically much drier than expected and sometimes covered by desert-like terrain.

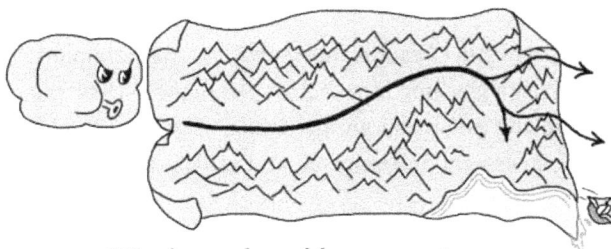

Winds are shaped by mountains . . .   *and drop their rain when climbing them.*

We saw several examples of how rain is caused by rising air. When air cools, its ability to hold water lowers. The consequences to this can be observed on many hot and humid summer days when the late afternoon approaches and the temperature begins to drop. As the air cools, a quick rain can develop releasing much of the day's evaporated water back to the ground. Sometimes when a pocket of cold air is pushed into an area of warmer air, the warm air is

---

[126] This is not the only cause for water currents. In fact, many of the arguments made about wind also apply to water currents.

rapidly pushed upward, resulting in strong rains. This is called a **cold front**.[127] In contrast, when warm air pockets encroach upon areas of cold air, the warm air gradually pushes upward in the process, resulting in prolonged slow rains. This is a **warm front**.

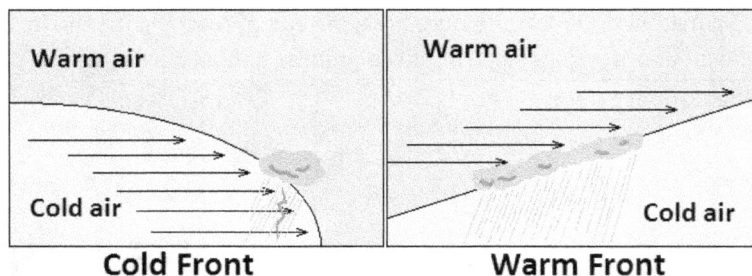

| Cold Front | Warm Front |

The point of all this weather hocus-pocus is that wind gets its energy from the Sun and travels all around the globe in predictable but uneven patterns due to 1) the placement of water and land, 2) the Moon and 3) the rotation of the Earth. These winds drag moisture which gets released any time warm air cools. The rain that results causes deviation from the otherwise simple "dry Tropics of Cancer and Capricorn" principle I talked about previously. When all is said and done, the complexity of the wind currents leads to some regions receiving predictably high levels of rainfall and others very little. Since water is a critical component in life, the difference in moisture levels leads to striking contrasts between one ecosystem and another. In the next unit, we'll be looking a little deeper into that particular topic, but for now let's shift back to our discussion of the Sun's energy, focusing on the part trapped by photosynthesis.

Energy and Food Chains

Plants require light in order to do photosynthesis. On the other hand, light makes heat, so the areas of the Earth that receive the most light are often the warmest places as well. As such, photosynthesis and warmth usually go hand-in-hand, assuming enough moisture is present.

Like moisture, warmth tends to get moved around by wind. Despite this tendency, the dominant effect on temperature is latitude. At the poles, the sunlight barely makes a glancing blow across the planet. Near the equator, the sunlight beats down with all its fury.[128] So, the closer to the poles you go, the colder it gets. The closer to the equator you go,[129] the hotter it gets. Since photosynthesis follows light and warmth, you usually see a steady increase in photosynthesis as you move toward the equator.

Regardless of where a plant happens to be growing, it captures energy from the Sun and converts it to chemical form. Photosynthesis binds water and carbon dioxide together to form sugar molecules. Clumping all the Earth's photosynthesis together, the total stored energy is called the **gross primary productivity** of the biosphere.

---

[127] Cold fronts are typically characterized by thunderstorms. The lightning that results is due to the separation of charge that develops between the sky and ground. It's more common in cold fronts because of the rapid manner in which the air rises, causing static electricity to build between all the dust, water particles and other molecules caught in the path. The details are outside the scope of this book.

[128] Not that I blame it. It must be lonely and frustrating to be a star. Everyone is always watching you, but no one wants to get close enough to really know you. Poor star.

[129] . . .or more precisely: the closer you go to the place where the Sun is shining directly overhead . . .

Part of this energy is immediately used up by plants during their own respiration.[130] The part left over is the **net primary productivity**, which is basically just a fancy way of saying "spare sugar." Plants take the excess sugar and make bigger molecules, like cellulose, oils and so forth. Animals eat the plants and make other molecules out of them as well. All this stuff eventually gets put into even larger things, such as trees, animals, cardboard, door-to-door salesmen, infomercials and so forth. It is all due to the net primary productivity – the excess energy that plants capture from the Sun.

Where does most of the net primary productivity occur? Not surprisingly, most of it comes from the rainforests, where the plant biomass is incredibly dense. Another large chunk comes from the oceans, specifically in the first meter or so of water. Why only the first meter? Well, water absorbs light quite readily, so the intensity of light drops considerably after the first meter or so. The deeper you go, the darker it tends to get and the less efficient photosynthesis is. Most of the photosynthesis in the oceans is done by cyanobacteria, which are present in high concentrations near the surface. The oceans cover the majority of the Earth. Still, despite such a massive surface area and huge numbers of cyanobacteria, surprisingly less photosynthesis occurs in the oceans than you would think. The reason is that many nutrients, such as phosphorus, are often the limiting factors in biotic growth. Occasionally human activity results in huge amounts of nutrients being dumped into the water. When this happens, you sometimes get temporary overgrowths of cyanobacteria, algae, dinoflagellates and other ocean critters. While this may sound like a good idea, these critters often release poisons into the water that kill fish.

Anyway, after rain forests and oceans, the next biggest "photosynthesizers" (in order of most to least) are the expansive broadleaf forests, boreal forests (taiga), savannas and then grasslands. Finally, places like deserts and the tundra bring up the rear. All these places are described in the next unit.

Regardless of where the primary productivity occurs, the products of photosynthesis serve as the base of energy from which all other organisms draw. This energy percolates through ecosystems in the form of energetic chemicals from plants to herbivores to carnivores in what is commonly referred to as a **food chain**. Each ecosystem has a number of food chains that crisscross their way from plants to carnivores in a **food web**. All along each chain are the decomposers. Decomposers are the fungi, bacteria, worms, bill collectors and other bottom-feeders that consume dead biological material. These organisms break down organic molecules into smaller ones, freeing a lot of CO_2, NH_3 and other simple molecules in the process. We'll return to these molecules shortly.[131]

At each link in the chain (a link being an individual), large amounts of energy is lost as heat. In fact, so much heat is lost along a food chain, that fewer and fewer individuals can ultimately be supported as you go higher in the food web - er, well, as a first approximation that is. So, whereas a geographic area might support 10,000 plants, say, these plants might only support 500 herbivores, which in turn might only support 10 carnivores. This scale-down effect is a common phenomenon, so much so that energy flow in ecosystems is often depicted using pyramid models that are divided into levels called **trophic levels**. The bottom of the pyramid is comprised of the first trophic level, occupied by the energy within **primary producers** (autotrophs). The second level, smaller than the first, is comprised of the energy within **primary consumers** (the herbivores). The third level, smaller still, is comprised of the energy within **secondary consumers** (the first level of carnivores). Many ecosystems have a fourth and fifth level,

---

[130] After all, do not forget that plants breathe too! They don't do photosynthesis to be nice.
[131] That sounded ominous. I mean we'll be discussing these molecules soon.

representing the carnivores that feed off other carnivores. Each of these levels becomes incredibly tiny, so much so that they aren't usually shown in the pyramid. The decomposers are also not shown because they feed off all levels.

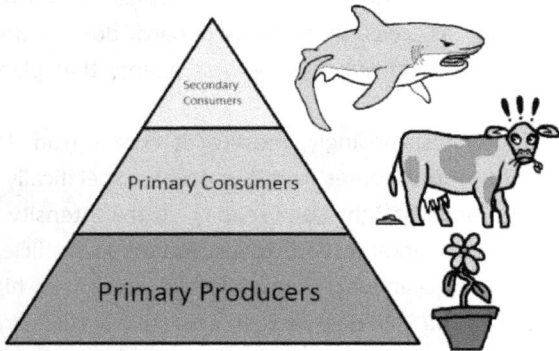

Let's clarify a bit. It is important to note that food web pyramids such as this do not necessarily represent the number of individuals – despite my example of 10,000 plants, 500 herbivores and so forth. Each level represents the total *energy* of those kinds of organisms. Why isn't the number of organisms a good measurement for the total energy? Well, consider an ecosystem comprised of one massive tree fed upon by forty thousand insects. In that case, there would be 1 plant, and 40,000 herbivores. That doesn't make a pyramid! If we take into account the total energy of the plant and the total energy of the bugs, the pyramid is restored.

Whereas these kinds of pyramids are not necessarily good representations of the total numbers of individuals within a given trophic level, they *can* be used to represent the total amount of biomass. If we return to our example of the big tree plagued by the swarm of insects, it should be readily clear that the tree could have a mass of several thousand kilograms, whereas the insects might have a combined mass of only a few hundred. So, a diagram representing the total biomass at each trophic level would be pyramid-shaped.

It's also important to note that many organisms can occupy multiple trophic levels. Consider humans. We eat plants, herbivores and many carnivores. Actually, an evening in front of the television will provide plenty of evidence that humans eat just about anything you put in front of them, including their last chance at dignity. As such, we occupy multiple trophic levels in the pyramid. In return, not much eats us, except for decomposers. It's sad; the only things that regularly feed on us are the organisms that have a taste for rotten stuff. Nature has us pegged all right.

Nutrient Recycling

We've seen how temperature and moisture cycles around the Earth, and we've also seen how energy trickles through an ecosystem. Now, we turn our attention to how basic molecules and atoms cycle around as well. In terms of matter, the Earth is a closed system. Until the second half of the 20th century, not much of anything ever left the Earth and not much entered either. The consequence of this is that life must recycle the materials at its disposal. If it didn't, it would run out. All the oxygen, carbon, nitrogen, phosphorus and other basics materials must be put back after use.

The Water Cycle

Water rolls downhill whenever possible. When rain falls, it quickly makes its way into streams, lakes, rivers and finally into the oceans. At any point in this process, the Sun heats the water causing it to gradually evaporate. A small amount evaporates before it reaches the ocean. A larger fraction evaporates once it gets to the ocean.

Not all the water runs downhill. Some of it drips down through the soil and rocks. Plants pick up a lot of it as well. Once it gets into a plant, it is slowly drawn into the trunks and branches, and then eventually into the leaves. When it gets to the leaves, it either gets put into glucose via photosynthesis or it gets released as "plant sweat" through a process called **transpiration**. This process is studied more in detail in chapters or courses on botany. The highly anticipated blockbuster sequels to this book cover it in fascinating and heart-pounding detail...

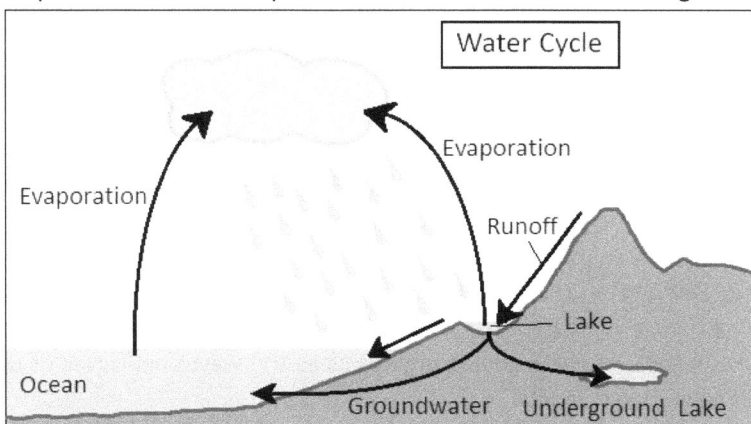

Anyway, the water that is not picked up by plants either dribbles into streams and oceans via underground routes, or gets stuck in deep underground lakes. You might think that all the water would eventually get trapped underground, but the deep earth can only hold so much moisture before it gets saturated. As such, we're in no danger of losing our water to the underground. Sorry Mole People, but you can't have all our H_2O!

The Carbon Cycle

The carbon cycle refers to the movement of carbon from the atmosphere to animal products and then back into the atmosphere. Based on CO_2's great stability and physical characteristics, as well as from the observation on other planets, we know that primitive Earth had a great deal of CO_2 in the air. Since the advent of photosynthesis, a lot of this carbon has been pulled from the sky by plants, specifically by the Calvin Cycle.[132] When plants die, they collapse and soon become buried by other plants that grow or fall on top. Over time, the plant material builds up. Some plants are eaten by animals, and the carbohydrates then get converted into various fats. When the animals die, they too tend to result in a build-up of dead material. Of course, not all the carbon gets stuck in dead bodies; a lot of it is returned to CO_2 by respiration.

Either way, imagine a world in which dead material is never broken down. The carbon in the atmosphere would slowly get captured and stuck in layer after layer of dead bodies. The CO_2 in the air would gradually drop until there wasn't enough for plants. Then they wouldn't be able to make their cell walls and would start to die. Then the

---

[132] Yikes! Old material!

animals would begin to die. Soon, you'd have a planet full of dead bodies.[133] Clearly, while some carbon *does* get stuck in the ground, we need to recycle as much as possible or life will stop.

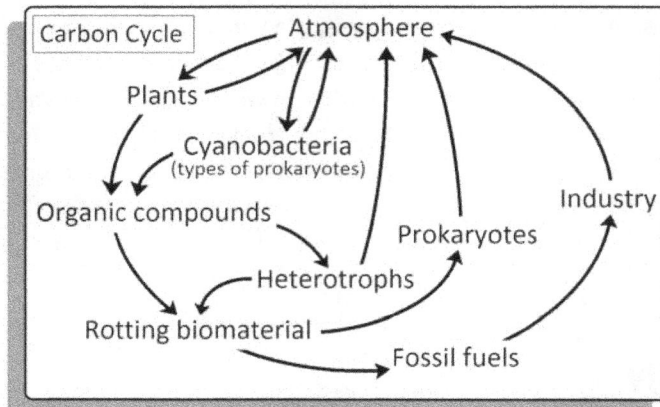

How do dead bodies get broken down? It's a complex topic, but the primary movers and shakers of this process are bacteria.[134] In fact – now that we're on the topic of bacteria – bacteria occupy two important points of the carbon cycle. First the cyanobacteria work to capture the CO_2 via photosynthesis. Along with plants, they fix a great deal of the CO_2 that is pulled from the sky. Second, a great deal of the remaining bacteria work to break down the dead, consuming the material and oxidizing the carbon back into CO_2, which then returns to the atmosphere. While not all the prokaryotes are involved, you can think of these simple organisms as the waste managers of the world. Without them, the process of life wouldn't have gotten very far.

You'll notice from the carbon cycle image above that humans play an important role through industry. We pull oil and other hydrocarbons from the ground (all of which are dead organic material from eons past) and burn it. In doing so, this releases the buried carbon back into the air. You might think that this is a good thing. After all, we wouldn't want it all to get stuck in the ground, right? Well, our role in the process is actually more destabilizing than helpful. When bacteria do it, they release CO_2 very carefully. When we do it, we release all manner of carbon compounds and soot into the air, rapidly changing the quality of the atmosphere. It is believed that these compounds have gradually begun to change the base temperature of the Earth (i.e. **global warming**), leading to more severe weather patterns and the melting of the polar ice caps.[135] Also, if the countless species on the planet do not have proper time to adapt to this rapid change in atmosphere, it may result in mass extinction.

Lastly, it's important to realize that we burn very *old* dead organic materials, not the new dead stuff. Despite our industrial habits, without bacteria we'd still find our world piling up with the dead.

---

[133] Based on the current social trends, some argue that we've already reached that mark. I'm waiting for "America's Top Dead-guy" to air. Finally a show with a real pulse!
[134] And, yes, worms and other icky things are involved too!
[135] It's important to understand that global warming will lead to severe weather patterns, not necessarily just warmer average temperatures. Increases in storms and blizzards, ironically, are a symptom of this process.

The Nitrogen Cycle

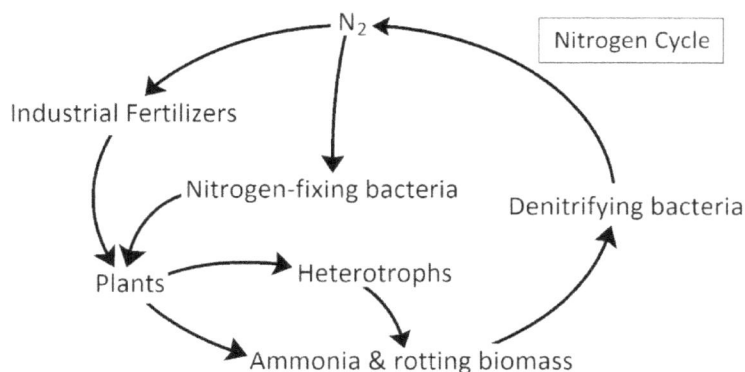

Like carbon, a great deal of the nitrogen on primitive planets is usually trapped in their atmospheres, this time in the form of N_2. The reason is that N_2 is very stubbornly stable, and as such, not many of Earth's organisms are capable of pulling it from the air and recombining it into organic molecules. Bacteria are among the very few, and without them the other organisms would have a very hard time getting nitrogen. If you'll recall from earlier studies of biology, nitrogen is a pivotal component of amino acids and nucleic acids. If organisms do not have access to nitrogen, they cannot create their proteins or replicate/repair their DNA. Just like the carbon cycle, things would come to a screeching halt if bacteria weren't around to yank nitrogen from the air. This important ability is called **nitrogen fixing** (made possible by the enzyme **nitrogenase**), and those that can do it are called **nitrogen fixers**. On the other end of the spectrum are the **denitrifying bacteria**. These are the critters that return nitrogen back to the air by converting it back into N_2.

Oxygen Cycling

If you remember from previous studies, oxygen is happiest when it hogs the electrons from other atoms. As such, molecules like CO_2 and H_2O (water) are very stable because the oxygen atoms in them are immensely happy.[136] Conversely, O_2 is a *miserable* molecule, because neither of the oxygen atoms gets to hog electrons from the other. Left to themselves, all of these molecules would eventually react with the surrounding atoms to form more CO_2 and water. Our world would have no O_2 left to breathe! In addition to the automatic reactivity of oxygen, most organisms use it in respiration, returning it to CO_2 and H_2O.

---

[136] It may seem silly to say this, but atoms are not technically capable of happiness. I am merely describing it in a way that will hopefully be memorable. The physical and chemical reactions of reality take place because of differences in energy and chaos; emotional desires have no role, apart from those caused by humans and other misinformed organisms.

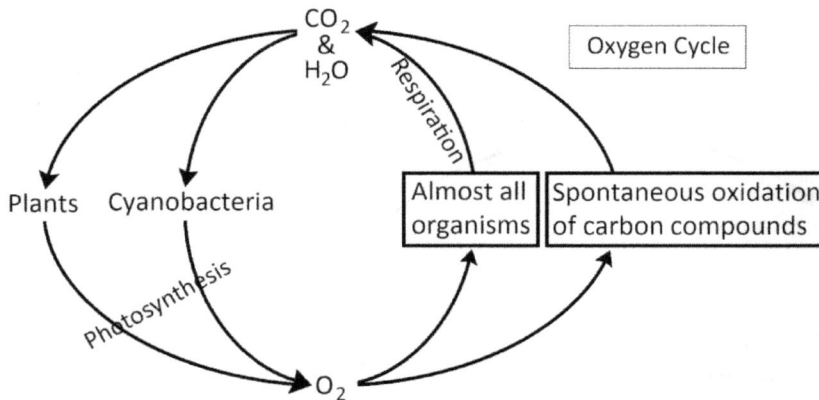

Oxygen Cycle

Prokaryotes (mostly cyanobacteria again) and plants play a major role in the cycling of O_2 by "fixing" it during the photosynthetic process. In fact, it is widely believed that cyanobacteria-like organisms were responsible for Earth's transition to an O_2-high atmosphere (see Topic Text D).

The Phosphorus Cycle

The primary reservoir for phosphorus is rocks. Billions of years of erosion have slowly etched it out from mountains and other rocky areas. Each time it rains a small amount of phosphorus trickles down into the streams and oceans. Once in the water, it is taken up by plants and animals where it gradually gets cycled through the marine and land ecosystems. Some of this phosphorus ultimately sinks to the bottom of the water or settles into the soil on land; it is then compacted by layers that form on top of it. Through geological time, these layers eventually get moved around by the many forces that act upon the Earth's land masses. Whenever new mountains are pushed up from below, the phosphorus that was compacted gets exposed again.

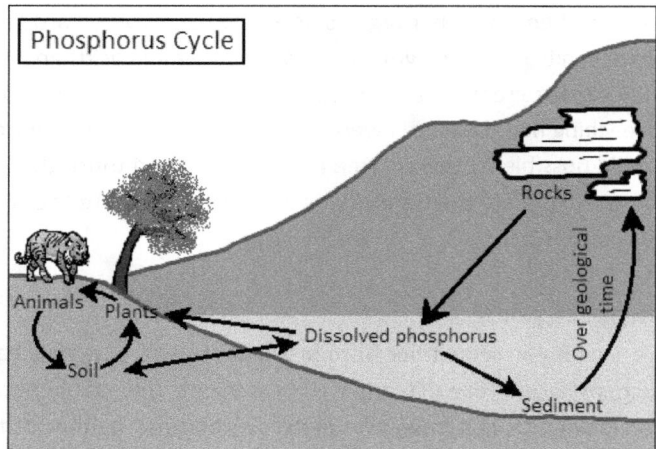

Phosphorus is often one of the many limiting nutrients in marine environments. It's one of the nutrients I mentioned before that sometimes get released when humans bumble around and drop stuff in the water. When this happens, large algal and dinoflagellate blooms develop, using up the oxygen and releasing toxins. This kills lots of fish and other marine life. Further, the drop in oxygen changes the chemical reactions that occur in the muck at the bottom of the water, which promotes the release of toxic gasses. Now that I think of it, this gives me a strong clue as to why my roommate's room smells the way it does, but that's a topic for a different day.

Novel Chemicals

In the modern world (far away from my apartment), new chemicals are being created at an unprecedented pace. They are created for so many things, such as novel industrials solvents, pharmaceutical chemicals, precursors to new plastics, pesticides and countless other things. Sadly, the rate at which these things are being created is outrunning our ability to safely test their effects on the environment. Sometimes these chemicals directly harm the ecosystem. Sometimes, the chemicals are harmless, but the ecosystem changes them into more harmful forms. I guess that's the price of technology. We might wipe out the planet in our attempt to find the perfect anti-pimple cream, but darn-it at least we'll look good doing it!

New chemicals can damage an ecosystem at any level. It can kill microscopic organisms, plants, insects, or even larger animals. The ways in which they do damage is a huge topic, and we could fill entire libraries with the details. A few simple facts bubble to the surface though. For instance, the damage that a novel chemical can do to its environment is not necessarily related to the amount released. One chemical might be many times more harmful than another, so that even tiny amounts will have devastating effects. In addition, we need to consider the phenomenon of **biological magnification**. In this process, a chemical that is released in trace quantities can sometimes be found in concentrations many *hundreds or thousands* of times higher inside some animals. How does this happen? Well, at the smallest level, bacteria and algae will take up a few molecules here and there. Many of these bacteria will be eaten by larger organisms, like protists or bugs, say. By doing so, the protists and bugs will acquire all the chemicals that the bacteria picked up, resulting in a small amplification in the concentration of the chemical. As larger organisms eat the smaller critters, the chemical gets amplified even further. By the time large animals enter into the food chain, the chemical has been greatly amplified. The higher the concentration within an organism, the greater is the chance that it will have an effect on its health. This is concerning to us, because the chemical can build inside the food that we eat – such as in fish, say. When we eat the fish, we get all the chemical magnification that happens to be there. Here is an illustration of the phenomenon.

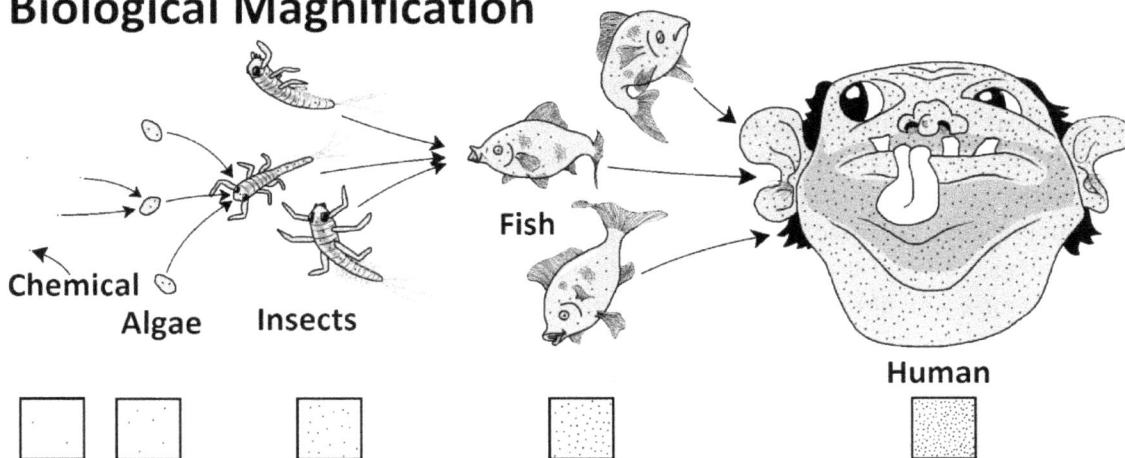

A classic example of biological amplification is DDT. DDT and its relative chemicals were used as insecticides in the agricultural industry, as well as controls for the spread of insect-borne diseases like malaria. Unfortunately, the chemical gathered in local streams and lakes, where it built up in concentration within fish. These were eaten by a number of birds and larger predators. DDT is believed to have resulted in the near-extinction of the American bald eagle, to name a few of the species affected. It is now banned for almost all but a few small-scale uses.

New chemicals can also alter the non-biological components of our environment as well. For example, fluorocarbons are used in a variety of processes, from cleaning and degreasing, to aerosol propellants, to refrigeration. Their use has dropped considerably, however, after it was discovered that they readily react with components of the ozone layer. The ozone layer is very important in the process of life because it blocks many of the harmful rays of sunlight. Continued use could result in a weakened ozone layer and subject practically all organisms to much higher potency of solar radiation. I suppose it is ok, I guess, because we'll soon have a cream for that too; all we have to do is whip up a few new chemicals.

The Soil

Now that we've looked at the wind, water and nutrient cycling, let's have a look at the basic characteristics of soil. We spoke briefly about soil when we talked about biological succession on beaches and newly exposed rocky terrain. Soil is critically important in the processes of life. First and foremost, it acts as a substrate to which plants anchor themselves. Without it, plants would just blow away, roll around and be subject to a great amount of physical damage.[137] Second, the soil acts as an important reservoir for nutrients and water. It keeps these things from quickly flowing away into the oceans.

Not all soils are equal. Soils have biological components and non-biological components. As discussed in ecological succession, the biological components of soil are the dead pieces of previous organisms. The more of this kind of material there is, the more the soil will tend to be able to support plant life, which accounts for the slow progression of succession. Pioneer plants will take root in sand and then gradually lay down dead material, which gradually increases the sand's life-supporting properties. More plants move in and so forth.

The non-biological components are typically comprised of variably-sized particles made from silicon- or aluminum-based compounds. Soils that have really small particles are typically referred to as **clay**. Mid-sized particles make up **silt**, and large particles comprise sand. The size and composition of the inorganic particles also affect its ability to hold water. For instance, in clay the tiny particles bind water and nutrients very tightly, making it too stingy for many plants to survive. In addition, these particles clump together into large globs, and new water flows off of them too quickly for plants to grab. When water rapidly flows away from soil, it tends to drag precious nutrients with it, called **nutrient leeching**. If particle size gets too big, such as in sands, the water percolates through it rapidly, resulting in the same problem. Mid-sized particles seem more ideal for holding water and nutrients, although this alone is a gross over-simplification of the various qualities of soils. We aren't even scratching the surface and could easily fill up an entire course with dirt. Trust me; it's a very messy subject to dig through, but don't let it soil your opinion of the topic... Ok, I'll stop before I run aground.

Despite soil qualities, the process of nutrient leeching is also affected by the acidity of water entering it. For example, when acid rain falls on a soil, the H^+ ions will displace other cations, such as Ca^{2+}, Mg^{2+}, Fe^{2+}, K^+, Mn^{2+} and so forth from the soil's inorganic particles. Once dislodged, these compounds flow away and are no longer available for future generations of plants. In addition, the H+ ions can dislodge toxic ions such as Al^{3+}, which can kill plants and

---

[137] In case you are wondering, tumble-weeds are already dead.

354

animals. Generally speaking, small amounts of acid rain are natural, but human activities have increased its severity by releasing compounds such as sulfur dioxide and nitrogen oxides into the air. These react with water to produce acidic chemicals that fall with rain. The evidence of its effects can be seen in the deterioration of many of our forests, and the damage is becoming more and more obvious.

Human activities can affect soil in other ways as well. For instance, poor farming practice can lead to a different type of nutrient leeching. Crops take up the nutrients as they grow, and then we pull the plants from the ground and ship them away. As a result, the nutrients that were normally recycled locally (soil to plant and back again), now get transported to your neighborhood's supermarket shelf. The next generation of plants doesn't grow as well because most of the Mg^{2+}, say, is now sitting in "aisle three" next to the cheesy poofs, a tofu turkey and a plastic dinosaur-shaped neon-green sandwich cutter. To battle this, modern farmers often replace the leeched nutrients by spreading fertilizers and animal dung on their fields. In the latter respect, I'm quite glad that 'organic' produce is available at my local grocer. After all, only the finest veneer of cow poo will do for me, thank you.

Pulling crops away from fields isn't the only way in which poor farming practice can wreck a local ecosystem. Sometimes, poorly planned irrigation can destroy a field as well. For instance, if standing fresh water is introduced on top of soil that is sitting above a lot of salt – such as in farmlands near sea level – the newly introduced fresh water can pull salt up from the ground. As the water evaporates, the salt begins to accumulate on the surface, radically changing the life-sustaining qualities of the field. There is some evidence to suggest that many of the fertile lands of Ancient Middle East have been destroyed by this kind of irrigation. The solution to this appears to be slower irrigation – less fresh water standing around – although this remains to be proven.

Lastly, removal of vegetation can speed nutrient leeching in a way unrelated to the direct removal of nutrients that go with the plants. The roots of the killed plants die, and as they die they lose their ability to hold the soil together. When this happens, water can flow away from the area more quickly, resulting in nutrient leeching and rapid erosion. If you don't like a hill, remove all the plants on it and watch what starts to happen.

That just about wraps up the first half of our look at biogeography. Let's have a short break and pick it up again in the next unit. See you then.

Summary

In this chapter, we learned about some basic concepts in biogeography.

- The Earth gets almost all its energy from the Sun. Sunlight heats the Earth, and much of this heat is lost to space. The remainder is trapped as potential energy in the form of new chemical bonds, altered weather, changes in the environment and so forth.
- The equatorial region of the Earth receives the most intense sunlight. This causes air and water to rise up. As the air rises, the water comes back down, resulting in lots of rain near the equator. The dry air moves north and south, dropping around the Tropics of Cancer and Capricorn. These latter areas tend to be very dry.
- Wind patterns are caused by a variety of phenomenon. Sea breezes occur because of differences in temperature between land and sea. During the day, the land is warmer than the sea, and rising land air drags

cooler sea air toward shore. During the night, the warmer sea air rises, dragging cooler land air away from shore.

- Heated air tends to have higher pressure than cold air. All other things being equal, high pressure air moves toward low pressure air, in the form of winds.
- The Coriolis Effect causes wind to travel in circular patterns due to the rotation of the Earth. This happens because wind at one latitude has a different rotational speed than the Earth at a different latitude. So, when wind travels away from the equator, it deflects east. When it travels toward the equator, it deflects west. The combination of the two leads to circular wind, as is seen in hurricanes.
- The Moon influences wind by pulling air toward the side of the Earth closest to it. Tides result from the same kind of process, and water currents are similarly influenced by it as well.
- Mountains affect the direction of wind flow by providing a physical barrier, deflecting air flow or forcing air higher into the atmosphere. When wind tries to cross a mountain chain, the increase in altitude causes the air to cool, which forces it to drop its water on the near side of the mountain. The far side tends to have a dry environment as a result.
- Cold air rushing into a warm region will cause warm air to rapidly rise, leading to thunderstorms. This is called a cold front.
- Warm air rushing into a cold region will slowly rise up in altitude, leading to prolonged rains. This is called a warm front.
- The closer to the equator you go, the more photosynthesis tends to be done. This is because of the intensity of light in that region. As you progress toward the poles, less light is available, so the amount of photosynthesis drops off.
- All the energy captured by photosynthesis is called the gross primary productivity of the Earth. Part of this is used for respiration. The part that remains is the net primary productivity, which is the amount available to the remaining organisms. The bulk of life draws its energy from the net primary productivity.
- Most photosynthesis occurs in the rainforests and oceans. Next are the deciduous forests, boreal forests, savannas and grasslands. Lastly, deserts and the tundra do the least.
- Energy flowing through an ecosystem is usually described as a food chain. There are many food chains within an ecosystem, and all these together comprise the food web.
- At the very bottom of the food web are the primary producers – the autotrophs. These have the greatest amount of energy in the ecosystem. Next are the primary consumers – the herbivores – that live by consuming autotrophs. These individuals make up the second energy level (or second trophic level) of the ecosystem. Last are the carnivores, which live by eating the herbivores. This is the third (and sometimes fourth and fifth) trophic level. Collectively, all the trophic levels form the shape of a pyramid, where most energy is at the bottom and progressively less is at the top.
- The pyramid model can also be used to show the total biomass of each trophic level. The primary producers occupy the most mass, followed by the primary consumers, then secondary consumers and so forth. Some species occupy multiple levels of the pyramid model, depending on the different ways in which they acquire their energy.
- The decomposers are not represented in the pyramid model, because they feed on all organisms.

356

- The atoms and molecules used in the reactions of life must be continually recycled because the Earth is a closed system. Almost nothing gets in or out.
- Water cycles back and forth between the atmosphere and the ground via evaporation and rain. Water flows downhill into streams, rivers and lakes, eventually going to the ocean. Water absorbed by plants gets drawn up into leaves where it gets put into glucose molecules or "sweated" by plant transpiration. Water can also soak into the ground and find its way to oceans and lakes via underground routes or get stuck in underground lakes.
- Carbon is pulled from the air by photosynthesis. It is released by respiration. Most of the carbon in dead biological material is released by decomposers, such as worms and bacteria. The carbon that is not freed gets stuck in deep layers of soil and eventually turns into coal and oil over geological time. Recently, humans have begun to release this trapped carbon through industrial combustion, and the rapid release has started to destabilize the atmosphere.
- Nitrogen is pulled from the air by nitrogen fixing bacteria using enzymes such as nitrogenase. The resulting nitrogen compounds are taken up by plants and animals. Eventually, these compounds are converted back into N_2 and released into the air by denitrifying bacteria. Recently, humans have begun pulling nitrogen from the air using industrial methods.
- The oxygen cycle starts with oxygen bound within water and carbon dioxide. These are incorporated into organic molecules during photosynthesis. In addition, some oxygen is released as O_2 during this process as well. This O_2 is then used during respiration to release water and carbon dioxide, completing the cycle.
- The phosphorus cycle starts with phosphorus in exposed rocks. Water erodes the rocks, releasing the mineral into streams, rivers and other waterways. Once dissolved in water, phosphorus is taken up by plants and animals. It is also deposited in the soil. As the soil gets deeper and deeper, the phosphorus gets compacted into newly forming rocks, and over geological periods it gets moved around by the large forces that work upon the planet. Eventually, the rocks get exposed again, starting the cycle over.
- Humans create a great deal of new compounds, few of which get fully tested for their effects on the environment.
  - Some compounds harm the environment directly. Other compounds are harmless until the environment changes them into a more harmful form.
  - Some compounds require great quantities in order to cause harm. Others do damage in small amounts.
  - Some compounds affect the physical environment, such as fluorocarbons vs. the ozone layer. Other compounds are poisonous to some species in the environment.
  - Chemicals can increase in concentration within a food chain via biological magnification. DDT is a good example and has resulted in the near extinction of the American Bald Eagle.
- Soil is mainly comprised of two parts: dead organic matter and inorganic particles.
  - Inorganic soil particles consist of silicon or aluminum compounds. The particles are divided arbitrarily into three sizes. The small particles make up clay. The medium particles make up silt, and the large particles make up sand. Clay tends to hold water too tightly. Sand tends to hold water too

poorly. They are poor substrates for plant growth. Silt soils typically hold water much better and serve as better substrates for plants.

- High water runoff can cause nutrient leeching by dragging away many of the minerals within the soil.
- Acid rain can accentuate nutrient leeching by dislodging cations from inorganic particles. Examples include K^+, Mn^{2+}, Ca^{2+} and Fe^{2+}. Once dislodged, they can get washed away and become inaccessible to plants and animals.
- Crop harvesting can leech nutrients from fields because the nutrients are carried away with the crops. To correct this, farmers frequently apply fertilizers to their fields.
- Standing water from irrigation can pull salt from the ground if the farmland is too close to sea level. The salt gathers when the water evaporates and eventually makes the soil no longer able to support plant growth.

Unit 43

Biogeography II

Land Biomes

The various physical parameters that exist around our planet – from temperature and rainfall, wind and humidity, latitude, mountains and soil quality – have given rise to a number of ecosystem templates, called biomes. There are a number of these biomes on our planet, some becoming less common and others more so. They include places like the **tundra**, **taiga**, **temperate deciduous forests, rainforests, grasslands, deserts** and coffee shops.[138] We'll discuss many of these here.

---

[138] Ok. I'm kidding about the coffee shops, but in all honesty, I wouldn't be surprised if biologists of the distant future included 'cities' as one of the Earth's biomes – or *only* biome, sadly. After all, humans are a result of natural evolution. That means we *are* natural, so shouldn't we include our behavior as an extension of nature? I mean, a beaver's dam radically changes its environment, but we consider it to be natural, right? Perhaps in the very distant future when we are long gone there will be a small entry in the galactic record of species like this: '*Homo sapiens* – an extinct bipedal species with a natural tendency to build coffee shops every fifteen feet. They became extinct when a beaver built a dam and flooded the last remaining farmland, devoted of course to coffee beans. The last remaining specimens of their species were found encased in a shell of hardened anti-pimple cream.'

Tundra

The tundra is a circumglobal strip of area just south of the artic in northern Canada and Russia. It has a permanently frozen layer of subsoil. As a result, the tundra has a difficult time sustaining large plants and is occupied mainly by mosses, lichens, perennial plants and small grasses. The summers are warm enough to allow the flowering plants to bloom, transforming the region into a vibrant display of colors. In addition, these few warm months melt the very top layer of frozen soil, making the ground mushy. The tundra is dotted by lots of lakes, ponds, and marshes, and is host to a small number of critters. Some of the representative animals are arctic wolves, foxes and rabbit as well as reindeer and caribou.

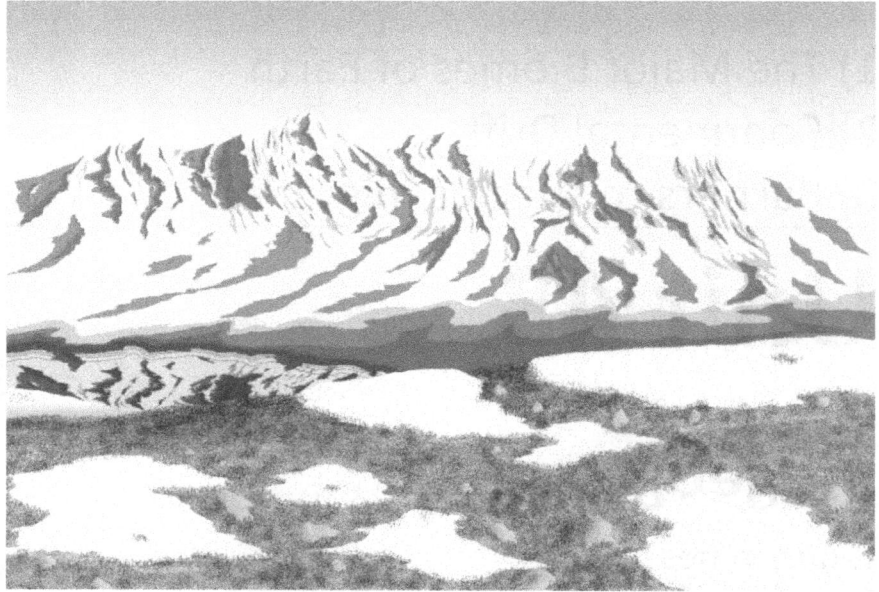

Taiga (Boreal forest)

The taiga (or **boreal forest**) is the region south of the tundra. It too is ring shaped, extending around the globe. Unlike the tundra, its subsoil is not permanently frozen, thawing for parts of the year to allow more vegetation to take hold and grow. Conifer trees are among the most dominant vegetation there. The region also supports a larger number of animal species than the tundra, including moose, black bears, wolverines, porcupines, wolves and lynx. Also like the tundra, the taiga is dotted by numerous ponds, lakes, streams and marshes. And no, the taiga is not related to cats.

Deciduous Forests

The tundra and taiga are both circumglobal rings, centered on the poles. Their main characteristic is their latitudes; they are so close to the poles that their soil is frozen for at least part of the year. As you move further away from the poles, the soil becomes less frozen and more hospitable. This yields ecosystems capable of supporting much larger vegetation. In areas where moisture is readily available, deciduous forests grow. They have a lot of hardwood (broadleaf) trees, such as elm, birch, maple and oak, among countless others. Deciduous forests offer a large number of vertically stratified micro-ecosystems as well, which support a variety of smaller plants and animals. The total number of plant and animal species is much larger than the tundra and taiga, encroaching upon a size too large to list. The most represented are mammals, birds and insects. The least represented are the reptiles and amphibians, although a few can be found here and there. Common animals include foxes, bears, deer, squirrels, humans and other rodents.[139]

Grasslands and Shrublands

In temperate areas where the moisture is not sufficient to support forests, grasslands and shrublands develop. The grasslands consist of fields of herbs and grasses, with the occasional island of trees here and there. They are often characterized by warm-cold or wet-dry annual cycles, depending on the location. Because of the cold and/or dry parts of the year, large vegetation has a hard time taking hold. As a result, grasslands are less complex than most of the other biomes. They support a number of rodents and are often occupied by a few large herbivores. The grasslands of North America, for instance, used to be home to many herds of buffalo, and the grasslands of Africa (also called savannas) are often the home of many zebra, gazelles and elephants.

The term **shrublands** (or **scrubland**) refers to an area where the dominant vegetation consists of bushes, scrubs and small trees, like a field of giant green and prickly cotton balls. Some of them receive higher levels of yearly

---

[139] Humans are not considered rodents, of course, but at times I wonder. . .

rainfall than grasslands. Others receive less and can be found predominantly around deserts. The factors leading to grassland versus shrubland include rainfall, temperature, soil content and so forth, but as a first degree of approximation you can think of shrublands as being a midway point between grasslands and temperate forests.

Tropical Rainforest

Tropical rainforests are like deciduous forests taken to the next level. These are some of the most complex ecosystems on the planet and can be found in the most moist and warm land areas. By far, most are contained within the tropical zone, i.e. south of the Tropic of Cancer and north of the Tropic of Capricorn. Other tropical rainforests can be found on island climates that receive abundant warm winds and moistures, as is common in many Pacific regions. Tropical forests can grow to be several hundred feet tall, allowing incredibly complex and rich vertical micro-ecosystems to develop. At the very top of tropical forests, the thick tree canopy blocks practically all sunlight from the areas below, creating a dark world of underbrush full of highly efficient broadleaf vegetation and a plethora of almost every imaginable critter. Reptiles, amphibians, birds, insects, mammals and whip-wielding archeologists are all represented. In addition, the thick growth of vegetation stifles the wind, yielding an icky soup of humid air. Altogether, these biology-gone-wild ecosystems fix between 10-20% of the Earth's total carbon. Sadly, rainforests are rapidly disappearing, owing to a species of highly pervasive coffee-shop-creating bipeds that occupy the rest of the globe. In the place of jungles, we are sure to get an entirely different kind of biome, perhaps a parking lot or shopping mall. Keep your fingers crossed.

Deserts

The last of the biomes that we will discuss are the deserts. Each is unique, but they all have one thing in common: low annual rainfall. As a result, they contain very little vegetation. Some deserts, such as the Sahara Desert, have so few vegetation that it is little more than sand dunes. Others, such as the Mohave Desert, have a variety of hardy, drought-resistant succulents, such as cacti. Still others have a brief wet period, during which fast growing annual plants germinate, produce seeds and die. The seeds lay dormant until the next season. Since deserts have very little vegetation, they have poor temperature moderation. The days are blistering hot and the nights sometimes quite frigid. Deserts support a number of different organisms, such as lizards, snakes, mice, coyotes, insects and various birds, but compared to other ecosystems, they are much less complex.

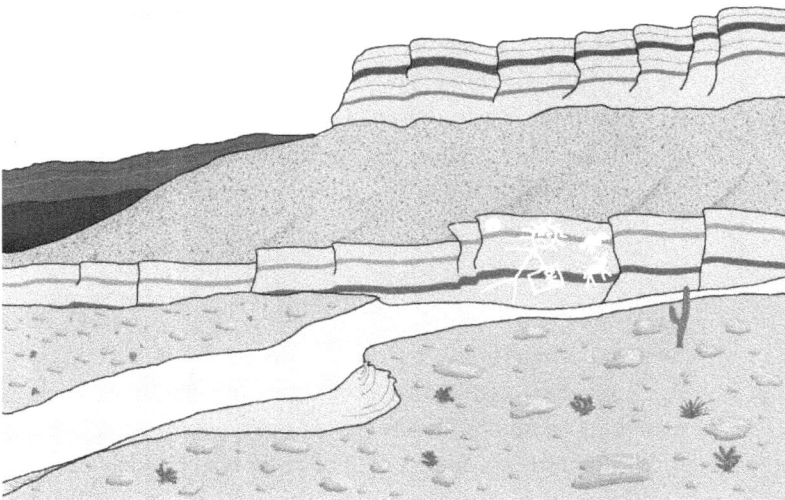

In the next image, we see the relative spread of the various biomes. Notice how the rainforests are largely contained within the strip between the two tropic lines and how many of the deserts fall roughly under the tropic lines. Further, notice how grasslands are often positioned between deserts and forests.

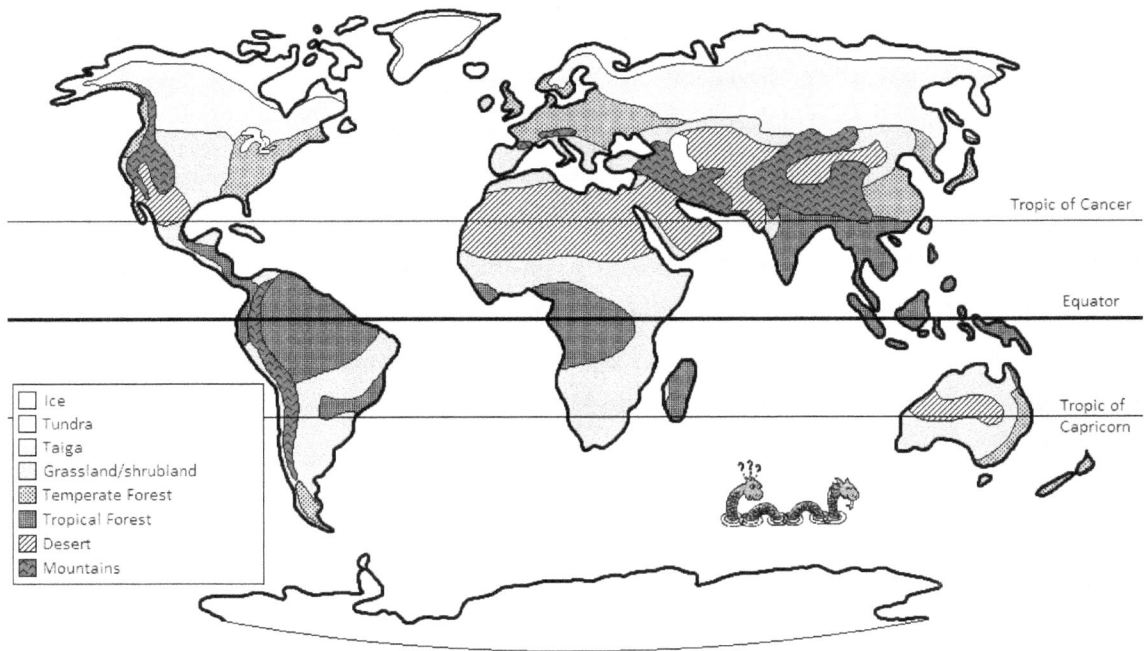

Tropic of Cancer

Equator

Tropic of Capricorn

☐ Ice
☐ Tundra
☐ Taiga
☐ Grassland/shrubland
▨ Temperate Forest
▧ Tropical Forest
▨ Desert
▨ Mountains

363

Effects of Altitude

Distance from the equator is one of the primary factors in determining the average temperature experienced by a location. Another important factor is altitude. The higher you go, the colder it gets. As a result, an increase in altitude often mimics the effect of increasing latitude. More explicitly, at the very tops of mountains you see regions that are equivalent to the Earth's icy poles. As you move down the mountain, you begin to see frozen tundra areas, then taiga, then finally a blend into whatever biome is at the bottom. If a mountain face is on the windward side, the water frequently dumped there will result in deciduous forests or rainforests at the bottom, (depending on the amount of rain and the temperature). On the opposite side, you'll typically see drier biomes, like grasslands or deserts. It all depends on the finite conditions.

Aquatic Ecosystems

Land biomes form largely as a result of moisture and temperature, but these two parameters are largely irrelevant under water. Forgoing the silliness of pondering moisture conditions, underwater communities typically experience very little variation in temperature. We discussed the reason for this before: water heats and cools slowly because of hydrogen bonds. As such, underwater ecosystems tend to form in response to other factors, such as water movement, salinity,[140] depth and so forth. How quickly the water moves, its depth, and the nature of its motion can result in a huge variety of unique environment, from shallow rapids, to deep and turgid ponds, to vast and dynamic oceanic bottoms.

When it comes to aquatic environments, we often divide them up in a variety of ways. For instance, the bottom part of the marine environment is called the **benthic zone**. This is where the bottom feeders are. The water above – and all things suspended in it – is considered part of the **pelagic zone**. When it comes to oceanic ecosystems, we sometimes divide them up based on how far they are from the shore. Those nearest to the shore in the areas pelted by the regular actions of the waves are in the **littoral zone**. Further out is the **neritic zone**, which extends to the drop off of the continental shelf. Ecosystems in the deep part of the ocean, beyond the drop off are in the **oceanic zone**.

Lastly, we can divide up aquatic ecosystems based on depth. The first 200 m is often referred to as the **photic zone**. "Photic" means 'light,' so the photic zone is the lighted part. The first 100 m in particular is called the **euphotic zone**, which means 'true light,' meaning that it's the brightest of the lighted part. Below 200 m is the **aphotic zone**, which just means it's the dark part.

---

[140] Salinity = how salty the water is.

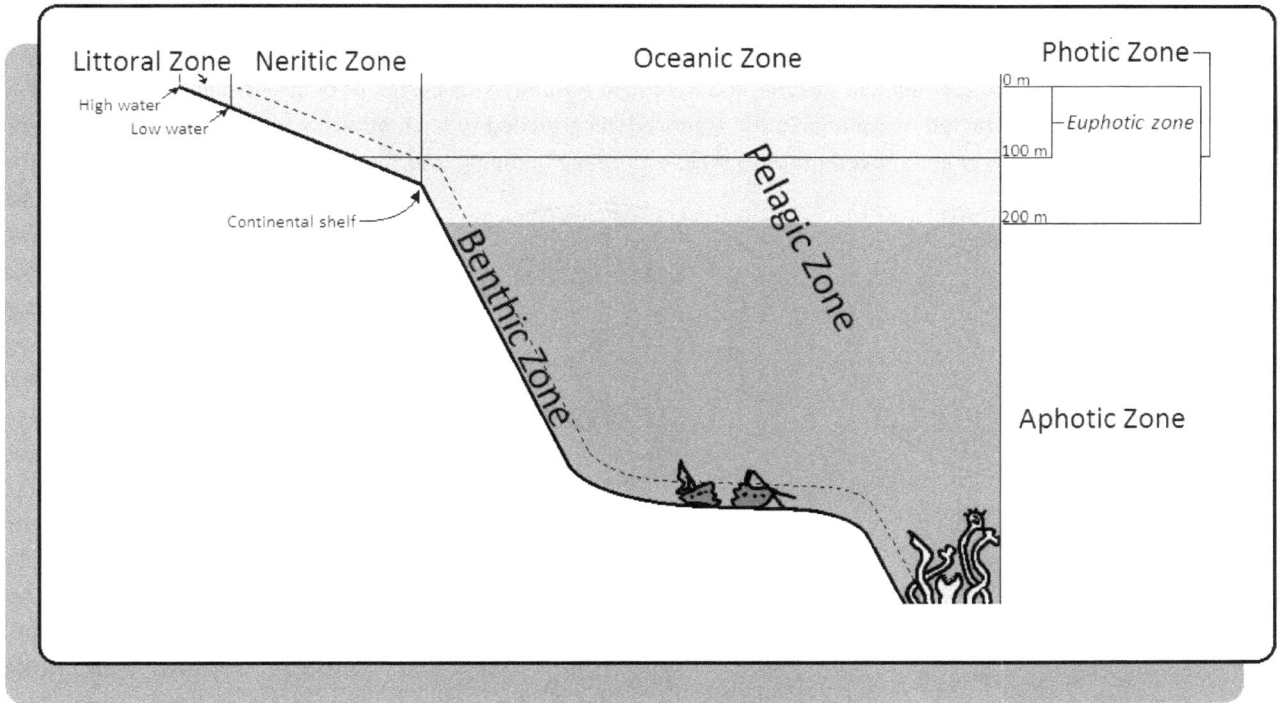

The bulk of the photosynthetic region of aquatic systems occurs mostly in the first meter of water. Most of the aquatic bottom communities (in the benthic zone) are comprised of organisms that 1) feed on floating protists, bacteria and other tiny critters, or 2) sift through dead materials that drift down from the photosynthetic layer above. The most grandiose of these ecosystems are the coral reefs, which are often referred to as the rainforests of the oceans. These colorful communities are created by **cnidarian** organisms that build their homes using calcium-based compounds. These form rocky but fragile structures upon which grow a huge variety of other organisms. The cnidarians feed off of zooplankton, floating nutrients and dead debris, although some also have symbiotic photosynthetic organisms called **zooxanthellae** living in or with them. Unfortunately, these ecosystems take a great deal of time to form but are easily destroyed. As a result, human activity has caused many of these underwater wonders to vanish.[141]

Primitive Earth

We've seen some of today's biomes and how they are related to the position of land and water, but how did the land get to where it is today? All evidence points to Earth being about 4.5 billion years old. As recently as 250 million years ago, all the contents were clumped together in one big continent called **Pangaea.**[142] This giant land mass began to break apart around 200 million years ago into two smaller land masses (both still bigger than any of today's continents). The two resulting continents were **Laurasia** in the north and **Gondwana** in the south. Laurasia eventually

---

[141] You can be sure that if we ever find a way to build a parking lot on the ocean floor, you can kiss these ecosystems goodbye.
[142] What it looked like prior to that is not entirely known.

gave rise to North America, Europe and most of Asia. Gondwana in turn gave rise to Africa, South America, India, Australia and Antarctica. This, of course, is putting it simply. Let's have a little deeper look at the details.

Based on what we can tell, Gondwana first split into a number of pieces prior to yielding the modern day continents. One piece contained Africa and South America (still attached to each other), which retained a connection to Laurasia. India broke away and moved north as well, eventually bumping into the southern part of Laurasia. In the process, the India landmass pushed up the Himalayan Mountains.[143] South America then split from Africa and soon rejoined with the southern-most part of Gondwana that comprised Antarctica and Australia. These three continents were connected to each other before South America and Australia finally broke away, drifting isolated for a long time. Eventually South America collided with North America via the small land bridge where Central America currently is.

Eventually Laurasia divided as well. North America broke off from what was to become Europe and began drifting west. Before doing so, it existed for a while in connection with Africa. Granted, all of this is probably as clear as mud to you, so I've included the image to the left to approximate what these continents looked like.

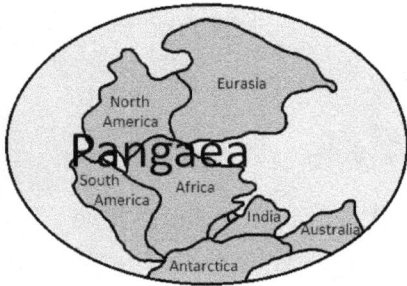
~250 million years ago

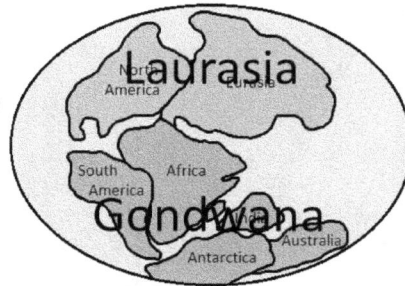
~200 million years ago

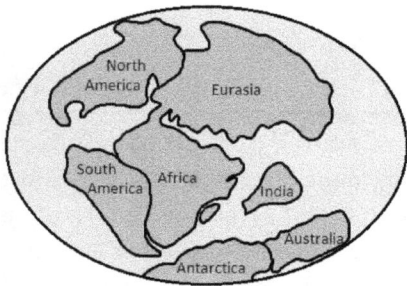
~150 million years ago

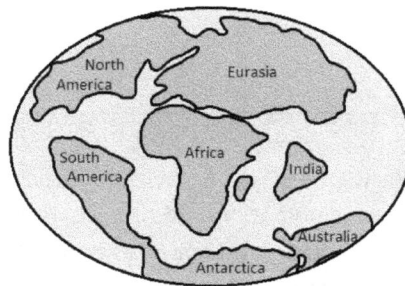
~50 million years ago

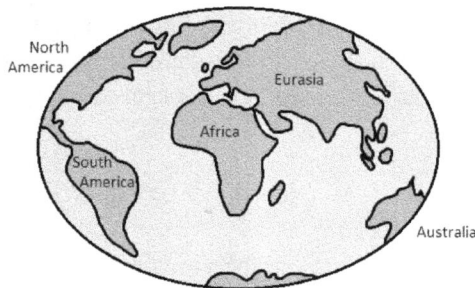
Present Day

---

[143] Like what happens if two rugs are pushed against each other.

366

So how do we know all of this? Well, the entire story is rather long and complex, but the pieces were put together based on a number of observations. First, the shapes of many of today's continents look like they fit together, the most obvious one being the facing borders of South America and Africa. This was the first clue. Second is the fossil record. For instance, large fossils found in both South America and Africa of the same species could only be possible if the two continents were once connected. Also, the fossil records for many continents show plants and animals from biomes that do not match with the current biomes occupying those areas. Antarctica, for example, has many fossil records that indicate it once existed in a much warmer climate. Lastly, geological deposits from one continent commonly match those of other continents. Once again, this could only be true if the two continents were once connected.

How could continents move apart like this? Well, the prominent theory to explain this is the **plate tectonic theory**, which basically says that the continents are floating around atop the liquid molten core of the Earth. Most of us are not used to the idea of floating rocks, so "floating continents" may sound absurd. The important point to remember, however, is that rocks are much less dense than molten metals. As the metals melt and sink toward the *very hot* center of the Earth, they push up the rocks, which end up floating on the surface. As such, the continents drift around like pieces of wood bumping into each other, only they move so slowly that it takes millions of years to notice the difference. Once in a while, we get a reminder of their slow activity in the form of earthquakes[144] and volcanos. Apart from these things, the Earth is rather silent on the matter.

At any rate, several important outcomes have resulted from the continents moving and forming as they have. First, those continents that were the most isolated, such as Australia, have had the greatest amount of evolutionary deviation. This is the reason for Australia's bizarre line-up of species found only there (excluding the animals capable of flying or swimming there). South America has also experienced quite a bit of time isolated from the others . . . after splitting off from Antarctica during its brief connection to it. During its isolation, it too evolved a number of unique species. On the other hand, South America and Australia have a large number of marsupials on them. Antarctica has fossils of marsupials. From this, we can conclude that marsupials evolved when these three continents were still connected. In the modern world, North America has a few marsupials as well. They arrived here by migration *after* South America became physically attached via Central America. Out of all the continental bodies, the chunk that had Europe, North America, most of Asia and (for a long time) Africa were connected for what appears to be the longest time out of any of the great land masses. As a result, many of the fossil records and modern species are similar across these three continents. Throughout that time, the only real barriers to migration were the developing Himalayas, the Mediterranean Ocean and the early Sahara Desert. Despite these barriers, many of the species are similar to each other.

Speciation across Barriers

Throughout evolutionary history, species have migrated from place to place. We saw an example of this when I mentioned how marsupials migrated into North America from South America. Here, we'll have a look at the general principles that describe the phenomenon of species migration. For simplicity, we'll break it down into three considerations.

---

[144] Earthquakes are believed to occur as a result of two land masses scraping against each other.

Physical Access

Before a species can migrate into a new ecosystem, it must have physical access to the new region. Oceans, mountains, impassible land and so forth can keep one species from entering into a new home. On the other hand, a physical barrier for one species may not necessarily represent a barrier for another. A chain link fence might stop my neighbors' rotten kids from getting into my yard, but it doesn't stop a bird . . . or their annoying yapping dog that constantly crawls under it. An open field will not stop a caribou but it's sure to stop a fish.

Throughout evolutionary time, oceans have been perhaps the strongest of all barriers, resulting in large differences between the species of one continent and another. Still, bodies of water are not entirely impassable, especially to small animals. Naturally forming rafts made of logs and other refuse are not uncommon, usually after storms. Any small animal caught on them can drift for hundreds of miles before finding a new home elsewhere. Sometimes, a good storm can whip through an area and carry small animals such as insects away on the wind. This is commonly how insects will colonize small islands or distant lands. At other times, an animal can catch a ride on a more mobile organism, such as a bird, a fish, an herbivore or whatever, and can be deposited far from its original home. Bacteria, fungi, protists, bugs and countless others species can migrate this way. In fact, many plant species rely quite heavily on other organisms for their dispersal. They pack their seeds into fruit, the fruit is eaten, and the seeds survive in the digestive tract until it is dropped off with the rest of the feces. During this process, the animal that ate the fruit carries the seed to some distant location.

Physical barriers change with time, sometimes rapidly and sometimes imperceptibly slow. When they do so, they often force the species they bar to move with them. If the bed of a north-south river slowly drifts west over several hundreds of years, the species of organisms stuck on the west side will be forced to keep moving west. If an ocean between two continents grows wider, the species on either side have no choice but to get farther apart.

Anatomical/Physiological Compatibility

Even when a physical barrier is not present, speciation is not guaranteed to succeed. The migrating species must be physiologically adaptable to a new region. For example, suppose one species of tree frog exists within a region of rainforest that contains an abundant amount of a specific food type. Then, suppose that several members of the tree frog species attempt to migrate to the far end of the rainforest where the same food has a slightly higher level of a toxin. The presence of toxin may cause the frog to be unable to thrive in that location. As such, no physical barrier stopped them; they just weren't physiologically adapted to the types of food found in the far end of the forest.

Physiological compatibility extends to much more than just food. There may be differences in temperature, moisture, light and so forth in the new region. Even more importantly, the new region may have different predators or other species of tree frog that compete fiercely with the newcomers. Sometimes the new environment might be perfect for the new species, except for sporadic frosts that come through. Even a brief weather abnormality might be enough to make a new environment impossible to live in. It would be like human astronauts finding an Earth-like planet that is perfect for habitation, except for the one day out of the year when the planet heats up to 3,000 degrees Celsius. Such a temporary *inconvenience* would stop our attempts to colonize the new planet . . . or in the very least make it exceptionally hard.

As you hopefully can see, any number of factors may prevent the spread of a species from one region to another, and collectively these restrictions are known as the **Law of Tolerance**, which basically just says that the species will be limited by its ability to tolerate subtle differences between the original and new niches.

Other Environmental Factors

There are a number of other environmental factors limiting the spread of species as well. One is the size of the new environment. For instance, small environments often carry smaller odds of colonization success than larger regions. The reason is arguably due to the limitation of resources. Consider a bunch of spiders blown away by a storm to land on a distant island. Let's take two cases. In the first case, let's assume the island is only thirty square meters in area. In the second case, let's say the island is thirty *thousand* square meters. The islands are otherwise identical. The spiders landing on the first island are less likely to survive because they have access to fewer resources.

Another ecological factor in species migration is complexity of the new environment. The more complex an environment is the greater are the odds of finding hiding places, nesting sites, novel food sources and other materials you might need to survive. Consider spiders landing on an island that is perfectly flat with little more than grass on it. They will be less likely to survive than similar spiders landing on an island of the same size occupied by a mountain and a rainforest.

The last ecological factor that is important in species migrate is the distance between the original ecosystem and the new ecosystem. The farther it is, the less likely the species will be able to get a foothold. Conversely, the closer the new environment is, the more members of your species will invade . . . as will members of *other* species as well. This aspect of speciation is often studied by examining islands close to shore. What we often find is that the rate of appearance of new species on such islands (or **rate of speciation**) depends heavily on the distance from the island to shore. Islands near the shore have a high rate of new species appearing during any given year. The farther away the island is the less new species will appear in the same given year.

On the other hand, the more species arrive on the island, the higher the competition and – consequently – the higher the **rate of extinction**. This latter aspect is heavily dependent on the size of the island. Small islands are battlegrounds with high rates of extinction. Large islands are more stable, with less species dying out during any given year.

Tying these two principles together, we find that at any given point in time, the total number of species that an island can support tends to equilibrate at a number in which the rate of speciation equals the rate of extinction.

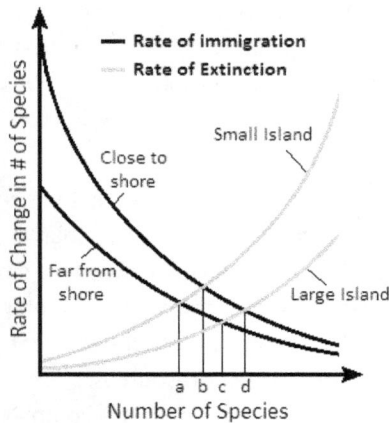

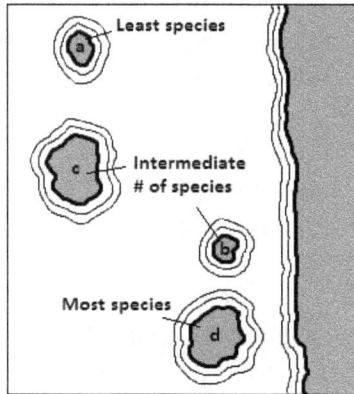

So an island that is, say, one mile offshore might be able to support 100 species because the rate of species migration is large, whereas the same island 50 miles offshore might only be able to support 30 species. In the latter case, the rate of new incoming species is far less, but so is the rate of extinction.

Summary

In this chapter, we learned about some of the basic biomes, the formation of today's continents and the basics of species migration.

- A biome is an ecological template. The major ones include: the tundra, the taiga, grasslands, deserts, deciduous forests and tropical rainforests.
- The tundra is a circular band-like region just south of the North Pole. It has permanently frozen soil and contains small plants and a few animals.
- The taiga is south of the tundra and is frozen for a large part of the year. It is dominated by coniferous trees. It contains wolves, moose and a variety of small animals and birds. Its other name is the boreal forest.
- Grasslands are regions that are warm enough to support forests but receive insufficient rainfall to grow them. They are occupied by grasses, small shrubs and pockets of trees. They are home to small rodents and herds of grazing herbivores.
- Deciduous forests are ecological regions that are sufficiently warm and moist enough to house a permanent collection of deciduous (leaf-shedding) trees and a small amount of undergrowth. They contain vertical stratification and have greater numbers of species than grasslands.
- Tropical rainforests have rich collections of dense vegetation and towering trees, with dark and moist undergrowth. They are primarily found within the tropical belt and have a rich number of species from

practically all the major animal groups. The tropical rainforests are rapidly disappearing due to deforestation by humans.

- Altitude often mimics the effects of latitude. The tops of tall mountains have similar ecological characteristics as the Earth's poles. The further down the mountain, the environment changes from tundra-like, to taiga-like, and then to a blend of ecology with whatever biome exists at the base of the mountain. The windward sides of mountains usually have forest biomes at the base, while the opposite side commonly has deserts at the base.
- Within aquatic ecosystems, factors such as water movement, salt content and lighting help to establish the ecological setting, of which there are huge variations from one point on Earth to another. We divide these regions very broadly in the following ways:
  - Bottom part vs. watery part: Benthic zone (bottom) and Pelagic zone (watery part).
  - Shore vs. near shore vs. distant ocean: Littoral zone (shore under the waves), Neritic zone (shallow ocean out to the continental shelf) and Oceanic zone (deep ocean beyond continental shelf).
  - Water depth: Photic zone (first 200 m of depth, i.e. lighted part, with first 100 m called the euphotic zone) and Aphotic zone (depths beyond 200 m).
- Most photosynthesis occurs in the top one meter.
- Most of the organisms in the benthic zone (sea floor) survive by eating the dead materials that drift down from above.
- The coral reefs are a fragile ecosystem formed by the action of cnidarians, which create calcium-based structures as homes. The other critters in coral reefs live in, among or around these calcium structures.
- One of the first recognizable continents was Pangaea which existed at least until around 250 million years ago. This separated into two clumps of land: Laurasia in the north and Gondwana in the south. Each of these eventually fragmented into smaller land masses:
  - Laurasia -> North America and Eurasia (Europe and Asia)
  - Gondwana -> South America, Africa, India, Australia and Antarctica.
- During the slow fragmentation of the continents, the following continental collections resulted in important biological patterns:
  - South America, Antarctica and Australia were once connected for a long time and gave rise to marsupials. A few of these migrated to North America once a land connection was finally established with South America.
  - Australia has since existed for a long time in isolation. This gave rise to some of Australia's very unique animals.
  - North America, Europe, Africa and Asia were connected for a very long time, which resulted in a large array of species that are common between these major land masses.
- The continents move because of plate tectonics, a phenomenon in which the land masses float atop the Earth's molten core and slowly move and bump against each other. Earthquakes and volcanoes are consequences of this.
- Species migrate from one region to another if all the following conditions are met:

- There is physical access to a new region (where the meaning of physical access varies depending on the species and their mode of transportation).
- The species has anatomical/physiological compatibility with the new region. (See Law of Tolerance, below)
- The new region is capable of supporting more organisms.

- The Law of Tolerance states that each individual in a region must be able to tolerate all the factors that make up the region in question. If the organism is unable to tolerate even one factor within that region, the organism will be unable to prosper there.
- By a first degree of approximation, the rate of speciation and extinction is determined by a number of simple factors:
  - The rate of speciation between two regions depends on the distance between the regions. The closer a region is to another region, the greater the number of species that will immigrate between the two.
  - The rate of extinction within a region depends on the size of the region and the total number of species currently present. The larger the region, the greater the resources and the lower the extinction rate will be. The greater the current number of species, the higher will be the competition, leading to higher extinction rates.
  - An equilibrium in the number of species will be established when the rate of speciation equals the rate of extinction. At that point, the types of species might change over time, but the total number of species will remain constant.

Index

B

Bacteria (Domain), 17
bacteriophage, 238
basal body, 130, *131*
base, 37
belt desmosome, 109, *110*, 127
benthic zone, **364**, 365, 371
beta tubulin, *129*
beta-carbon of amino acid, *60*
beta-sheet, 64, *65*
bicarbonate, 25
biogeographic, 283
biogeography, **343**, 355
biological magnification, **353**, 357
biome, **359**, 362, 364, 370, 371
biosphere, 13, 18, **312**, 320, 332, 343, 346
blood platelet, 219
boreal forest. *See* taiga
bottleneck effect, *292*, *See* founder effect
buffer, 39
butane, *148*

C

cadherin, 109
calcium hydroxylapatite (in bone), 288
Calvin Cycle, **142**, 144
    step 1, 142
    step 2, 142
    step 3, 142
    step 4, 143
    summary, 145
cAMP, 94, *See* cyclic adenosine monophosphate
cancer, **94**, 106, 124, 135, 211, 214, 220, 292
capsid (viral), 203
carbohydrate, *164*
carbon cycle, 349–50
carboxyl group, *44*
carotenoid, 138
carrier protein, 95
cat and the stick, 13
catenin, *110*
cdc, **220**, 221

cell, 12
cell coat. *See* glycocalyx
cell cycle, 211, **216**, 220
cell membrane, **92**, 109, 110, 111, 115, *123*
    roles, 92
cell wall, **103**, 111, *112*, 115
    plant, *104*
    primary, plant, 104
    secondary, plant, 104
cell-division-cycle protein. *See* cdc
cellular clock, 220
cellulose, *51*, 52, 104
Central Dogma of Biology, 184, **186**, 192, 202, 216
centriole, *117*, **129**, 130
centromere, **217**, 218, 225, 229
centrosome, *129*, 218
cGMP, 94, *See* cyclic guanosine monophosphate
channel protein, **94**, *122*
chemical reaction, 27
chiasmata, 226
chitin, *51*, 105
chlorophyll, 137, *138*
chloroplast, 115, *117*, 124, 125, **126**, *139*, *141*, 169
cholesterol, *43*
chromatin, *117*, 118, *119*
chromosome, 115, 116, *117*, **118**, 119, 174, 182, 206, 217,
    224, 225, 229, 232, 236, 241, 242, 246, 247, 252, 258,
    259, 260, 262, 263, 269, 275
    eukaryotic, *174*
    prokaryotic, *174*
cilia, 131
citrate synthase, 164
citric acid, **164**, *165*
citric acid cycle, 126, 152, 158, 162, **163**, 164, 240
    summary, 168
class (taxonomy), 16
clathrin, 100
climax community, 327
cnidarian, 365
CoA. *See* coenzyme A
codon, **192**, 193, 194, 195, 197, 209, 210, 240
coenocytic cell, 219
coenzyme, 82
coenzyme A, 82

triplet (codon). *See* codon
tRNA, 186, 192, *194*, 195, 196, 197, 199
trophic levels, **347**, 348, 356
tryptophan, *61*
tundra, 347, 356, 359, ***360***, 361, 364, 370, 371
turgor pressure, 103
tyrosine, *61*
    genetic code, 193

U

ultraviolet light, ***136***, 213
Uncertainty Principle, 20
Uniformity of Space-Time, 8
unsaturated fatty acid, 55
uridine, **78**, *186*, 187
UV. *See* ultraviolet light
UV mutation, *213*

V

vacuole, *103*, *112*, *117*
valence shell, 24
valine, *61*, *62*
    genetic code, 193
Verhulst equation (in footnote), 316
vesicle, **100**, 121, 122
vinculin, *110*
viruses, **202**, 213, 238, 284
    by nucleic acid type, 204

    possible origins, 205
    retro, 205, 213
    reverse transcriptase, 205

W

warm front, 346
water, *31*, *34*
water cycle, 348–49
wax, 54
what fire is, 148
why ice floats, 34
wildtype, 241
wind patterns, 344–46
Wobble Hypothesis, 194

X

X chromosome, 217, 225, **245**, 246, 247, 252
X-ray, ***136***, 213

Y

Y chromosome, **245**, 246, 247

Z

zooxanthellae, 365

www.ingramcontent.com/pod-product-compliance
Lightning Source LLC
Chambersburg PA
CBHW080710220326
41598CB00033B/5366